High-Voltage Engineering

ELECTRICAL ENGINEERING AND ELECTRONICS

A Series of Reference Books and Textbooks

EXECUTIVE EDITORS

Marlin O. Thurston
Department of
Electrical Engineering
The Ohio State University
Columbus, Ohio

William Middendorf
Department of
Electrical and Computer Engineering
University of Cincinnati
Cincinnati, Ohio

EDITORIAL BOARD

Maurice Bellanger
Télécommunications, Radioélectriques, et
 Téléphoniques (TRT)
Le Plessis-Robinson, France

J. Lewis Blackburn
Bothell, Washington

Sing T. Bow
Department of Electrical Engineering
The Pennsylvania State University
University Park, Pennsylvania

Norman B. Fuqua
Reliability Analysis Center
Griffiss Air Force Base, New York

Charles A. Harper
Westinghouse Electric Corporation
 and Technology Seminars, Inc.
Timonium, Maryland

Naim A. Kheir
Department of Electrical and
 Systems Engineering
Oakland University
Rochester, Michigan

Lionel M. Levinson
General Electric Company
Schenectady, New York

V. Rajagopalan
Department of Engineering
Université du Québec
 à Trois-Rivières
Trois-Rivières, Quebec, Canada

Earl Swartzlander
TRW Defense Systems Group
Redondo Beach, California

Spyros G. Tzafestas
Department of Electrical Engineering
National Technical University
 of Athens
Athens, Greece

Sakae Yamamura
Central Research Institute of
 the Electric Power Industry
Tokyo, Japan

1. Rational Fault Analysis, edited by Richard Saeks and S. R. Liberty
2. Nonparametric Methods in Communications, edited by P. Papantoni-Kazakos and Dimitri Kazakos
3. Interactive Pattern Recognition, Yi-tzuu Chien
4. Solid-State Electronics, Lawrence E. Murr
5. Electronic, Magnetic, and Thermal Properties of Solid Materials, Klaus Schröder
6. Magnetic-Bubble Memory Technology, Hsu Chang
7. Transformer and Inductor Design Handbook, Colonel Wm. T. McLyman
8. Electromagnetics: Classical and Modern Theory and Applications, Samuel Seely and Alexander D. Poularikas
9. One-Dimensional Digital Signal Processing, Chi-Tsong Chen
10. Interconnected Dynamical Systems, Raymond A. DeCarlo and Richard Saeks
11. Modern Digital Control Systems, Raymond G. Jacquot
12. Hybrid Circuit Design and Manufacture, Roydn D. Jones
13. Magnetic Core Selection for Transformers and Inductors: A User's Guide to Practice and Specification, Colonel Wm. T. McLyman
14. Static and Rotating Electromagnetic Devices, Richard H. Engelmann
15. Energy-Efficient Electric Motors: Selection and Application, John C. Andreas
16. Electromagnetic Compossibility, Heinz M. Schlicke
17. Electronics: Models, Analysis, and Systems, James G. Gottling
18. Digital Filter Design Handbook, Fred J. Taylor
19. Multivariable Control: An Introduction, P. K. Sinha
20. Flexible Circuits: Design and Applications, Steve Gurley, with contributions by Carl A. Edstrom, Jr., Ray D. Greenway, and William P. Kelly
21. Circuit Interruption: Theory and Techniques, Thomas E. Browne, Jr.
22. Switch Mode Power Conversion: Basic Theory and Design, K. Kit Sum
23. Pattern Recognition: Applications to Large Data-Set Problems, Sing-Tze Bow
24. Custom-Specific Integrated Circuits: Design and Fabrication, Stanley L. Hurst
25. Digital Circuits: Logic and Design, Ronald C. Emery
26. Large-Scale Control Systems: Theories and Techniques, Magdi S. Mahmoud, Mohamed F. Hassan, and Mohamed G. Darwish
27. Microprocessor Software Project Management, Eli T. Fathi and Cedric V. W. Armstrong (Sponsored by Ontario Centre for Microelectronics)
28. Low Frequency Electromagnetic Design, Michael P. Perry
29. Multidimensional Systems: Techniques and Applications, edited by Spyros G. Tzafestas
30. AC Motors for High-Performance Applications: Analysis and Control, Sakae Yamamura

31. Ceramic Materials for Electronics: Processing, Properties, and Applications, *edited by Relva C. Buchanan*
32. Microcomputer Bus Structures and Bus Interface Design, *Arthur L. Dexter*
33. End User's Guide to Innovative Flexible Circuit Packaging, *Jay J. Miniet*
34. Reliability Engineering for Electronic Design, *Norman B. Fuqua*
35. Design Fundamentals for Low-Voltage Distribution and Control, *Frank W. Kussy and Jack L. Warren*
36. Encapsulation of Electronic Devices and Components, *Edward R. Salmon*
37. Protective Relaying: Principles and Applications, *J. Lewis Blackburn*
38. Testing Active and Passive Electronic Components, *Richard F. Powell*
39. Adaptive Control Systems: Techniques and Applications, *V. V. Chalam*
40. Computer-Aided Analysis of Power Electronic Systems, *Venkatachari Rajagopalan*
41. Integrated Circuit Quality and Reliability, *Eugene R. Hnatek*
42. Systolic Signal Processing Systems, *edited by Earl E. Swartzlander, Jr.*
43. Adaptive Digital Filters and Signal Analysis, *Maurice G. Bellanger*
44. Electronic Ceramics: Properties, Configuration, and Applications, *edited by Lionel M. Levinson*
45. Computer Systems Engineering Management, *Robert S. Alford*
46. Systems Modeling and Computer Simulation, *edited by Naim A. Kheir*
47. Rigid-Flex Printed Wiring Design for Production Readiness, *Walter S. Rigling*
48. Analog Methods for Computer-Aided Circuit Analysis and Diagnosis, *edited by Takao Ozawa*
49. Transformer and Inductor Design Handbook, Second Edition, Revised and Expanded, *Colonel Wm. T. McLyman*
50. Power System Grounding and Transients: An Introduction, *A. P. Sakis Meliopoulos*
51. Signal Processing Handbook, *edited by C. H. Chen*
52. Electronic Product Design for Automated Manufacturing, *H. Richard Stillwell*
53. Dynamic Models and Discrete Event Simulation, *William Delaney and Erminia Vaccari*
54. FET Technology and Application: An Introduction, *Edwin S. Oxner*
55. Digital Speech Processing, Synthesis, and Recognition, *Sadaoki Furui*
56. VLSI RISC Architecture and Organization, *Stephen B. Furber*
57. Surface Mount and Related Technologies, *Gerald Ginsberg*
58. Uninterruptible Power Supplies: Power Conditioners for Critical Equipment, *David C. Griffith*
59. Polyphase Induction Motors: Analysis, Design, and Application, *Paul L. Cochran*

60. Battery Technology Handbook, *edited by H. A. Kiehne*
61. Network Modeling, Simulation, and Analysis, *edited by Ricardo F. Garzia and Mario R. Garzia*
62. Linear Circuits, Systems and Signal Processing: Advanced Theory and Applications, *edited by Nobuo Nagai*
63. High-Voltage Engineering: Theory and Practice, *edited by M. Khalifa*
64. Large-Scale Systems Control and Decision Making, *edited by Hiroyuki Tamura and Tsuneo Yoshikawa*

Additional Volumes in Preparation

Distributed Computer Control for Industrial Automation, *edited by D. Popovic and Vijay P. Bhatkar*

Industrial Power Distribution and Illuminating Systems, *Kao Chen*

Computer-Aided Analysis of Active Circuits, *Adrian Ioinovici*

Mathematical Modeling and Simulation for Electronic Design, *edited by Ricardo F. Garzia and Mario R. Garzia*

Electrical Engineering-Electronics Software

1. Transformer and Inductor Design Software for the IBM PC, *Colonel Wm. T. McLyman*
2. Transformer and Inductor Design Software for the Macintosh, *Colonel Wm. T. McLyman*
3. Digital Filter Design Software for the IBM PC, *Fred J. Taylor and Thanos Stouraitis*

High-Voltage Engineering

Theory and Practice

edited by

M. Khalifa
Cairo University
Giza, Egypt

Marcel Dekker, Inc. • New York and Basel

Library of Congress Cataloging-in-Publication Data

High-voltage Engineering.

(Electrical engineering and electronics ; 63)
Includes bibliographical references.
1. Electric engineering. 2. High voltages. I. Khalifa,
Mohamed   . II. Series
TK153.H47 1990   621.319'13
ISBN 0-8247-8128-7 (alk. paper)

This book is printed on acid-free paper.

Copyright © 1990 by MARCEL DEKKER, INC. All Rights Reserved

Neither this book nor any part may be reproduced or transmitted in any form or by any means, electronic or mechanical, including photocopying, microfilming, and recording, or by any information storage and retrieval system, without permission in writing from the publisher.

MARCEL DEKKER, INC.
270 Madison Avenue, New York, New York 10016

Current printing (last digit):
10 9 8 7 6 5 4 3 2 1

PRINTED IN THE UNITED STATES OF AMERICA

" . . . and say 'O my Lord increase me in knowledge..."

The Sublime Quran (20:114)

Foreword

Although high voltages have existed in nature since creation, they were not harnessed by human beings until the beginning of the twentieth century. They have long been in use for power transmission and many other applications in the fields of electric power, electronics, physics, medicine, and many others.

The advances in high-voltage engineering during the last few decades are staggering, especially with regard to our present deep insights into discharge phenomena, development of new materials for electrical insulation, and new and versatile applications, as well as the gigantic leaps in utilized voltages. The world record for high-voltage in power transmission now stands at about 1 million volts, whereas 50 years ago it was only about 300 kV.

The history of high-voltage engineering education at the University of Cairo goes back to 1936, long before that at many other institutions. At that time we trained undergraduate students in our high-voltage laboratory at 40 kV ac, reaching 250 kV in 1940, and 1400-kV impulse in 1951.

With the well-established tradition of high-voltage engineering at the university level, coupled with the international experience of my colleagues as both research leaders and consultants to industry, this book ably bridges the gap between research in the laboratory and practice in industry and power utilities. This book will be of considerable interest to experienced engineers and students alike. Two particularly valuable aspects of the book are its comprehensive approach and the copious bibliographies that appear at the end of chapters, which will enable the reader to pursue research in further detail on any point discussed in the book.

So gigantic and rapid have been the advances in many branches of electrical and electronics engineering that the tendency has been for individual engineers to specialize in a particular branch. One important attribute of a good engineer is breadth of view. This book will be invaluable in assisting specialists to maintain contact with other branches in their field.

It is my great pleasure to write the foreword for such a notable contribution to the literature in this important multidisciplinary field of study.

Mahmoud A. B. El-Koshairy
Consultant, Ministry of Electricity and Energy
Cairo, Egypt

Preface

The need for electrical and electronics engineers to be well trained in high-voltage techniques and to be familiar with related basic theories and concepts is being felt more keenly today than ever before in both industrial and developing countries. The range of high-voltage applications is no longer confined to high-voltage laboratories and substations.

Such topics as electrical insulation and its breakdown under various stresses have direct applications in industries where electrical and electronic equipment is being manufactured. Knowledge of the techniques for generating high-voltage dc, impulse, and ac low and high frequencies is essential for engineers in charge of the design and manufacture of high-voltage power supplies. Such equipment has a wide range of applications, extending from aerospace electronics, radio transmitting stations, and particle accelerators used in the medical and nuclear physics fields, down to television receivers. The subject of gas discharges is another example, as they are encountered not only as noisy corona on HV and EHV overhead power lines and as troublesome arcs in switches, but also in electrostatic precipitators, ozone production plants, gas-discharge lamps, arc furnaces, plasma torches, and ion-implantation equipment, to give just a few examples.

All or selected areas of this vast field of high-voltage engineering represent essential ingredients in the information and training that should be acquired by all electrical and electronics engineers. Included are engineers responsible for the design and operation of power transmission and distribution systems at all voltage levels, those responsible for the design and manufacture of electrical and electronic equipment, those in charge of the design and installation of industrial plants and radio transmitting stations, and engineers and scientists involved in the design and installation of particle accelerators in medical and nuclear research. Not to be overlooked are the staff of high-voltage research centers and university faculty active in research, education, and training in fields of high-voltage engineering.

Reference works and textbooks in the field of high-voltage engineering have been rather scarce, generally being limited to monographs on specific topics such as switchgear, corona or arc discharge, and high-voltage laboratory techniques, or, in effect, research reports on specific EHV projects. There is still a great demand for a book covering most, if not all, of the salient topics of high-voltage engineering in one volume. This book would be of considerable assistance to engineers in power utilities, electrical industries, and several other fields.

The topics covered include gas discharge, insulating materials, system earthing, overvoltage and insulation coordination, and high-voltage equipment and testing techniques. One volume to cover all these and related topics is bound to be limited in its coverage of each, but no apology is needed for the absence of detailed analyses on some topics. However, salient subjects are treated to the appropriate depth, reflecting the contributors' research and industrial experience recognized internationally. Readers interested in more information about any of the topics discussed will find a comprehensive list of references at the end of each chapter.

This book does not overlook essential basic theories and techniques but still covers up-to-date practice in major areas of EHV insulation coordination, HVDC line insulation problems, gas-insulated systems, electric field and corona computations, and solid-state circuit breakers. The book requires no prerequisites other than the physics and mathematics courses normally taught to undergraduates in electrical and electronics engineering.

Following the introduction, the book is divided into two parts. Part I comprises Chapters 2 through 8 and provides the theoretical basics in high-voltage engineering; Part II (Chapters 9 through 18) is devoted to common high-voltage equipment, some important design problems, and high-voltage testing techniques.

Chapter 2 is a rigorous but clear discussion of methods of field computation and mapping. In Chapters 3 and 4 the physical phenomena acting in different types of gas discharges are reviewed. In Chapters 5 and 6 corona and arc discharges are described in some detail. Liquid and solid insulating materials are surveyed briefly from the point of view of electrical engineers in Chapters 7 and 8.

In the first two chapters of Part II (Chapters 9 and 10), the principles of design of high-voltage busbars are discussed, together with their insulation and ampacity, whether they are of conventional air-insulated type or the metal-clad GIS types now widely used at the HV and EHV levels. The various types of circuit breakers and cables are discussed in Chapters 11 and 12, including mention of solid-state breakers and superconducting cables.

Preface

Chapters 13 through 15 are assigned to a treatment of power system grounding, external and internal overvoltages imposed on system insulation, and techniques adopted for insulation coordination. The last three chapters focus on the area of insulation testing, covering the topics of high-voltage generation, measurements, and standard specifications.

It is hoped that electrical and electronics engineers reading this book will find it rewarding and stimulating, and that it will enrich their knowledge of several segments of the field of high-voltage engineering, which continues to grow in importance and scope.

Many persons have contributed to this book in various ways, and their efforts are most sincerely appreciated.

The contributors and I are especially grateful to Professor Dr. M. A. B. El-Koshairy, the pioneer of high-voltage engineering in Egypt and, in fact, in the whole Middle East. He was the first to establish a high-voltage laboratory and to teach high-voltage engineering at Cairo University in the early 1940s. He was also the first to initiate the planning and design of Egypt's unified power network with its trunk line of 500kV in the 1960s.

We are also indebted to our colleagues for their keen interest, helpful suggestions, constructive criticism, and wholehearted cooperation. We appreciate the valuable help we have received from our numerous graduate students over the years at Cairo, Assiut, and other universities in Egypt, the United States of America, Canada, Jordan, and Saudi Arabia.

We also appreciate the courtesy of permission, freely granted by firms, organizations, and friends, to reproduce many illustrations included in this book, as well as the enthusiastic and sincere cooperation from the team at Marcel Dekker, Inc.

I am also profoundly indebted to my late parents and to my brothers and sisters for their enthusiastic encouragement. My brothers Abdel-Halim and Ezzat's unfailing and selfless support and advice with every step of education and training through the years are gratefully acknowledged.

M. Khalifa

Contents

Foreword *Mahmoud A. B. El-Koshairy*	v
Preface	vii
Contributors	xix

1. **Introduction**   1
   M. Khalifa

1.1	High-Voltage Lines	1
1.2	Power Networks	3
1.3	Insulation of Electrical Equipment	6
	References	6

Part I

2. **Electric Fields**   11
   M. Abdel-Salam

2.1	Introduction	11
2.2	Analytical Calculation of Space-Charge-Free Fields	11
2.3	Experimental Analogs for Space-Charge-Free Fields	21
2.4	Numerical Computation of Space-Charge-Free Fields	25
2.5	Analytical Calculations of Fields with Space Charges	51
2.6	Numerical Computation of Fields with Space Charges	51
2.7	Electric Stress Control and Optimization	59
	References	61

3. **Ionization and Deionization Processes in Gases**   65
   M. Abdel-Salam

   3.1   Introduction   65
   3.2   Kinetic Theory of Gases   65
   3.3   Behavior of a Gaseous Dielectric in an Electric Field   70
   3.4   Ion-Generation Processes   72
   3.5   Deionization Processes   83
   3.6   Gas-Discharge Parameters as Influenced by Atmospheric Conditions   87
         References   90

4. **Electrical Breakdown of Gases**   93
   M. Abdel-Salam

   4.1   Introduction   93
   4.2   Pre-Breakdown Phenomena in Gases   93
   4.3   Breakdown in Steady Uniform Fields   97
   4.4   Breakdown in Electronegative Gases and Their Mixtures   106
   4.5   Effects of Gas Parameters   109
   4.6   Breakdown in Nonuniform DC Fields   109
   4.7   Breakdown in Nonuniform AC Fields   112
   4.8   Breakdown Under Impulse Voltages   112
   4.9   High-Frequency Breakdown   117
         References   120

5. **The Corona Discharge**   123
   M. Khalifa

   5.1   Introduction   123
   5.2   Mechanism of Corona Discharge   123
   5.3   The Corona Onset Level   133
   5.4   Corona Power Loss   136
   5.5   Corona Noise   139
   5.6   Some Industrial Applications of Corona   140
         References   142

6. **The Arc Discharge**   145
   M. Khalifa

   6.1   Introduction   145
   6.2   Arcs in Circuit Breakers   145
   6.3   Regions of the Arc   147

Contents xiii

6.4	Energy Balance in a Steady Arc	150
6.5	Steady-State Arc Characteristics	152
6.6	Magnetic Phenomena in Arcs	154
6.7	Dynamic Arc Characteristics	154
6.8	The Arc as a Circuit Element	156
6.9	Arc Interruption	158
6.10	Arc Erosion	161
6.11	Applications	162
	References	163

7. **Insulating Liquids** 165
   M. Khalifa

7.1	Introduction	165
7.2	Types of Oils	165
7.3	Physical Properties	168
7.4	Chemical Properties	169
7.5	Electrical Properties	171
7.6	Theories of Dielectric Breakdown	174
7.7	Factors Influencing the Dielectric Strength of Insulating Liquids	179
7.8	Aging	182
7.9	Tests on Insulating Liquids	183
7.10	Reconditioning of Insulating Liquids	184
	References	185

8. **Solid Insulating Materials** 189
   M. Khalifa

8.1	Introduction	189
8.2	Electrical Properties	189
8.3	Thermal, Mechanical, and Chemical Properties	202
8.4	Dielectric Breakdown	205
8.5	Insulating Materials	213
8.6	Active Dielectrics	214
	References	215

Part II

9. **High-Voltage Busbars** 219
   A. El-Morshedy

9.1	Introduction	219
9.2	Busbar Arrangements	220
9.3	Rigid Bus Versus Strain Bus	225

9.4	Bus Conductor Materials	225
9.5	Busbar Clearances	226
9.6	Thermal Rating	231
9.7	Mechanical Stresses on Busbar Conductors	236
9.8	Insulators	239
	References	241

10. Gas-Insulated Switchgear 243
H. Anis

10.1	Introduction	243
10.2	Chemical and Physical Properties of SF_6	243
10.3	Layout of Gas-Insulated Switchgear	245
10.4	Components of GIS	251
10.5	Compressed Gas-Insulated Cables	256
10.6	Dimensioning of Compressed Gas Enclosure	258
10.7	Factors Affecting Insulation Strength	261
10.8	Electromagnetic Compatibility in GIS Substations	270
10.9	On-Site Testing	270
10.10	Maintenance	272
	References	273

11. Circuit Breaking 275
R. Radwan

11.1	Introduction	275
11.2	Arc Interruption	275
11.3	Circuit Breaker Rated Quantities	276
11.4	Switched Currents and Circuits	279
11.5	Rate of Rise of Restriking Voltage	287
11.6	Types of Circuit Breakers	288
	References	304

12. High-Voltage Cables 305
M. Abdel-Salam

12.1	Introduction	305
12.2	Cable Construction	306
12.3	Types of Cables	310
12.4	Cable Insulation Characteristics	314
12.5	Sheath Phenomena	315
12.6	Armor Loss	319
12.7	Cable Constants	319
12.8	Electric Fields in Cable Insulation	320
12.9	Current-Carrying Capacity	321

Contents xv

12.10	Jointing and Terminating High-Voltage Cables		323
12.11	Locating Faults in Cables		324
12.12	Superconducting Cables		326
12.13	Cable Testing		329
	References		329

13. Grounding Systems 331
A. El-Morshedy

13.1	Introduction	331
13.2	Resistance of Grounding Systems	332
13.3	Impulse Impedance of Grounding Systems	341
13.4	Principles of Design of Substation Grounding System	343
13.5	Neutral Grounding	346
13.6	Grounding of Gas-Insulated Substations	347
13.7	Transmission-Line Tower Grounding	348
13.8	Safety Ground	349
13.9	Measurement of Ground Resistance	350
13.10	Measurement of Earth Resistivity	352
13.11	Ground Resistance Meters	353
	References	354

14. Overvoltages on Power Systems 357
H. Anis

14.1	Introduction	357
14.2	Types of Overvoltages	357
14.3	Lightning Overvoltages	357
14.4	Switching Overvoltages	360
14.5	Temporary Overvoltages	364
14.6	Traveling Waves	366
14.7	Control of Overvoltages	374
14.8	Statistical Characteristics of Overvoltages	377
	References	379

15. Insulation Coordination 381
H. Anis

15.1	Introduction	381
15.2	Insulation Performance Under Voltage Stresses	382
15.3	Flashover of External Gaps in Parallel	388
15.4	Principles of Insulation Coordination	392
15.5	Overvoltage Protective Devices	398

15.6		Insulation Coordination of Gas-Insulated Substations	403
		References	408

16. High-Voltage Generation — 409
M. Abdel-Salam

16.1	Introduction	409
16.2	Generation of High Alternating Voltages	410
16.3	Generation of High DC Voltages	416
16.4	Generation of Impulse Voltages	428
16.5	Generation of Switching Impulses	440
16.6	Generation of Impulse Currents	444
	References	446

17. High-Voltage Measurements — 449
H. Anis

17.1	Introduction	449
17.2	Sphere Gaps	449
17.3	Peak Voltmeters	458
17.4	Electrostatic Voltmeters	459
17.5	Voltage Transformers	461
17.6	Ammeter in Series with a High Impedance	463
17.7	Potential Dividers for AC and DC	463
17.8	Dividers for Impulse Voltages	469
17.9	The Measuring Cable	474
17.10	Measurement of Transient Currents	477
17.11	Measurement of Electric Field	480
17.12	Sources of Error in High-Voltage Measurements	485
	References	486

18. Testing Techniques — 487
R. Radwan and A. El-Morshedy

18.1	Introduction	487
18.2	Classification of Tests	488
18.3	Test Voltages	488
18.4	Tests with Impulse Currents	494
18.5	Safety Precautions in the Laboratory	495
18.6	Nondestructive Testing	497
18.7	Circuit Breaker Testing	505
18.8	Cable Testing	507
18.9	Testing of Power Transformers	508

18.10	Testing of Surge Arresters	509
18.11	Testing of Insulators	510
	References	511

Index 513

Contributors

M. Abdel-Salam, M.Sc., Ph.D., MIEE, SMIEEE, A.V.Fel. Professor, Electrical Department, Faculty of Engineering, Assiut University, Assiut, Egypt

H. Anis, M.Eng., Ph.D., SMIEEE Professor of H.V. Engineering, Electrical Engineering Department, Faculty of Engineering, Cairo University, Giza, Egypt

A. El-Morshedy, M.Sc., Ph.D., MIEEE Professor of H.V. Engineering, Electrical Engineering Department, Faculty of Engineering, Cairo University, Giza, Egypt

M. Khalifa, M.Sc., Ph.D., FIEE, FIEEE Professor of H.V. Engineering, Electrical Engineering Department, Faculty of Engineering, Cairo University, Giza, Egypt

R. Radwan, Ph.D., MIEEE Professor of H.V. Engineering, Electrical Engineering Department, Faculty of Engineering, Cairo University, Giza, Egypt

High-Voltage Engineering

1
Introduction

M. KHALIFA *Cairo University, Giza, Egypt*

It was about a century ago when the first high-voltage transmission line ever built was used to deliver about 100 kW at 2 kV over a distance of 20 km near Cerchi in Italy. The load was exclusively lighting of public buildings. Since then very many other applications for electric energy have been discovered. At home, in industry and commerce, in hospitals, and on the road, the applications of electric energy are just too numerous to count. Its greatest advantage is that it is very easily transmittable and convertible to other forms of energy. In almost every country, electric energy is now being utilized in everyday life at an ever-increasing rate. The average annual consumption of electric energy per capita is a recognized criterion for a country's degree of technical development and standard of living. The annual rate of increase in electric energy consumption in North America and Europe is now about 2 to 4%. In developing countries the rate is much higher.

To meet the ever-increasing demand for electricity, larger and larger power stations are being planned, built, and commissioned for efficient utilization of water power, conventional fuel, and nuclear fuel. It is now quite common to find 1000 MW being generated in a single power station (Fig. 1.1) (Chrobak and Pollak, 1986). Such power stations are designed to be appropriately located in terms of fuel handling and cooling-water supply, often at long distances from areas of concentrated electricity consumption.

1.1 High-Voltage Lines

The heavier the power to be transmitted and the longer the transmission distance, the higher the line voltage will have to be. This is determined by several technical and economical factors, including efficiency, voltage drop, system stability, and number of circuits necessary for a secure supply of energy. For example, to transmit

Fig. 1.1 Nuclear power station. (Courtesy of Aesa Brown Boveri AG, Switzerland.)

10 to 100 GW of power over a distance of about 500 km, the suitable line voltage could be either 750 kV ac or ±600 kV dc.

Therefore, with the ever-growing need to transmit greater amounts of power longer distances, transmission lines with higher and higher voltages have been built in many countries over the years. Figure 1.2 shows the world record voltages for ac and dc lines. Ultrahigh voltage (UHV) of about 1500 kV ac may be in service before long (Thorén, 1983).

Introduction

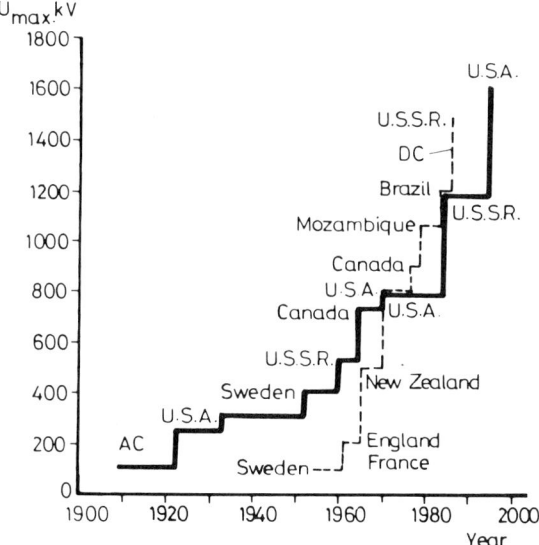

Fig. 1.2 The highest voltage used for ac and dc transmission in the world, showing the increase over time. The dc voltages are between the poles of bipolar lines. [From Thorén (1983).]

In certain situations it may be preferable to use dc rather than ac for power transmission. DC transmission lines are more economical to build, suffer fewer voltage drops and losses, are most suitable for interconnecting asynchronous networks, and do not impose an extra burden on the switchgear of connected ac networks. On the other hand, they do not carry reactive power, and their terminal inverter stations produce harmonics that have to be carefully filtered.

1.2 POWER NETWORKS

Rather than having every power station or small group of power stations look after its own loads, tie lines connect power stations into an integrated network over an entire country or a group of neighboring countries. These interconnections greatly improve the reliability and economy of the system and make available the benefits of diversity among the peaks of the component loads. The extra-high voltage (EHV) grid covering Western Europe operates at 400 kV ac and is interconnected with that of Eastern Europe at substations in Austria (Putz, 1987). The United States and Canada are interconnected at voltage levels of about 750 kV ac and ±400 kV dc. The partial schematic of a typical power network in Figure 1.3 shows the main components of an outdoor 500-kV substation; Figure 1.4 shows a partial view (Bethmann and Boehle, 1985).

As electrical engineers are aware, the design, installation, and operation of high-voltage equipment involve numerous technical problems. For instance, a higher overhead line voltage involves larger separations among power conductors and clearances to their tower structures and to ground. As shown in Figure 1.5, an air gap has to be increased three to four times if its withstand votlage is to be doubled from 1 million volts to 2 million volts. This probably entails a huge increase in the cost of the structure and of the system insulation in general.

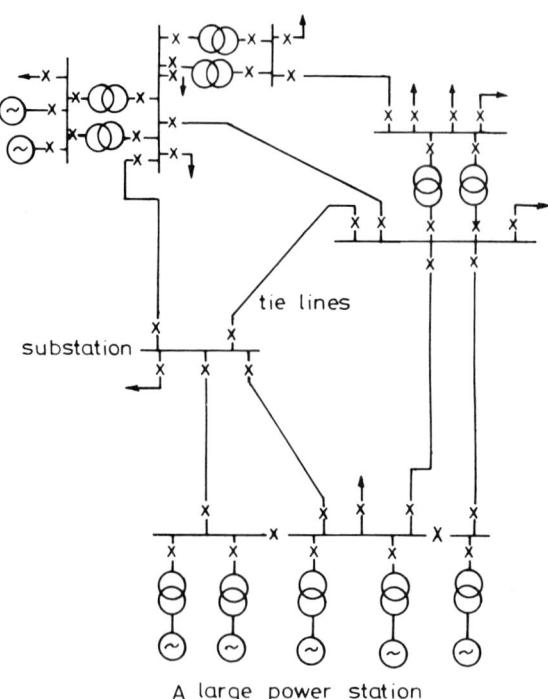

Fig. 1.3 Schematic of a power network showing busbars, transformers, generators, breakers, and feeders.

Fig. 1.4 A 500-kV outdoor substation. (Courtesy of South Wales Switchgear.)

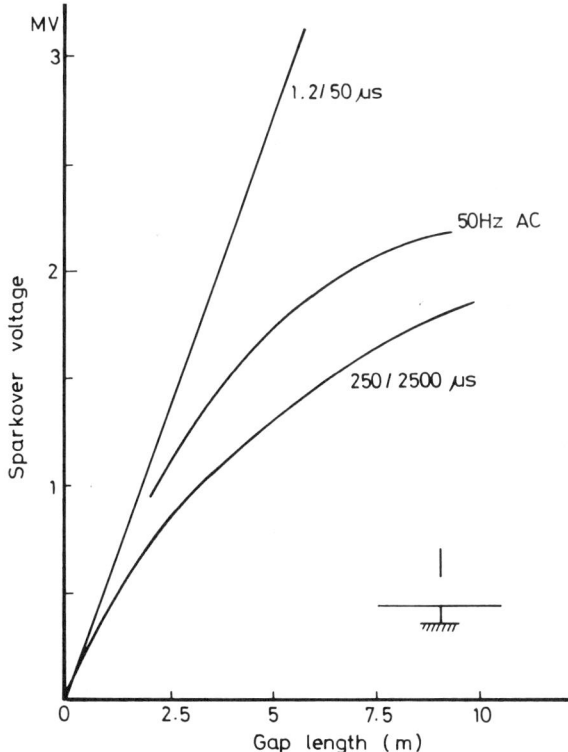

Fig. 1.5 Breakdown voltages of a rod-to-plane gap under ac, positive lightning, and switching impulse. [From Meek and Craggs (1978).]

1.3 INSULATION OF ELECTRICAL EQUIPMENT

Equipment insulation is stressed both by continuous normal voltage and by relatively slow dynamic and fast transient overvoltages. For insulation to withstand the various stresses, it has to be carefully designed, coordinated, and properly tested. The test voltages have to exceed the normal rated voltages by some factor of safety. The lightning-impulse test voltage is about 10 times the rated ac voltage for low-voltage equipment. The corresponding ratio is only about 3 for UHV equipment (IEC, 1976), for the sake of system economy.

To achieve an economically and technically sound design for equipment insulation, adequate information needs to be acquired. This applies to equipment in power as well as electronics systems, no matter what the size. It applies equally to components and to devices. We should be able to calculate and measure system overvoltages. We should also be able to translate these voltage magnitudes and durations to electrical stresses across the vulnerable parts of equipment insulation and accordingly, to dimension them. This involves computing and, in some cases, measuring electric fields. Additional necessary information includes the physics of breakdown of solid, liquid, and gaseous insulating materials, and the design, service performance, and testing of insulating components.

Other areas of high-voltage engineering include the design of high-voltage power supplies for various aerospace and ground applications; high-voltage high-frequence sources for radio transmission; design, operation, and testing problems for HV, EHV, and UHF overhead lines, cables, and other equipment, and their insulation coordination; various industrial, medical, and other applications of gas discharge; and other high-voltage phenomena. The field of high-voltage engineering covers all these areas. In the following chapters we deal with the salient theoretical bases of high-voltage engineering and correlate them with the many issues encountered in practice.

REFERENCES

Bethmann, J. and Boehle, B. (1985). *Brown Boveri Review*, 72: 464–471.

Chrobak, E. and Pollak, M. (1986). *Brown Boveri Review*, 73: 224–230.

IEC (1976). *Insulation Coordination*, Publication 71, International Electrotechnical Commission, Geneva.

Meek, J. and Craggs, J. (1978). *Electrical Breakdown of Gases*, John Wiley & Sons, Inc., New York.
Putz, W. (1987), *Electra, 114* (October):21–24.
Thorén, B. (1983). *ASEA Journal, 56*(3):4–7.

Part I

2
Electric Fields

M. ABDEL-SALAM *Assiut University, Assiut, Egypt*

2.1 INTRODUCTION

Knowledge of electric fields is necessary in numerous applications in the design and operation of electrical and electronic equipment. To name just a few of these, it is necessary:

1. For the design of insulation and for assessing electrical stresses in high-voltage sources, machine windings, and cables as well as within electronic components.
2. In the study of gas discharges
3. In the design of UHV substations and the effects of electric fields in their vicinity
4. In industrial applications such as electrostatic filters and xerography

In this chapter we survey how the electric field is determined for various gap geometries. For space-charge-free fields, we seek solutions to Laplace's equation that satisfy the boundary conditions. In the presence of a space charge, a solution to Poisson's equation is obtained. In some gap geometries, the electric fields can simply be expressed analytically in a closed-form solution; in others, the electric field problem is complex because of the sophisticated boundary conditions, including media with different permittivities and conductivities. In such cases we must resort to experimental modeling or numerical techniques.

2.2 ANALYTICAL CALCULATION OF SPACE-CHARGE-FREE FIELDS

The general governing equation for space-charge-free fields is Laplace's equation,

$$\nabla^2 \phi = 0$$

where ϕ is the electrical potential. Laplace's equation has a closed-form solution for geometries with great symmetry and with one-dimensional fields, as explained in detail in books on electromagnetic fields (Shen and Kong, 1983).

2.2.1 Simple Geometries

Configurations consisting of spheres or cylinders occur frequently in high-voltage equipment and the electric field E can easily be calculated for many of these configurations. Table 2.1 contains formulas for the potential ϕ, field E, maximum field E_{max}, and field efficiency factor η:

$$\eta = \frac{E_{av}}{E_{max}} = \frac{V}{dE_{max}} \tag{2.1}$$

where $E_{av} = V/d$ (d is the electrode spacing and V is the applied voltage).

For other cases with electrode surfaces that could be expressed by analytical relations, the field calculation could be simplified by the well-known transformation of coordinates. For example, with two confocal paraboloidal electrodes, the field at any point spaced x from the tip of the inner paraboloid along its axis is given by

$$E = \frac{V}{\ln(F/f)} \frac{1}{x+f} \tag{2.2}$$

where f and F are the focal lengths of the inner and outer paraboloids, respectively. For a needle-plane gap with needle-tip radius $r \ll h$, the gap length

$$E(x) = \frac{2V}{\ln(4h/r)} \frac{h}{h(2x+r) - (x)^2} \tag{2.3}$$

2.2.2 Transmission-Line Conductors to Ground

Consider a multiconductor transmission line of very long conductors with potentials V_i and charges q_i per unit length (i = 1, ... , n) strung parallel to the ground plane (Fig. 2.1). The effect of ground is usually simulated by mirror images of the charges q_i. This guarantees that the potential at every point on the plane is zero.

Electric Fields

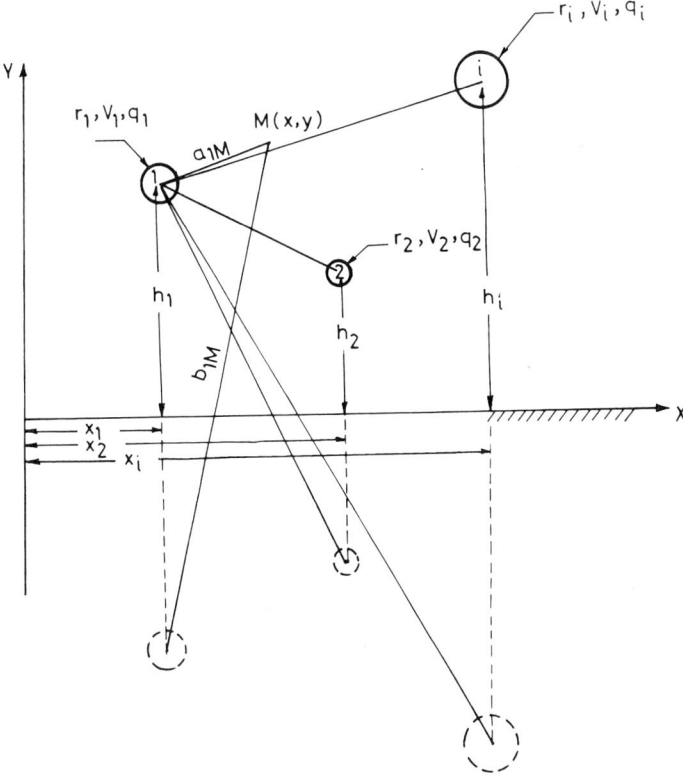

Fig. 2.1 Transmission-line conductors over the ground plane. h_i and r_i are the height above ground and radius of the ith conductor.

The potential at an arbitrary point M(x,y) due to conductor 1 and its image (Fig. 2.1) is given by

$$\phi_{M1} = \frac{q_1}{2\pi\xi_0} \ln \frac{b_{1M}}{a_{1M}} \qquad (2.4)$$

where a_{1M} and b_{1M} are the distances shown in Figure 2.1. Where the conductor separations and heights above ground are much larger than their diameters, a useful approximation is to consider that the line charge is located at the axis of each conductor.

Table 2.1 Potential, Field, Maximum Field, and Ratio η of Average to Maximum Fields

Configuration	Potential	E
Concentric spheres	$\phi(r) = \dfrac{a}{b}\dfrac{V}{r}(b-r)$	$E(r) = \dfrac{Vba}{r^2(b-a)}$
Coaxial cylinders	$\phi(r) = \dfrac{V \ln(b/r)}{\ln(b/a)}$	$E(r) = \dfrac{V}{r \ln(b/a)}$
Separated equal spheres	Two-dimensional field	Two-dimensional field
Equal parallel cylinders	Two-dimensional field	Two-dimensional field

E_{max}	$\eta = E_{av}/E_{max}$	Field of application
$E_{max} = \dfrac{Vb}{a(b-a)}$	$\dfrac{a}{b}$	Spherical capacitors, capacitance representation of the dome of a Van de Graaff generator and the structure of the room
$E_{max} = \dfrac{V}{a \ln(b/a)}$	$\dfrac{a \ln(b/a)}{b-a}$	Cable bushing and GIS
$E_{max} = \dfrac{V}{2R}$ if $d \gg R$	$\dfrac{2R}{d}$ if $\dfrac{d}{R} \gg 1$	Sphere gap for HV measurements, etc.
$E_{max} = \dfrac{V}{2R \ln[(d+2R)/R]}$	$\dfrac{2R \ln(d/R)}{d}$ if $\dfrac{d}{R} \gg 4$	Overhead transmission-line arrangements

The field at point M due to conductor 1 and its image is the gradient of ϕ_{M1}, that is,

$$E_{M,1} = -\frac{\partial \phi_{M1}}{\partial x}\bar{u}_x - \frac{\partial \phi_{M1}}{\partial y}\bar{u}_y$$

$$= \frac{q_1(x-x_1)}{2\pi\varepsilon_0}\left(\frac{1}{a_{1M}^2} - \frac{1}{b_{1M}^2}\right)\bar{u}_x$$

$$+ \frac{q_1}{2\pi\varepsilon_0}\left(\frac{y-h_1}{a_{1M}^2} - \frac{y+h_1}{b_{1M}^2}\right)\bar{u}_y \quad (2.5)$$

Thus the resultant potential at M due to all the transmission-line conductors is given by

$$\phi_M = \sum_{i=1}^{i=n} \frac{q_i}{2\pi\varepsilon_0} \ln \frac{b_{iM}}{a_{iM}} \quad (2.6)$$

Subsequently, the resultant field at M has the components

$$(E_x)_M = \sum_{i=1}^{i=n} \frac{q_i(x-x_i)}{2\pi\varepsilon_0}\left[\frac{1}{(x-x_i)^2 + (y-h_i)^2}\right.$$

$$\left. - \frac{1}{(x-x_i)^2 + (y+h_i)^2}\right] \quad (2.7)$$

$$(E_y)_M = \sum_{i=1}^{i=n} \frac{q_i}{2\pi\varepsilon_0}\left[\frac{y-h_i}{(x-x_i)^2 + (y-h_i)^2}\right.$$

$$\left. - \frac{y+h_i}{(x-x_i)^2 + (y+h_i)^2}\right] \quad (2.8)$$

If the point M is placed on the first conductor, then $\phi_M = V_1$, $b_{1M} = 2h_1$, $a_{1M} = r_1$, and equation (2.6) takes the form

$$V_1 = \frac{q_i}{2\pi\varepsilon_0} \ln \frac{2h_1}{r_1} + \frac{q_2}{2\pi\varepsilon_0} \ln \frac{b_{12}}{a_{12}} + \frac{q_3}{2\pi\varepsilon_0} \ln \frac{b_{13}}{a_{13}}$$

$$+ \cdots + \frac{q_n}{2\pi\varepsilon_0} \ln \frac{b_{1n}}{a_{1n}} \qquad (2.9)$$

For brevity, equation (2.9) may be written in the form

$$V_1 = q_1 P_{11} + q_2 P_{21} + q_3 P_{31} + \cdots$$

$$V_2 = q_1 P_{12} + q_2 P_{22} + q_3 P_{32} + \cdots$$

$$\vdots \qquad (2.10)$$

$$V_i = q_1 P_{1i} + q_2 P_{2i} + q_3 P_{3i} + \cdots$$

or

$$[V] = [P][q]$$

where $P_{ij} = (1/2\pi\varepsilon_0)\ln(b_{ij}/a_{ij})$ is the potential coefficient in meters per farad. It is evident that $P_{ij} = P_{ji}$ and that they are positive quantities, independent of the sign of charge and potential. Equations (2.10) make up the first set of Maxwell's equations.

Transmission Line with Single Conductor

For a monopolar dc line with a single conductor, equation (2.9) reduces to

$$V = qP = \frac{q}{2\pi\varepsilon_0} \ln \frac{2h}{r} \qquad (2.11)$$

It could easily be shown that the equipotentials around the conductor take the form of circles. The field at the conductor surface reaches its maximum value at the point facing the ground as expected and as could easily be evaluated.

$$E_{max} = \frac{V}{r \ln(2h/r)} \qquad (2.12)$$

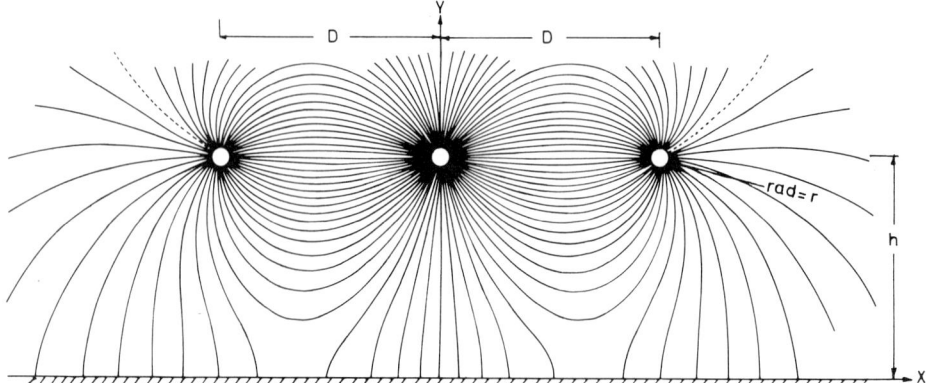

Fig. 2.2 Three-phase transmission line with a sketch of field lines at the instant when the center phase voltage is maximum.

The field strength at the conductor surface is decided primarily by the conductor radius and whether the conductors are bundled. For bipolar dc lines and for three-phase ac lines (Fig. 2.2), the spatial field variations could be calculated using equations (2.7) and (2.8). Knowledge of the maximum surface field for different conductor bundles is necessary for estimating the corona discharge phenomena on HV and EHV transmission lines.

Transmission Line with Bundle Conductors

In practical designs, the line conductor radius r, the separation d between the subconductors of a bundle, the separation D between phases, and their height h above ground would comply with the following relation:

$$r < d \ll D \text{ and } h \qquad (2.13)$$

The effects of the image charges could safely be neglected when evaluating the fields at the conductor surfaces. The magnitude and direction of the maximum conductor surface field could be evaluated using the line conductor charges and potential coefficients. These could, in turn, be evaluated in terms of the line voltage and geometry. For example, in case of a monopolar bundle 2, dc line of voltage V, equation (2.10) reduces to:

$$V = P_{11}q_1 + P_{12}q_2 = P_{21}q_1 + P_{22}q_2 \qquad (2.14)$$

Electric Fields

Table 2.2 Potential Coefficients of Equation (2.14) for a Bundle of Two Subconductors When Arranged Horizontally and Vertically

Horizontal bundle—2 (◄─o o─►)	Vertical bundle—2 $\begin{pmatrix} \uparrow \\ \circ \\ \circ \\ \downarrow \end{pmatrix}$
$P_{11} = \dfrac{1}{2\pi\epsilon_0} \ln \dfrac{2h}{r} = P_{22}$	$P_{11} = \dfrac{1}{2\pi\epsilon_0} \ln \dfrac{2(h + d/2)}{r}$
$P_{12} = \dfrac{1}{2\pi\epsilon_0} \ln \dfrac{\sqrt{4h^2 + d^2}}{d} = P_{21}$	$P_{12} = \dfrac{1}{2\pi\epsilon_0} \ln \dfrac{2h}{d} = P_{21}$
$\phantom{P_{12}} = \dfrac{1}{2\pi\epsilon_0} \ln \dfrac{2h}{d}$	if $h \gg d$
if $h \gg d$	$P_{22} = \dfrac{1}{2\pi\epsilon_0} \ln \dfrac{2(h - d/2)}{r}$
$q_1 = \dfrac{V}{P_{11} + P_{12}} = \dfrac{2\pi\epsilon_0 V}{\ln(2h/r) + \ln(2h/d)}$	$q_1 = \dfrac{(P_{22} - P_{12})V}{P_{11}P_{22} - P_{12}^2}$
$ = q_2$	$q_2 = \dfrac{(P_{11} - P_{12})V}{P_{11}P_{22} - P_{12}^2}$

The potential coefficients P_{11}, P_{12}, and P_{22} depend on the conductor arrangement in the bundle (Table 2.2). The maximum field strength E_{max} at either subconductor of the bundle is expressed as

$$E_{max} = \frac{V}{\ln(4h^2/rd)} \left(\frac{1}{r} + \frac{1}{d}\right) \qquad (2.15)$$

For three-phase ac or bipolar dc lines, the maximum surface field could be similarly calculated for bundles of two, three, and four subconductors (El-Arabaty et al., 1977).

Approximations

For quick estimation of the maximum surface field for overhead line conductors, some expressions were proposed by previous workers (e.g., Timascheff, 1975; Comber and Zaffanella, 1974). They suggested the following relation for the angular variation of the surface field E around the subconductor of a multiconductor bundle:

$$E_\theta = E_a \left[1 + \frac{2r}{C} (n - 1) \cos \theta \right] \qquad (2.16)$$

where E_a is the average field obtained from the well-known relation

$$E_a = \frac{q}{2\pi \varepsilon_0 r} \qquad (2.17)$$

with the parameters being defined as given in Figure 2.3.

2.2.3 Fields in Multidielectric Media

Usually a low-voltage or a high-voltage insulation system (e.g., a high-voltage bushing and a machine insulation) comprises more than one insulating material. The electric field distribution in a multidielectric insulation system can easily be explained in a parallel-plate capacitor with two layers of different permittivities ε_1 amd ε_2.

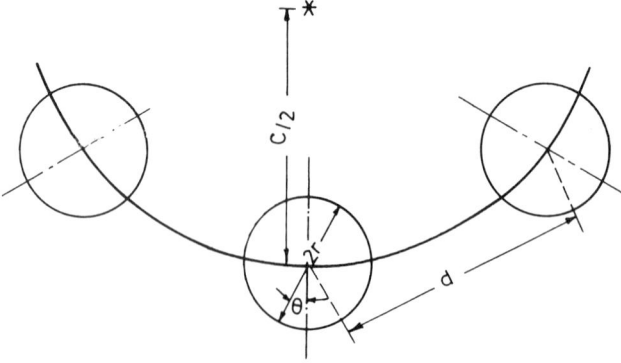

Fig. 2.3 Phase geometry of a bundle of n conductors.

Electric Fields

The electric displacement is the same in both layers; hence the electric field intensities E_1 and E_2 are related as $E_1/E_2 = \varepsilon_2/\varepsilon_1$. This means that partial replacement of the gas insulation with a solid material does not improve the dielectric strength of the whole. On the contrary, it becomes more heavily stressed, the higher the ε value of the inserted solid dielectric slab.

The constancy of the electric flux density at the interface of a multidielectric system makes it possible to increase the field uniformity by an appropriate choice of the dielectric permitivities. A typical example is the grading of insulation in coaxial cables.

Refraction Law of Electric Field

For ac voltage applications, free charges are absent at the interface between dielectrics and the polarization charges define the boundary conditions. Then the angle of incidence (θ_1) and refraction (θ_2) are related as follows:

$$\frac{\tan \theta_1}{\tan \theta_2} = \frac{E_{t1}/E_{n1}}{E_{t2}/E_{n2}} = \frac{E_{n2}}{E_{n1}} = \frac{\varepsilon_1}{\varepsilon_2} \qquad (2.18)$$

On the other hand, in dc voltage applications, accumulation of free charges at the interface takes place due to the differing conductivities of the materials (interfacial polarization). More is said about such cases in Section 2.5.

Equation (2.18) points out that the electric displacement flux lines penetrating from a dielectric of a high ε into one of a much lower ε are forced to leave the material nearly perpendicular to its surface. This means that the equipotential surfaces in the lower-permittivity dielectric are forced to be nearly parallel to the interface, and the dielectric of the much higher ε behaves almost like a conductor as $\varepsilon \longrightarrow \infty$.

2.3 EXPERIMENTAL ANALOGS FOR SPACE-CHARGE-FREE FIELDS

The infinitely long, smooth cylindrical conductors parallel to each other and to ground with separations much larger than their diameters represent overhead transmission lines. They are only special simple cases compared to such conductor arrangements as in multicore cables, gas-insulated switchgear, transformers, and machines. There, conductors take various shapes, their sizes are comparable to their separations, and more than one dielectric medium is involved.

For assessing the electric field distributions in such complex arrangements, sometimes in three dimensions, analytical methods are by no means suitable. Two other approaches are in use, experimental analogs and numerical techniques. The former is the subject of this section and the latter is covered in Section 2.4.

The potential distribution in conductive media in current equilibrium conditions satisfies Laplace's equation, the same as the electric fields in space-charge-free regions. This fact makes it possible to obtain solutions to many difficult electrostatic field problems by constructing an analogous potential distribution in a conductive medium, where the potential and field distributions could be measured directly. The conductors and insulation arrangements can be represented using an electrolytic tank, a sheet of semiconducting paper, or by a mesh of resistors.

2.3.1 Electrolytic Tank

This method has been widely used for decades. Equipotential boundaries are represented in the tank by specially formed sheets of metal. For example, a single dielectric problem such as a three-core cable may be represented using a flat tank as shown in Figure 2.4. Different permittivities are represented by electrolytes of different conductivities separated by special partitions. Otherwise, the tank base could be specially shaped. The conductance of the entire model is a scale model of the capacitance of the system being represented, care being taken to minimize the errors.

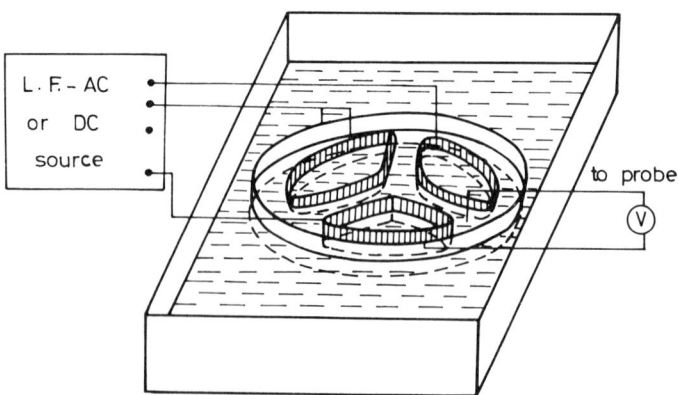

Fig. 2.4 Electrolytic tank model of a three-core cable represented at the instant when one core is at zero voltage, the same as the sheath.

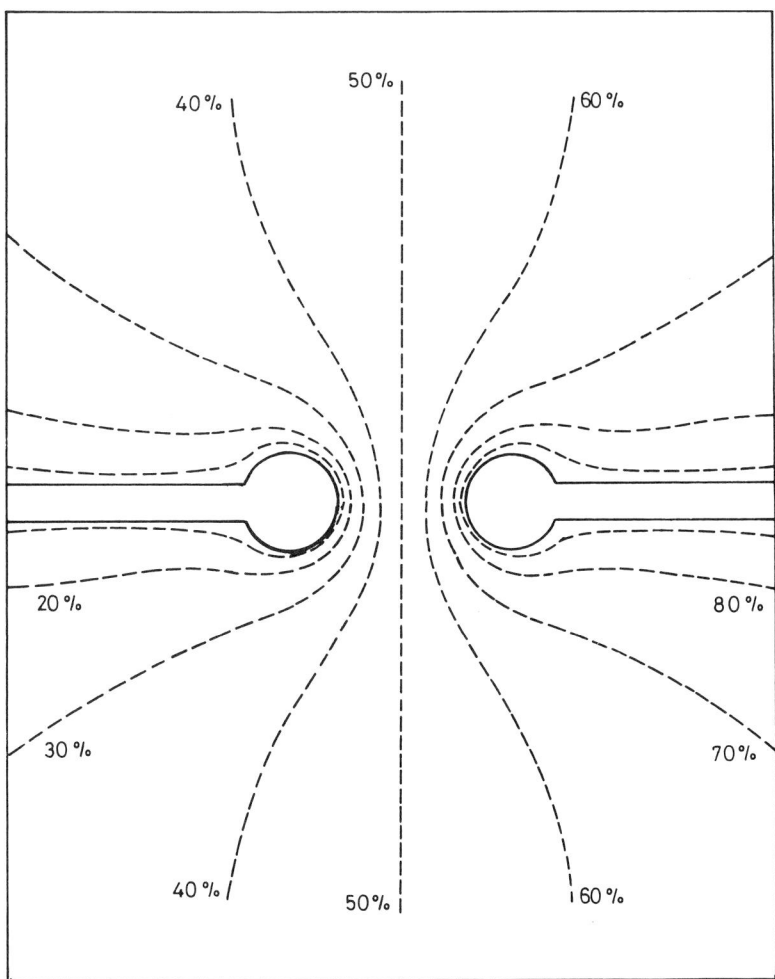

Fig. 2.5 Field plot between two spheres with shanks as plotted by a semiconducting-paper model.

2.3.2 Semiconducting Paper Analog

A somewhat less accurate but attractively simple alternative to the electrolyte is semiconductive paper. Its surface resistivity is of the order of 1 kΩ per square (Fig. 2.5). Errors in this method result among other things, from the nonhomogeneity of the paper resistivity, its dependence on the ambient humidity, and the contact resistance to electrodes.

2.3.3 Resistive-Mesh Analog

In this analog, the continuous field is replaced by a discrete set of points as depicted by a mesh of resistors (Fig. 2.6). Replacement of the continuous field by discrete resistors does, of course, introduce an error arising from the finite mesh size. This error may be reduced by reducing mesh size.

For the mesh point 0 of Figure 2.6, the continuity equation of current demands that

$$\frac{V_1 - V_0}{R} + \frac{V_2 - V_0}{R} + \frac{V_3 - V_0}{R} + \frac{V_4 - V_0}{R} = I_0 \qquad (2.19)$$

or

$$V_1 + V_2 + V_3 + V_4 - 4V_0 = I_0 R$$

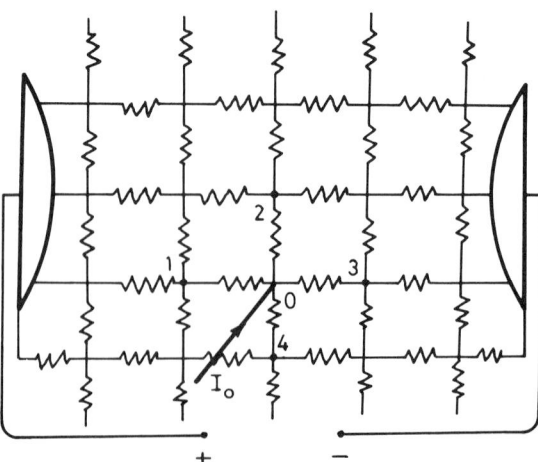

Fig. 2.6 Resistive mesh analog of the field pattern between two electrodes.

Electric Fields

Thus the resistive network analog can approximately represent Laplacian fields ($I_0 = 0$) as well as Poissonian fields ($I_0 \neq 0$). For $I_0 = 0$,

$$V = \frac{V_1 + V_2 + V_3 + V_4}{4} \tag{2.20}$$

Thus the potential at each mesh point is determined by the adjacent point potentials, which are in turn determined by others. Each potential value represents in fact an interpolation between those one step away from it, the same as in numerical calculations using finite differences (Section 2.4.4).

Due to discretization, simulation of electrostatic fields in the vicinity of curved surfaces of the electrodes is bound to be of reduced accuracy. Nevertheless, the analog could be used in three-dimensional field problems by extending the mesh grid in three dimentions. Time dependence may also be represented using impedance networks containing capacitors and inductors.

2.4 NUMERICAL COMPUTATION OF SPACE-CHARGE-FREE FIELDS

Several numerical techniques have recently been reported in the literature for solving Laplace's and Poisson's equations for the fields between complex electrode arrangements. Each numerical technique has advantages and drawbacks. Some techniques may complement others. They are discussed briefly below.

2.4.1 Successive Imaging Technique

The successive imaging technique is based on the concept of imaginary point or line charges located outside the region of field evaluation, but chosen such that their field within this region is identical to that of the induced charges on the electrode boundaries of the region. In two cases only is it possible to implement the successive imaging technique (Khalifa and Abdel-Salam, 1974a) when the region of field evaluation is a single-permittivity medium. The first case is the field between a long, thin wire with a charge q (C/m) and a parallel long, sizable conducting cylinder. The usual practice is to represent the cylinder by a line charge as an image within the cylinder, parallel to its axis and displaced from it by a distance:

$$\delta = \frac{r^2}{D} \tag{2.21}$$

The two charges on the wire and cylinder are equal and opposite. The field outside the cylinder could be computed in terms of the charges, which in turn are determined by the voltage applied to the cylinder.

The second case is that of a field between a point charge Q and a charged spherical conductor. The sphere may be represented by a point charge as an image equal to $-Qr/D$, located a distance $\delta = r^2/D$ away from its center. The imaging technique cannot be extended to multidielectric media or to three-dimensional fields.

Successive Imaging Among Cylinders

A system of parallel cylindrical conductors, as in gas-insulated switchgear (GIS) is used as an example to explain how the image charges are determined. For the system of conducting cylinders shown in Figure 2.7a, their respective potentials are V_1, V_2, . . . , V_4 and their surface charges are q_1, q_2, . . . , q_4. Of course, if they are placed near the ground or its equivalent, each conductor will also have its image (see Fig. 2.1).

Now, as discussed above, each cylinder would present an equal and opposite image for each of the charges existing outside it (Fig. 2.7b). The image locations are calculated according to equation (2.21). It is noticed that each of the n cylindrical conductors has (n − 1) images. If the ground is considered, the number of images inside each cylinder would be (2n − 1). These line charges are used to compute the field in the space *outside* the cylinders. The computations would not yield an equipotential surface exactly coinciding on each cylinder with its respective voltage. Therefore, to reduce errors in the results, a next-imaging process is performed.

At any step in the imaging process, each of the images in a conductor obtained by the preceding step is resolved into (2n − 1) new images. The largest distance of each of these (2n − 1) new images from the image it replaces can be used as a criterion to determine whether the successive imaging process should be continued. A large distance of this type indicates that further imaging is necessary to improve accuracy.

If the imaging process is continued uniformly for all the conductors, the number of images in any conductor after the mth imaging process will be $(2n - 1)^m$. The system of images starts to form a branching network in each conductor, as suggested in Figure 2.7c.

Let N line charges represent a system of N conductors (N = 2n) at various potentials. Suppose that the number of line charges found to represent the system after termination of the imaging process is M. Choosing N points, one on each conductor surface,

Electric Fields

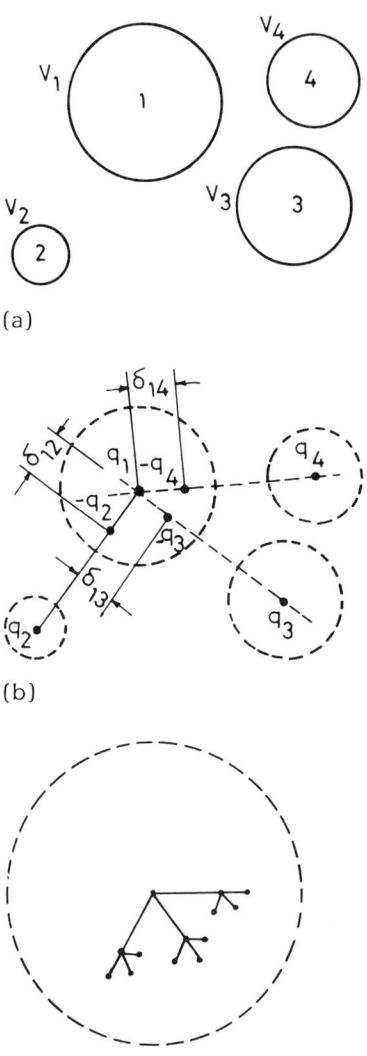

Fig. 2.7 (a) Group of conductors to be modeled; (b) first-order images (i.e., of charges q_2, q_3, and q_4 inside cylinder 1; (c) locating higher-order images.

to be the N points of known potentials, the following equation needs to be solved for the unknown charge density vector [q']:

$$[V] = [P'] [q'] \qquad (2.22)$$

where [V] is an N × 1 known vector of potentials of the N conductors, [P'] is an N × 1 known vector of potential-coefficient matrix, and [q'] is an M × 1 vector of all line charges used in the system. However, the M line charges representing the system are linearly related to the original N line charges by

$$[q'] = [G] [q] \qquad (2.23)$$

where [G] is an M × N matrix evaluated during the imaging process.

The electric potential and field may be calculated at any point outside the conductor boundaries as

$$\phi(a) = \sum_{i=1}^{i=M} P'_i q'_i \qquad (2.24)$$

$$E(a) = \frac{1}{2\pi\varepsilon_0} \sum_{i=1}^{i=M} \frac{q'_i}{r_{ia}} \bar{u}_{ia} \qquad (2.25)$$

where r_{ia} is the distance from q'_i to the point a and $\bar{u}_{ia}$ is a unit vector along the direction from q'_i to a. As a practical application, successive imaging technique has been used to calculate equipotential contours for an overhead transmission line with two circuits and two ground wires (Fig. 2.8).

Successive Imaging Between Spheres

Given two spheres located near, and isolated from, each other, with potentials V_1 and V_2, the same imaging process as that described above for parallel cylinders can easily be adopted here with n = 2, but only if the spheres are far from ground.

The advantages characterizing the successive imaging technique are the well-defined coordinates of the image charges, and the accuracy, which can be improved by increasing the number of successive images. Recently, successive imaging has been combined with the charge simulation technique to compute the electric field of monopolar, homopolar, and bipolar conductor bundles (Abdel-Salam and El-Mohandes, 1987). Details are given in Section 2.4.3.

Electric Fields 29

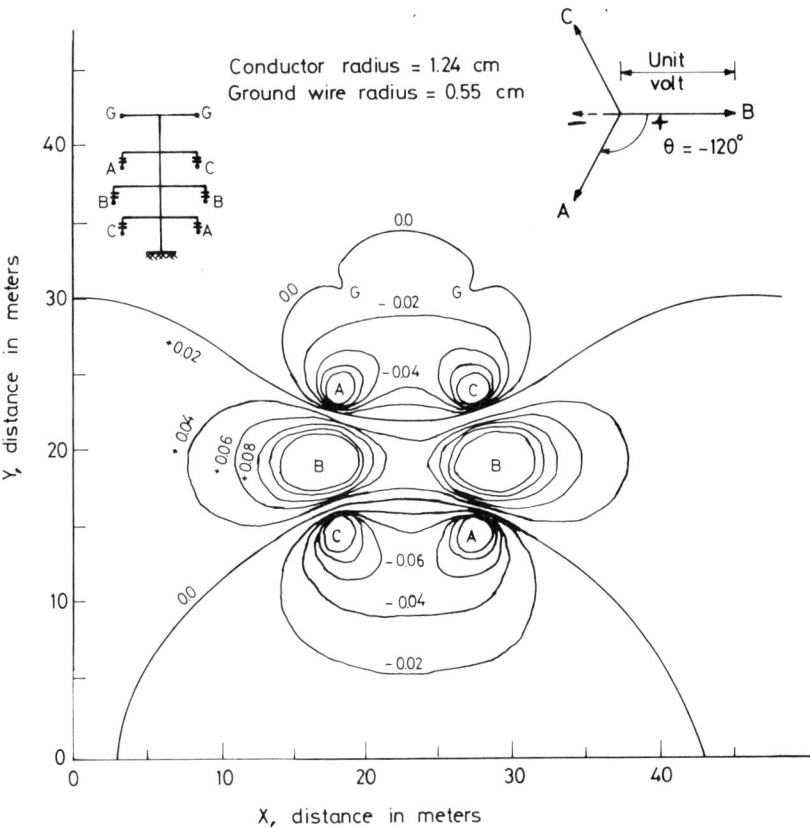

Fig. 2.8 Equipotential contours plotted for a double-circuit overhead test line using a successive-imaging technique.

2.4.2 The Dipole Method

The dipole method was developed by Thanassoulis and Comsa in 1971 to calculate the electric field in the vicinity of bundle conductors. It is an extension of the successive imaging technique described above (Abdel-Salam and El-Mohandes, 1989).

To establish this model for a bundle of n subconductors, each of them with (2n − 1) image charges, the effect of earth is included. According to equation (2.21), the n images inside each subconductor caused by the effect of ground can safely be assumed concentrated at the subconductor axis. They are of the same polarity as the subconductor itself. The other images, corresponding to the other members of the bundle, will have the opposite polarity, and their displacements from the subconductor axis are significant.

A dipole is composed of each of the negative charges together with one of the positive charges at the subconductor axis. The dipoles with the remaining axial charge can be used in computing the field distribution around the bundle. The accuracy of the method is satisfactory for practical separations exceeding 10 times the subconductor radius.

2.4.3 Charge-Simulation Technique

The charge-simulation technique can be employed successfully for the computation of an electric field between electrodes in a medium where one or more dielectrics are involved. The technique is described for single- and two-dielectric arrangements. The technique is discretization of the integral-equations technique explained in Section 2.4.8.

Single-Dielectric Arrangements

In this case the distributed charges on the surfaces of the stressed electrodes are replaced by a number (n) of fictitious discrete charges arranged inside the electrodes, that is, outside the space in which the field is to be computed (Singer et al., 1974). The number of charges and their coordinates are arbitrary. However, the larger the number of charges, the higher the possible accuracy of computation. The magnitudes of these discrete charges are to be such that under their integrated effect, each electrode has its assigned potential as checked at every contour point on its surface, that is,

$$\sum_{j=1}^{j=n} P_j q_j = V \qquad (2.26)$$

where q_j is the discrete charge and P_j is the associated potential coefficient.

Checking could be at a number m of contour points selected equal to or exceeding the number of discrete charges (n) and used for setting a group of simultaneous equations (Fig. 2.9):

$$[P][q] = [V] \qquad (2.10)$$

Thus, the unknown charges q_j could be evaluated.

Electric Fields 31

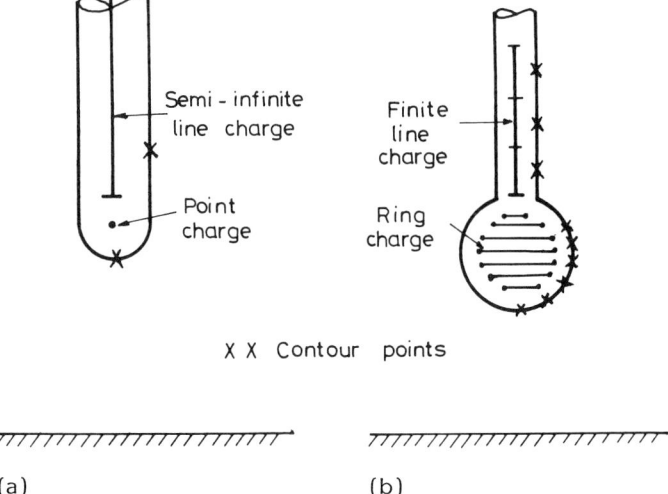

Fig. 2.9 Arrangement of simulation charges and contour points: (a) hemispherically capped rod, (b) sphere with cylindrical shank.

The charges simulating a given electrode could be chosen as point charges, ring charges, or as finite, semi-infinite, or infinite line charges. This choice should suit the shape of the electrode being simulated. For example, spherical electrodes can easily be simulated by point charges. For fields with axial symmetry, ring charges centered on the axis are an effective method of simulation. Finite and infinite line charges have commonly been used to simulate the charge on the surfaces of cylindrical conductors. Other charge shapes, such as oblate spheroid, prolate spheroid, axihyperbola, and elliptic cylinder, have been proposed. Shell and annular plate charges have also been suggested (Mukherjee and Roy, 1983).

In terms of the calculated charges q_j, the electric field at any point in space around the electrodes can be determined:

$$\bar{E} = E_x \bar{u}_x + E_y \bar{u}_y = \sum_{j=1}^{j=n} f_j q_j \qquad (2.27)$$

The potential and field coefficients for the different discrete charges are reported elsewhere (Singer et al., 1974).

For sufficient accuracy, effort should be devoted to choosing the proper number and location of simulation charges. The calculated potential cannot easily equal V at every point on the electrode surface under the effect of the simulation charges. Therefore, the magnitudes of the charges q_j are chosen such that the calculated potentials at a large number of points $m > n$ deviate only slightly from the actual potential V. The deviations are minimized by the least squares technique (Abdel-Salam and Ibrahim, 1977). This was found to improve considerably the stability of point matching. Indeed, the choice $m > n$ represents a step forward in improving the conventional charge-simulation technique. The improved technique showed better accuracy than the conventional one, even with a smaller number of simulation charges (Khalifa et al., 1975).

In another version of the charge-simulation technique, instead of the electrode charges that surface field values on the electrodes are first evaluated from the known voltages (V) of these electrodes by solving a set of simultaneous equations analogous to (2.10). The values of charges on the electrode surface are calculated from the field values there. The method has been applied (Foo and King, 1976) successfully to symmetrical and asymmetrical overhead bundled transmission lines. This version of the charge-simulation technique has the advantage of yielding surface fields directly with simple computations.

Two-Dielectric Arrangements

In this case dipoles are aligned by the electric field and compensate for each other in the bulk of each dielectric, leaving a net surface charge on the boundary between the dielectrics. This surface charge can be simulated by discrete charges on both sides of the dielectric boundary.

Figure 2.10 shows, for example, the cross section of a metal electrode with an applied voltage (V) meeting dielectrics I and II. At the electrode, there are n_e charges and contour points, with n_1 points on the interface between the electrode and dielectric I and $(n_e - n_1)$ points on the interface between the electrode and dielectric II. The n_e charges contribute to the potential and field in both dielectrics. At the dielectric boundary, there are n contour points with n discrete charges inside the surface of the dielectric I, contributing to the potential and field in the dielectric II, and vice versa. In total, there are $(n_e + n)$ contour points and $(n_e + 2n)$ discrete charges (Fig. 2.10).

The system of simultaneous equations required for determining the unknown discrete charges is formulated according to the following boundary conditions:

Electric Fields 33

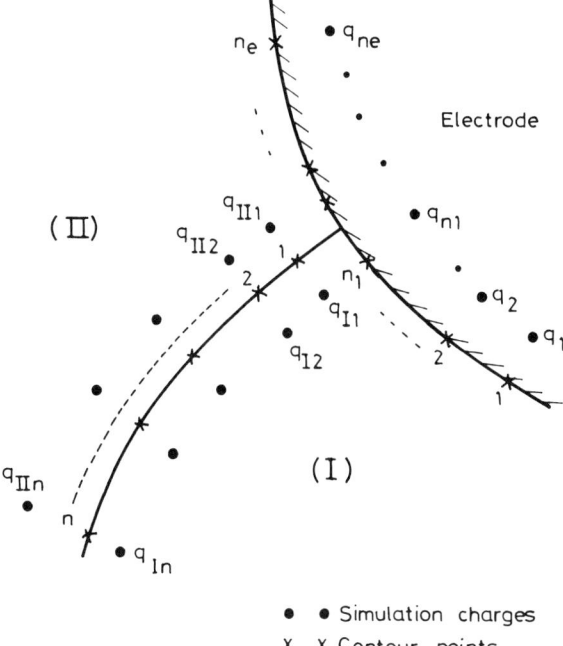

● ● Simulation charges
X X Contour points

Fig. 2.10 Discrete simulation charges at an electrode and at a dielectric boundary.

1. The potential at the n_1 contour points on the electrode/dielectric I interface must be equal to the applied voltage V:

$$\sum_{j=1}^{j=n_e} P_j q_j + \sum_{j=1}^{j=n} P_{IIj} q_{IIj} = V \qquad (2.28)$$

Similarly, the potential at the $(n_e - n_1)$ contour points on the electrode/dielectric II interface must be equal to V

2. The potential at each of the n contour points on the interface between the two eielectrics is the same whether seen from either side; that is,

$$\sum_{j=1}^{j=n} P_{Ij}q_{Ij} = \sum_{j=1}^{j=n} P_{IIj}q_{IIj} \qquad (2.29)$$

3. The displacement at the interface between the two dielectrics is continuous. Let the coefficient f'_j be defined as the contribution of the charge q_j to the component of the electric field normal to the dielectric interface. Then the continuity is expressed as

$$\varepsilon_I \left(\sum_{j=1}^{j=n_e} f'_j q_j + \sum_{j=1}^{j=n} f'_{IIj} q_{IIj} \right) = \varepsilon_{II} \left(\sum_{j=1}^{j=n_e} f'_j q_j + \sum_{j=1}^{j=n} f'_{Ij} q_{Ij} \right)$$
(2.30)

where ε_I and ε_{II} are the permittivities of the dielectrics I and II, respectively.

Thus (n_e + 2n) linear equations are formulated for the calculation of the same number of unknown discrete charges. In a two-dielectric arrangement, the contour points were chosen large compared with the number of discrete charges to achieve better matching of the boundary conditions (Abdel-Salam and Stanek, 1987b).

The charge-simulation technique has been applied successfully for calculating the electric field in and around practical suspension insulators (Khan and Alexander, 1982), and around overhead transmission lines (El-Arabaty et al., 1977). The simulation technique was also used for field calculation in arrangements including free potential electrodes, such as grading foils in condenser bushings and in pot heads of HV cables or suspended particles in gas-insulated systems (Abdel-Salam, 1987). The simulation technique was also attempted for computing electric fields, including surface and volume resistance of the insulation, such as pollution layers on insulation surfaces (Takuma et al., 1981b; Abdel-Salam and Stanek, 1987a; Singer, 1981).

With an increased number of dielectrics, solution of field problems using the charge-simulation technique becomes more and more complex. Therefore, the technique is limited to field calculations in arrangements with only one or two dielectrics. On the other hand, three-dimensional fields with and without symmetry can be handled without excessive computation (Fig. 2.11) (Chen and

Electric Fields

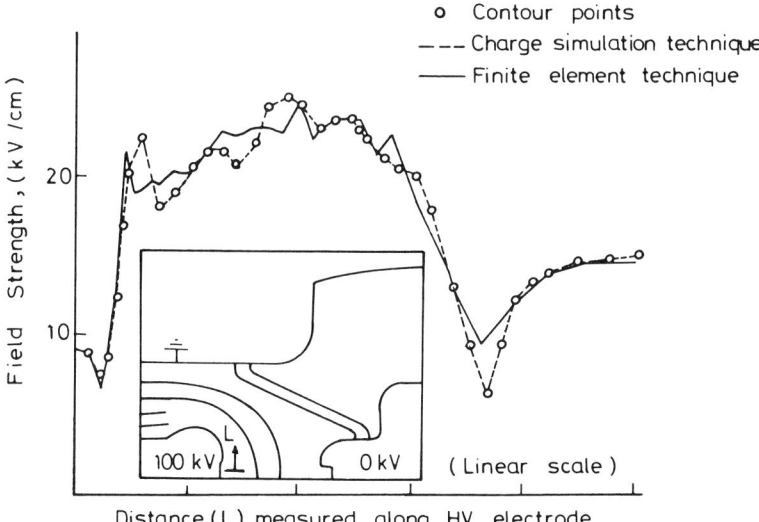

Fig. 2.11 Surface field calculations along the HV electrode and the dielectric spacer within a gas-insulated switchgear using the charge-simulation and finite-elements techniques. [From Okubo and Metz (1978).]

Pearmain, 1983). Another major advantage is the simple simulation of curved interfaces between dielectrics and conductors. The values of unknown charges satisfy the boundary conditions when their number and locations are chosen judiciously. Here experience plays an important role.

Abdel-Salam and El-Mohandes (1987) have proposed an efficient charge-simulation technique (ECST) for calculating electric fields around conductor bundles of EHV transmission lines. In the ECST, the number n_e of simulation charges and their coordinates are no longer arbitrary but rather, are dependent on the number of subconductors of the bundle and how they are arranged in space. The ECST was applied for bundles 2, 3, and 4 monopolar, homopolar, and bipolar dc transmission lines and showed high accuracy for the potential and field values calculated compared with results using the conventional technique. Moreover, the number of simulation charges involved with the ECST could thus be much reduced with a considerable saving of the computational time. As an example, for a bundle 2 of a monopolar dc line, using the ECST the number of charges could be reduced from 16 to 3 while accuracy was improved from 0.1% to 10^{-5}% (Abdel-Salam and El-Mohandes, 1987).

2.4.4 Finite-Difference Technique

The basis of this technique is the replacement of a continuous domain representing the entire space surrounding the high-voltage electrodes with a rectangular or polar grid of discrete "nodes" at which the value of unknown potential is to be computed. Thus we replace the derivatives describing Laplace's equation with "divided-difference" approximations obtained as functions of the nodal values.

The finite-difference technique is applicable for three-dimensional fields. However, a two-dimensional field is treated here for simplicity. Figure 2.12 shows a square grid with its sides parallel to the X or Y axis. All the elements have the same depth in the Z direction. Among the nodes of the grid, only nodes 0 to 4 will be of immediate interest, where nodes 1 to 4 surround node 0. Suppose that the potentials at these nodes are ϕ_1, ϕ_2, ϕ_3, ϕ_4, and ϕ_0. As the potential within the field region is continuous, it is possible to expand the potential at any point, such as node 0, using the well-known Taylor's series. If the series is terminated by neglecting the terms containing third and higher derivatives of the potential, then

$$\phi_1 = \phi_0 - \left(\frac{\partial \phi}{\partial x}\right) \Delta x + \frac{1}{2}\left(\frac{\partial^2 \phi}{\partial x^2}\right)(\Delta x)^2$$

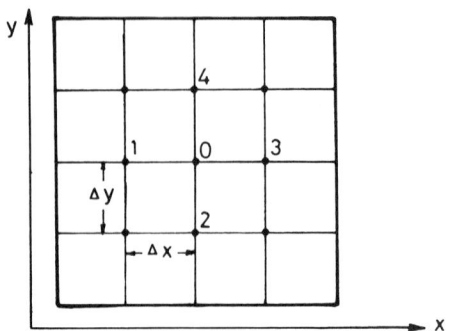

Fig. 2.12 Regular grid for finite-difference technique, indicating the node numbers.

Electric Fields

$$\phi_2 = \phi_0 - \left(\frac{\partial \phi}{\partial y}\right) \Delta y + \frac{1}{2} \left(\frac{\partial^2 \phi}{\partial y^2}\right) (\Delta y)^2 \qquad (2.31)$$

$$\phi_3 = \phi_0 + \left(\frac{\partial \phi}{\partial x}\right) \Delta x + \frac{1}{2} \left(\frac{\partial^2 \phi}{\partial x^2}\right) (\Delta x)^2$$

$$\phi_4 = \phi_0 + \left(\frac{\partial \phi}{\partial y}\right) \Delta y + \frac{1}{2} \left(\frac{\partial^2 \phi}{\partial y^2}\right) (\Delta y)^2$$

As Laplace's equation in Cartesian coordinates states that

$$\frac{\partial^2 \phi}{\partial x^2} + \frac{\partial^2 \phi}{\partial y^2} = 0$$

for a square grid, equation (2.31) yields

$$\phi_0 = \frac{1}{4} (\phi_1 + \phi_2 + \phi_3 + \phi_4) \qquad (2.32)$$

Equation (2.32) is the finite-difference form of Laplace's equation for a two-dimensional field.

The next step is to apply equation (2.32) for evaluating the potential at every node in the space between the electrodes. Of course, at the nodes located on the high-voltage electrode surface, $\phi = V$. This completes the set of equations relating the node potentials $[\phi]$ to the boundary coefficients $[B]$:

$$[\phi] [P] = [B] \qquad (2.33)$$

where [P] is an n × n matrix of the coefficients for n-unknown node potentials, $[\phi]$ is n × 1 column matrix of the unknown potentials, and [B] is an n × 1 column matrix of the sum of known potentials of the bounding electrodes. Equation (2.33) could be solved by the Gaussian elimination method for problems with a relatively small number of nodes (James et al., 1985). For complex cases, iterative schemes are most efficient in combination with successive overrelaxation methods.

There is no reason why this technique cannot be used for computing the field in multidielectric arrangements. Referring to Figure 2.12, assume the nodes 2, 0, and 4 to be at the interface between two dielectrics of permittivities ε_1 (containing node 1) and ε_2 containing node 3). At the interface, the potential and displacement

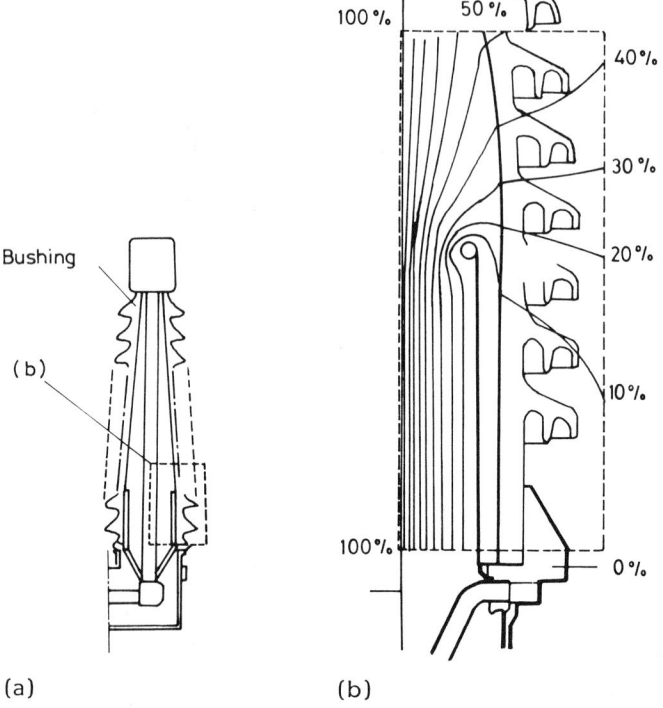

(a)   (b)

Fig. 2.13 Field computed by the finite-difference technique for a 300-kV gas-insulated system: (a) part of single-phase layout, showing area enlarged in (b); (b) plot of equipotentials at lower end of bushing. [From Scott et al. (1974).]

are continuous. Therefore, the resultant field may be composed by superposition of two Laplacian fields and equation (2.32) takes the form

$$\phi_0 = \frac{1}{4}\left(\phi_2 + \phi_4 + \frac{2\varepsilon_1}{\varepsilon_1 + \varepsilon_2}\phi_1 + \frac{2\varepsilon_2}{\varepsilon_1 + \varepsilon_2}\phi_3\right) \quad (2.34)$$

The finite-difference technique has thus been applied successfully for field calculation in switchgear design and development (Ryan et al., 1983). Figure 2.13 shows the computed field within a gas-insulated switchgear. The areas of field concentrations received detailed computations with finer grids. Potential values

Electric Fields 39

between the nodes are obtained by interpolation. The electric field components at any node are easily expressed in terms of the potential values at the nodes surrounding it.

Some difficulties are encountered in solving many problems using the finite-difference technique, and therefore its efficiency is considerably limited. A regular grid is not suitable for curved boundaries or interfaces because they intersect grid lines at points other than nodes. Therefore, the grid must be denser near the curved boundaries. The technique is well suited for use in field regions that are finite in space; otherwise, the number of nodes becomes very large and solution of equation (2.33) may be difficult even for machine computation.

2.4.5 Combined Charge-Simulation and Finite-Difference Technique

As noted in Section 2.4.3, the charge-simulation technique (CS) is well suited for computing fields in the vicinity of curved electrodes placed in open (unbounded) space with one or two dielectrics. On the other hand, the finite-difference technique (FD) is applicable to bounded space with multidielectrics. A combination of CS and FD can be utilized in some cases to obtain the advantages of both methods (Abdel-Salam and El-Mohandes, 1989).

To apply the combined method, the entire field is divided into two domains, where each technique is used separately (Fig. 2.14).

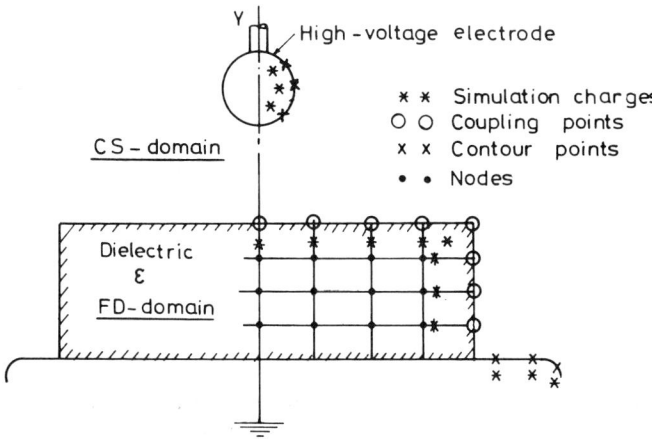

Fig. 2.14 Entire field in air and a dielectric divided into CS and FD domains. Being symmetrical, only one half is studied.

In the CS domain, the surface charges on the electrodes are represented by n_c unknown discrete charges with a set of contour points on the electrode surfaces. The FD domain is replaced by a rectangular grid whose n_g nodes are of unknown potentials. Naturally, equation (2.10) applies to the CS domain, and equation (2.33) applies to the FD domain. The coupling between the two domains is based on the condition of continuity of potential and displacement at the n_b coupling points along the CS-FD boundary.

While analyzing the CS domain, the FD domain is represented by a number n_b of discrete charges located inside it. Also, each of the n_b coupling points on the boundary between the two domains has its own potential to be evaluated in addition to the potentials of the n_g grid nodes. Therefore, the total number of unknowns in the entire space is $(n_c + 2n_b + n_g)$ to be evaluated by simultaneous solution of n_c equations (2.10) for the CS domain, n_g equations (2.33) for the FD domain, and $2n_b$ equations for the equality of potentials and continuity of displacements at the n_b coupling points.

2.4.6 Finite-Element Technique

According to this technique, the space in which the electric field is to be calculated is divided up into triangular elements as shown in Fig. 2.15. All the elements have the same depth in the Z direction, the same as in the FD technique (Zienkiewicz, 1977).

Solution of the electric field problems by the finite-element method is based on the fact, known from variational calculus, that Laplace's equation is satisfied when the total energy functional is minimal. If the permittivity of the medium is ε, the electrostatic energy functional F for a flat two-dimensional field is

$$F = \iint_A \frac{1}{2} \varepsilon \, (\mathrm{grad} \, \phi)^2 \, dA \qquad (2.35)$$

A represents the area scanned by the trigular elements.

The potentials ϕ at the different nodes are unknown variables and for minimum energy functionals,

$$\frac{\partial F}{\partial \phi} = 0 \qquad (2.36)$$

for all the potentials at the nodes (Silvester and Ferrari, 1983). It is obvious that when ϕ_1 is varied while all other potentials remain constant, only the energy in the adjoining triangles is affected. Therefore, in the partial derivative $\partial F / \partial \phi_1 = 0$, only the energies

Electric Fields 41

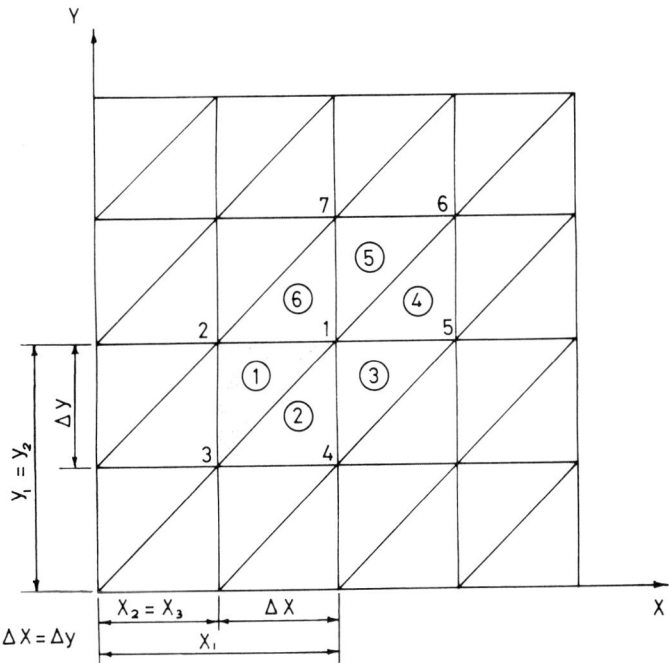

Fig. 2.15 Grid for the finite-element technique and arrangement of the elements.

in the six triangles numbered 1 to 6 in Figure 2.15 need to be considered. For the energy functional in the six triangles,

$$F = F_1 + F_2 + F_3 + F_4 + F_5 + F_6 \tag{2.37}$$

In triangle 1,

$$(\text{grad } \phi)^2 = (\text{grad } \phi)_x^2 + (\text{grad } \phi)_y^2 \tag{2.38}$$

Assuming a linear variation of the potential ϕ as a function of x and y within the triangle,

$$(\text{grad } \phi)^2 = \left(\frac{\phi_1 - \phi_2}{\Delta x}\right)^2 + \left(\frac{\phi_2 - \phi_3}{\Delta y}\right)^2 \tag{2.38a}$$

and the functional F_1 according to equation (2.35) becomes

$$F_1 = \frac{1}{2} \iint_A \varepsilon \, (\text{grad } \phi)^2 \, dA = \frac{1}{4} \varepsilon \left[\left(\frac{\phi_1 - \phi_2}{\Delta x} \right)^2 + \left(\frac{\phi_2 - \phi_3}{\Delta y} \right)^2 \right] \Delta x \, \Delta y$$

and the partial derivative takes the form

$$\frac{\partial F_1}{\partial \phi_1} = \frac{1}{2} \varepsilon \frac{\phi_1 - \phi_2}{(\Delta x)^2} \Delta x \, \Delta y \qquad (2.39)$$

For the other triangles, similar equations could be obtained. For $\Delta x = \Delta y$, the total derivative becomes (Fig. 2.15)

$$\frac{\partial F}{\partial \phi_1} = \frac{1}{2} \varepsilon \, [8\phi_1 - 2(\phi_2 + \phi_4 + \phi_5 + \phi_7)] \qquad (2.40)$$

Thus

$$\phi_1 = \frac{1}{4}(\phi_2 + \phi_4 + \phi_5 + \phi_7)$$

which is the same as the finite-difference solution (Section 2.4.4).

With a typical triangular element (Fig. 2.15), the potential is approximated by the polynomial

$$\phi = a + bx + cy + a_1 x^2 + b_1 xy + c_1 y^2 + \cdots \qquad (2.41)$$

Any increase in accuracy by including higher-order terms is outweighed by the increase in computation time and storage requirements. Most of the computations are satisfactorily based on a first-order approximation. Thus the actual continuous potential distribution over the XY plane is replaced by a piecewise-planar approximation.

The coefficients a, b, and c in equation (2.41) may be found from the three independent simultaneous equations obtained by constraining the potential to assume vertex values ϕ_1, ϕ_2, and ϕ_3 at the three vertices of coordinates shown in Figure 2.15:

Electric Fields

$$\begin{bmatrix} \phi_1 \\ \phi_2 \\ \phi_3 \end{bmatrix} = \begin{bmatrix} 1 & x_1 & y_1 \\ 1 & x_2 & y_2 \\ 1 & x_3 & y_3 \end{bmatrix} \begin{bmatrix} a \\ b \\ c \end{bmatrix} \quad (2.42)$$

That is,

$$\phi = \sum_{m=1}^{m=3} \phi_m \alpha_m(x,y) \quad (2.43)$$

where α_m is a linear function of x and y, easily evaluated. It can easily be verified that the variables α_m are interpolatory on the three vertices of the triangle; that is,

$$\alpha_m(x_n, y_n) = \begin{cases} 0 & m \neq n \\ 1 & m = n \end{cases}$$

where (x_n, y_n) are the x and y coordinates of the vertices. The potential gradient may be found from equation (2.43) as

$$\text{grad } \phi = \sum_{m=1}^{m=3} \phi_m \nabla \alpha_m \quad (2.44)$$

So according to equation (2.35), the energy of this element is

$$F = \frac{1}{2} \sum_{m=1}^{m=3} \sum_{n=1}^{n=3} \phi_m \iint_A \varepsilon \nabla \alpha_m \nabla \alpha_n \, dA \, \phi_n \quad (2.45)$$

where the dielectric within the element is assumed isotropic and the permittivity ε is constant. However, a solution is also possible if ε changes from element to element. For brevity, define the elements

$$S_{mn} = \iint_A \varepsilon \nabla \alpha_m \nabla \alpha_n \, dA$$

of the asymmetrical matrix [S] for the hatched triangle (Fig. 2.15).

$$[S] = \begin{bmatrix} S_{11} & S_{12} & S_{13} \\ S_{21} & S_{22} & S_{23} \\ S_{31} & S_{32} & S_{33} \end{bmatrix} \qquad (2.46)$$

The matrix [S[is the well-known "stiffness matrix" of the element under consideration, as it contains the sensitivity of the energy functional with respect to the potentials.

Equations (2.45) and (2.46) yield

$$F = \frac{1}{2} [\phi]^{tr}[S][\phi] \qquad (2.47)$$

where $[\phi]$ is the column vector of vertex potential values and "tr" denotes transposition.

The minimization of the energy functional for the specific element is expressed as

$$\frac{\partial F}{\partial [\phi]} = [S][\phi] \qquad (2.48)$$

Any composite triangular-element assembly may be built up one triangle at a time. It is therefore sufficient to consider how continuity is enforced when one triangular element is added to an already existing assembly.

For minimum energy functionals, and with properly arranged nodes, equations (2.35) and (2.48) yield

$$\begin{bmatrix} [S]_{ff} & [S]_{fb} \end{bmatrix} \begin{bmatrix} [\phi]_f \\ \\ [\phi]_b \end{bmatrix} \qquad (2.49)$$

where $[\phi]_b$ are the bound (known) potentials of the nodes lying on the electrode surfaces, and $[\phi]_f$ are the free (unknown) potentials of the other nodes. This equation can be rearranged so that the potential at every point in the space between the electrodes could be evaluated in terms of the electrode voltages. Thus

$$[\phi]_f = -[S]_{ff}^{-1}[S]_{fb}[\phi]_b = [T]^{-1}[\phi]_b \qquad (2.50)$$

Electric Fields 45

The solution of Laplace equation takes the form of a set of nodal potential values as obtained from equation (2.50).

By proper numbering of the nodes, the matrix [T] takes the form of a narrow diagonal band of symmetrical structure. The order of [T] can reach several thousands to cover an entire two-dimensional field. However, its bandwidth is about 5 to 15. Fortunately, techniques have been developed for efficiently solving such sparse matrix equations which can significantly decrease the computation time (Jennings, 1977).

Figure 2.11 shows a comparison of the finite-element (FE) and charge-simulation (CS) techniques for calculating the electric field at the contacts within a gas-insulated switchgear. The small deviation between the field values obtained by the two techniques is attributed to two reasons. The first is that with the CS technique the electrode surface is not exactly equipotential. The second reason is the discretization of the field space into finite triangular elements when using the FE technique. We should add that the FE method needs more computer storage and time than does the CS technique. On the other hand, the accuracy of calculation is improved with finer elements.

As in the FD technique, the electric field components at any node can be expressed in terms of the potential values at the nodes surrounding it. The precision of field calculation using the finite-element method increases significantly with the number of elements.

Similar to the FD technique, the FE technique is suitable for bounded regions. Many physical problems of interest, however, are only partly bounded (i.e., a portion of the boundary is at infinity). This difficulty is usually surmounted by assuming an artificial boundary, with an assumed potential, to be located far enough from the region of interest as to minimize the resulting errors. The price paid for doing this is that Laplace's equation must be solved for the entire larger region. This adds to the computation time. Some alternatives to this technique have recently been reported (Cendes and Hamann, 1981).

The flexibility of the FE technique is its greatest advantage with respect to the traditional FD technique. Elements can have various shapes and can easily be adapted to any shape of boundary and interface geometries. Further, its set of equations usually has a symmetric positive-definite matrix of coefficients. This property is not ensured in the FD technique. A third advantage of the FE technique is claimed to be simpler programming because of the easier introduction of boundary conditions. In general, it is more powerful for a wide range of problems with complicated geometries. On the other hand, it involves tedious preparations for the input data, most of which is a description of the element grid topology. For large problems, the time for grid preparation and editing becomes prohibitive. Some of these difficulties have recently been alleviated by the development of grid generation algorithms, adaptive grid refinement algorithms, and grid optimization (Shephard and Gallagher, 1979).

2.4.7 Combined Charge-Simulation and Finite-Element Technique

A comparison of the finite-element (FE) and charge-simulation (CS) techniques shows that each has advantages and disadvantages (Okubo et al., 1982; Tan and Steinbigler, 1985; Steinbigler, 1979). A combination of FE and CS can yield the advantages of both techniques, similar to the case of coupling CS with FD (Section 2.4.5) (Abdel-Salam and El-Mohandes, 1989). To apply the combined method, the space is divided into two domains—the FE and CS domains—which are treated separately. The coupling of the FE and CS domains is based on the fact that the potential and field strength must be continuous along the boundary between the domains (i.e., at all the boundary points). As an application of the combined technique, the axisymmetric three-dimensional field of a high-voltage bushing was computed (Fig. 2.16). The figure shows the configuration of the bushing and the computed equipotential surfaces, which speaks for itself.

2.4.8 Integral-Equations Technique

The integral-equations technique is applicable for evaluating the electric potential and field in a space composed of conducting and dielectric regions (Fig. 2.17). All the surfaces of the problem under study are subdivided into a number of segments. The charge density is assumed uniform on each of these segments, but of an unknown magnitude. The charge density $\sigma(\overline{\Omega})$ is assumed $\sigma_{12}(\overline{\Omega})$, $\sigma_{13}(\overline{\Omega})$, and $\sigma_{23}(\overline{\Omega})$ on the subsurfaces A_{12}, A_{13}, and A_{23}, respectively. $\overline{\Omega}$ represents the location of the segmental charge. The unit vectors u_{ni} are inward normal to the boundary surface A_i of the ith region ($A_i = A_{ij} + A_{ik}$).

By using Green's theorem, Laplace's equation can be expressed in surface integral form subject to the boundary conditions

$$\phi_1 = \phi_2 = V \text{ on the interface } A_{12}$$

$$\phi_1 = \phi_3 = V \text{ on } A_{13}$$

Electric Fields 47

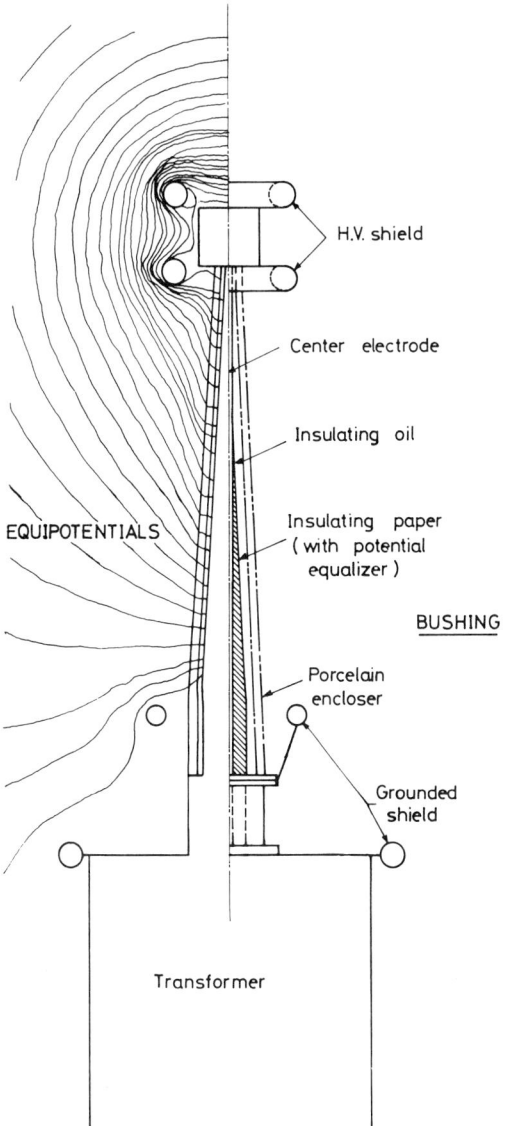

Fig. 2.16 Calculation of three-dimensional axisymmetrical field of a high-volume bushing using a combined charge-simulation/finite-element technique. [From Okubo et al. (1982).]

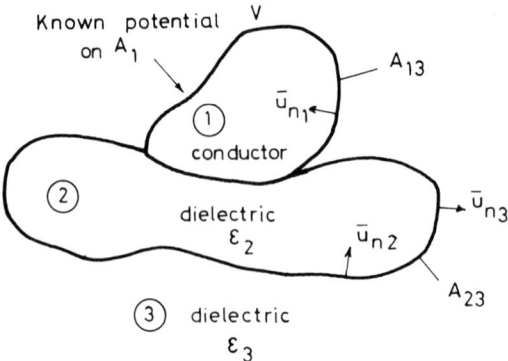

Fig. 2.17 System consisting of a conductor and two dielectrics.

$$\phi_2 = \phi_3 \text{ on } A_{23}$$

$\phi_3 = 0$ on the infinite boundary of the third region

$$\frac{\partial \phi_1}{\partial u_{n1}} = 0 \text{ on the boundary surface } A_1 \text{ (conducting surface)}$$

$$\varepsilon_2 \frac{\partial \phi_2}{\partial u_{n2}} = -\varepsilon_3 \frac{\partial \phi_3}{\partial u_{n3}} \text{ on } A_{23}$$

The solution of Laplace's equation is obtained by formulating the following two equations. One equation results from the condition that the potential must equal V on the conductor boundaries. The potential at any contour point on A_1 is equal to the sum of potential contributed by all surface charges that is, by polarization charge on the interface between dielectrics and by both the free and polarization charges on the conductor-dielectric interfaces. As each dielectric is assumed homogeneous, there is no volume polarization charge. Hence

$$\iint_A \sigma(\overline{\Omega}) G(\overline{P},\overline{\Omega}) \, dA = V \qquad \overline{P} \text{ on } A_1 \qquad (2.51)$$

where A is the collection of subsurfaces A_{12}, A_{13}, and A_{23}. $\overline{P}$ is the location of the contour point. $G(\overline{P},\overline{\Omega}) = 1/4\pi(\overline{P} - \overline{\Omega})$ is a

Electric Fields

generating function that satisfies $-\nabla_x^2 G(\overline{P},\overline{\Omega}) = \delta(\overline{P} - \overline{\Omega})$, and δ is the Dirac delta.

The second equation results from the condition that the displacement is continuous at the interface A_{23} between dielectrics; that is,

$$\frac{\varepsilon_2 + \varepsilon_3}{2} \sigma(\overline{P}) = (\varepsilon_3 - \varepsilon_2) \iint_A \sigma(\overline{\Omega}) \frac{\partial G(\overline{P},\overline{\Omega})}{\partial u_{n3}} dA \qquad \overline{P} \text{ on } A_{23} \qquad (2.52)$$

Therefore, each integral in equations (2.51) and (2.52) is reduced to a sum of terms, each being a constant depending on the system geometry multiplied by an unknown charge density $\sigma(\overline{\Omega})$. The constants for any problem can be evaluated using analytical and numerical integration schemes (Singer, 1984). By applying equations (2.51) and (2.52) at several contour points, a set of linear equations are obtained that can easily be solved by standard techniques. Once the charge densities are known, the potential and field at any point can be determined (Daffe and Olsen, 1979).

Equations (2.51) and (2.52) can be applied to regions composed of several conductors and several dielectrics. In this case A_1 becomes the collection of all conductor-dielectric interfaces, A_{23} becomes the collection of all dielectric-dielectric interfaces (Singer, 1984).

The integral-equations technique is appealing because the potential must be computed only at the required points. Another attractive quality is that it involves fewer unknown quantities than do the FD and FE techniques, thus considerably less computer time. It may be argued that the equations describing the charge-simulation technique could be obtained by discretizing the integral equations explained above. That is true, but it would result in a larger number of linear equations to solve than the integral-equations solution, in which the sources are treated as surface charges.

2.4.9 Monte Carlo Technique

Laplace's equation for the electric potential is solved here by numerically simulating random walks of a fictitious particle beginning from a point chosen in the space for which the potential distribution is to be evaluated. To illustrate the application of this technique, a conductor at a potential of +V centered in a rectangular duct at zero potential will be considered (Fig. 2.18). Let the space between the conductor and the duct be ruled off in a Cartesian grid, as for the finite-difference technique. Choose at random either of the letters x or y and either of the signs + or -, and move the particle one

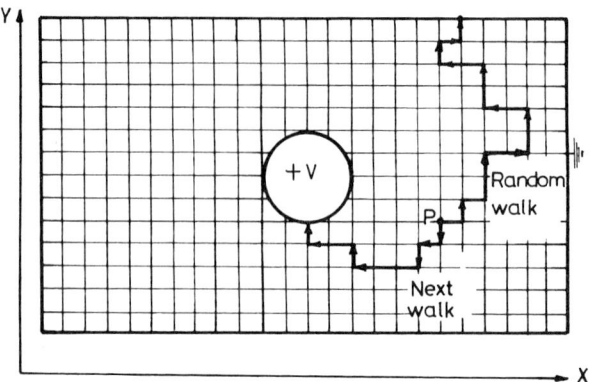

Fig. 2.18   Random walks in the Monte Carlo technique.

step from the chosen point P in the direction thus identified (say, +x). Again choose at random one of the four possible directions and move the particle ahead one more step. The path traced out by this sequence of random directions is called a "random walk" (Fig. 2.18). Eventually, this walk would reach either the conductor or the duct wall and would thus be terminated. This random walk of the particle is repeated a fairly large number of times (n). Then the potential at P is defined,

$$V_p = \lim_{n \to \infty} \frac{\Sigma V_n}{n} \qquad (2.53)$$

Each V_n will equal V or 0, depending on whether the corresponding walk terminates on the conductor or on the duct wall (Fig. 2.18). The rate of convergence of V_p to the correct value is proportional to $\sqrt{n}$, so many random walks are necessary if a fine grid is chosen and reasonable accuracy is required. The sequence of instructions that is used to govern the motion of the fictitious particle is derived from a sequence of random numbers. Methods for generating these numbers are described elsewhere.

As was true for the FD and FE techniques, it is assumed that the region between boundaries is circumscribed. Again, however, there are techniques that can be used to apply the Monte Carlo technique to unbounded regions. The most interesting characteristic of Monte Carlo solutions is that the potential can be computed for each point at a time. Neither a large array of potentials need to be stored, nor do a large number of simultaneous equations need to be solved. Despite this advantage, Monte Carlo techniques have been of only limited use for one- or two-dimensional problems because the time needed to calculate the potential at each point is not small. However, they are very appealing for solving three-dimensional problems, where they tend to supersede other techniques (Anis and Abd-Allah, 1989).

2.5 ANALYTICAL CALCULATIONS OF FIELDS WITH SPACE CHARGES

In high-voltage engineering, space charge is mainly developed by corona discharges at highly stressed electrodes. At alternating voltages, the space charge is constrained in the vicinity of the stressed electrodes and oscillates with periodic reversal of the electric field. This increases the difficulty of field calculation (Abdel-Salam et al., 1984; Abdel-Salam and Shamloul, 1988). On the other hand, the space charges created by dc corona fill the entire interelectrode space. Analytical calculation of electric fields with space charges created by dc corona is possible for simple configurations such as concentric spheres and coaxial cylinders.

2.6 NUMERICAL COMPUTATION OF FIELDS WITH SPACE CHARGES

In more complex electrode geometries such as wire-plane gaps, analytical calulcation of the electric field is extremely difficult, due to the nonlinear nature of space-charge generation by corona. A relatively simple case is that of unipolar space charge (e.g., unipolar dc lines). The equations in this case are as follows:

$$\nabla \cdot \overline{E} = \frac{\rho}{\varepsilon} \qquad \overline{J} = k\rho\overline{E} \qquad \nabla \cdot \overline{J} = 0 \qquad \overline{E} = -\nabla\Phi \qquad (2.54)$$

where $\overline{E}$ and $\overline{J}$ are the electric field and current density vectors at any point in space. The first is Poisson's equation, the second defines the relationship between $\overline{J}$ and $\overline{E}$, the third is the continuity equation for the ions, and the fourth is the electric field in terms of the potential Φ. ρ is the ion charge density and k is the ion mobility.

Similarly, bipolar dc ionized fields are described by the following equations:

$$\nabla \cdot \overline{E} = \frac{\rho_+ - \rho_-}{\varepsilon} \qquad \overline{J}_+ = k_+\rho_+\overline{E} \qquad \overline{J}_- = k_-\rho_-\overline{E}$$

$$\overline{J} = \overline{J}_+ + \overline{J}_- \qquad (2.55)$$

$$\nabla \cdot \overline{J}_+ = -\frac{\xi \rho_+ \rho_-}{e} \qquad \nabla \cdot \overline{J}_- = \frac{\xi \rho_+ \rho_-}{e} \qquad E = -\nabla \Phi$$

where the subscripts + and − correspond to positive and negative ions. ξ denotes the recombination coefficient for ions and e is the electronic charge.

Most attempts in the literature resort to simplifying assumptions (Khalifa and Abdel-Salam, 1974b; Abdel-Salam et al., 1982). The two most important such assumptions are:

1. The space charge affects only the magnitude, not the direction of the electric field. This assumption was originally suggested by Deutsch in 1933.
2. The surface field at the coronating electrode remains constant at its value corresponding to corona onset.

Both assumptions compensate for the lack of adequate boundary conditions, and thus equations (2.54) and (2.55) could be solved along a number of flux lines of the imposed electric field. Iterative methods were used for the conditions to be closely fulfilled at the electrodes.

Deutsch's assumption continued to be a basis until Abdel-Salam and Khalifa in 1974 showed that its use was not a necessity. They based their computation of the resultant local field on a physical understanding of the problem as a superposition of the applied electric field and the space charge. Also, the surface field of conductors in corona could be calculated (Khalifa and Abdel-Salam, 1973) and measured. It was found to decrease significantly with corona intensity. Computation may follow the finite-element or charge-simulation technique either separately or combined with the method of characteristics or the method of residues, as shown below.

2.6.1 Finite-Element Technique

By using some additional assumptions, Takuma et al. (1981a) succeeded in applying the finite-element method to a field problem with space charge in the case of wire-plane gaps. Instead of assumption 2, they postulated a constant space-charge density around the wire periphery, at a value to be determined experimentally. This added some empiricism to the analysis.

More recently, Abdel-Salam et al. (1983) applied the finite-element method for analyzing the electric field with monopolar space charge in wire-plane gaps. They could eliminate both Deutsch's assumption (1) and that of constant surface field intensity (2). They thus could compute the field in the interelectrode space without empiricism. From equation (2.54), we get

Electric Fields

$$-\nabla \cdot (\varepsilon \nabla \Phi) = \rho \tag{2.56}$$

$$-\nabla \cdot (k\rho \nabla \Phi) = 0 \tag{2.57}$$

Abdel-Salam et al. considered the potential Φ as the sum of two components, the applied electrostatic value ϕ_s and the elemental potential caused by the space charge ϕ_{sc}, that is,

$$\Phi = \phi_s + \phi_{sc} \tag{2.58}$$

The elemental potential ϕ_{sc} is bound to be zero at both electrodes, where $\Phi = \phi_s$. The potential ϕ_s was computed for the contour points by the charge-simulation technique (CS) as explained in Section 2.4.3. Now, to find ϕ_{sc}, substitute equation (2.58) into equations (2.56) and (2.57) to get

$$\nabla(\varepsilon \nabla \phi_{sc}) = -\rho \tag{2.56a}$$

$$\nabla(k\rho \nabla \phi_{sc}) = 0 \tag{2.57a}$$

or in a general form,

$$\nabla(\gamma \nabla \phi_{sc}) = \beta \tag{2.59}$$

where γ is equal to ε and $k\rho$ in equations (2.56a) and (2.57a), respectively, and β is equal to $-\rho$ and zero in equations (2.56a) and (2.57a), respectively.

In wire-plane gaps, and according to the finite-element technique, the inter-electrode space is an unbounded region and is usually defined by an artificial boundary as explained in Section 2.4.5. The inter-electrode space is divided into triangular elements, thus forming a grid, and the potential ϕ_{sc} within each element was approximated as a linear function of space coordinates.

To solve equations (2.56a) and (2.57a) in their general form (2.59), an energy functional given in equation (2.60) is used instead of equation (2.35) of the space-charge-free fields.

$$F = \iint_A \left\{ \frac{\gamma}{2} \left[\left(\frac{\partial \phi_{sc}}{\partial x} \right)^2 + \left(\frac{\partial \phi_{sc}}{\partial y} \right)^2 \right] - \beta \phi_{sc} \right\} dA \tag{2.60}$$

Representing the variation of ϕ_{sc} over each element in terms of the nodal values of ϕ_{sc} and imposing the minimization condition $\partial F/\partial \phi_{sc} = 0$ for the entire set of elements, we get a set of simultaneous equations for ϕ_{sc} at the nodes for a given ρ distribution. Solution of this set of equations gives the nodal values of ϕ_{sc}: [ϕ_{sc1}] and [ϕ_{sc2}], corresponding to equations (2.56a) and (2.57a), respectively. Iterations are used until the values of both ϕ_{sc1} and ϕ_{sc2} agree within an acceptable tolerance (Abdel-Salam et al., 1983). The relation for correcting the charge density ρ_i to ρ_n is

$$\rho_n = \rho_i \left(1 + \frac{f_{ac}}{2} \frac{\phi_{sc1} - \phi_{sc2}}{\phi_{sc1} + \phi_{sc2}} \right) \qquad (2.61)$$

where f_{ac} is an acceleration factor chosen equal to 0.5.

2.6.2 Finite-Element Technique Combined with the Method of Characteristics

A self-consistent description of the charge ρ and potential Φ structures of a monopolar dc ionized field was proposed by Davis and Hoburg (1986) based on simultaneous computation of both ρ and Φ. The technique employs iterative use of the finite-element method to compute potential and electric field distributions, and the method of characteristics to compute the charge distribution.

Use of the finite-element method to determine the distribution of Φ for a known charge density ρ in a region with permittivity ε is based on minimizing the energy functional:

$$F = \iint_A \left[\frac{1}{2} \varepsilon (\text{grad } \Phi)^2 - \rho \Phi \right] dA \qquad (2.62)$$

to obtain the solution of Poisson's equation ($\nabla^2 \Phi = \rho/\varepsilon$). It is noted that the energy functional given by equation (2.35) is obtained from the general equation (2.62) by setting $\rho = 0$ for space-charge-free fields. Once the Φ distribution is determined, the electric field $\overline{E}$ distribution is obtained simply.

According to the method of characteristics proposed by Davis and Hoburg (1986), the partial differential equations (2.54) that describe the effect of a known field pattern on the spatial distribution of ρ could be converted into an ordinary differential equation for ρ along specific space-time trajectories. Combination of equations (2.54) leads to a nonlinear partial differential equation governing the evolution of ρ:

$$\overline{E} \cdot \nabla \rho = -\frac{\rho^2}{\varepsilon} \qquad (2.63)$$

Along the "characteristic lines" defined by

$$\frac{d\bar{r}}{dt} = k_i \bar{E} \qquad (2.64)$$

equation (2.63) becomes

$$\frac{d\rho}{dt} = -\frac{k_i \rho^2}{\varepsilon} \qquad (2.65)$$

and yields the solution

$$\rho = \frac{1}{(1/\rho_0) + (k_i t/\varepsilon)} \qquad (2.66)$$

where ρ_0 is the charge density at the starting point of the characteristic line.

For a given starting point at the surface of the coronating electrode, the characteristic line is traced out by numerically integrating equation (2.64). At intersections with triangle boundaries, time t is determined, then used in equation (2.66) to compute ρ along the characteristic line.

2.6.3 Charge-Simulation Technique

In the application of the CS technique to space-charge-free fields approximate solutions to Laplace's equation are obtained by placing fictitious discrete charges outside the region in which the solution is sought (Section 2.4.3). This technique was adapted by Hornstein (1980) to include corona space charges. He concentrated them at some discrete points inside the region under study. Both sets of discrete charges have to be chosen to satisfy the boundary conditions of Poission's equation (2.54). These boundary conditions are given partially in terms of known electrode potentials and are obtained parially by assuming that the magnitudes of the electric field at the coronating electrodes remain constant at the onset value or vary in the manner computed by Khalifa and Abdel-Salam (1973). The contribution of these charges to the total electric field should satisfy the condition of current continuity.

To solve a given unipolar dc ionized field, first a space-charge-free field solution is found. This field represents conditions just prior to corona onset. Then, for a given value of corona current i_c, equipotential charge shells enclosing the wire and represented by discrete line charges are added to the volume as depicted

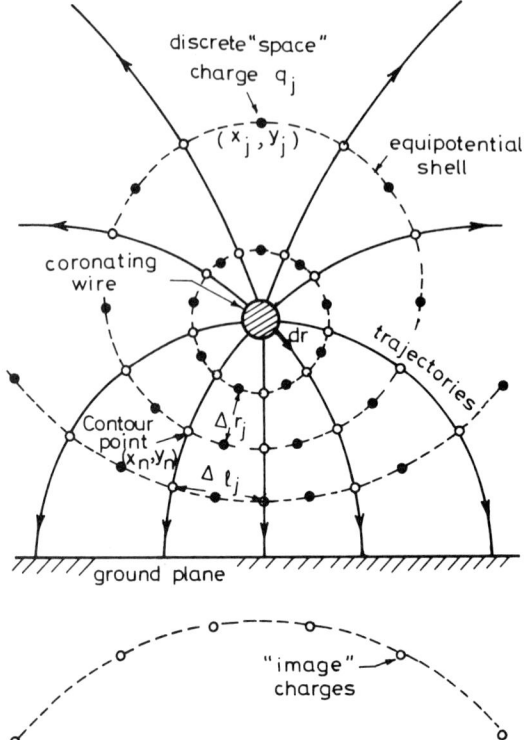

Fig. 2.19 Schematic representation of trajectories, discrete space charge, and equipotential contours for a wire-plane gap (not drawn to scale).

schematically in Figure 2.19 for a wire-plane geometry. Since these equipotential shells do not contribute to the surface field of the wire, it would remain at its onset level.

The equipotential charge shell nearest the wire is first located and simulated by M line charges uniformly distributed around its contour. Their images with respect to ground are also set. The M selected contour points are also chosen spread about the equipotential contour as shown in Figure 2.19 and located midway between the line charges. Under the constant levels of the equipotential contours, M equations are formulated as

$$\sum_{j=1}^{j=M} \frac{q_j}{2\pi \varepsilon_0} \ln \frac{1}{\bar{r}_n - \bar{r}_j} = \text{constant at each } \bar{r}_n \qquad (2.67)$$

Electric Fields 57

where $\bar{r}_n$ and $\bar{r}_j$ are, respectively, the location of the contour point and line charge q_j.

The values of q_j should also satisfy the discretized form of the current continuity equation:

$$\sum_{j=1}^{j=M} \frac{q_j}{\Delta r_j \, \Delta \ell_j} \, kE \, \Delta \ell_j = i_c \qquad (2.68)$$

where Δr_j and $\Delta \ell_j$ are distances defining an elemental volume with unit axial length around q_j as shown in Figure 2.19. The quantity $(q_j / \Delta r_j \, \Delta \ell_j)$ is the equivalent local space charge density ρ and i_c is the corona current per unit length of the wire with the applied voltage V. Equations (2.67) and (2.68) formulate M equations with M unknowns to be solved simultaneously.

After the first equipotential charge shell and its "image" charges are found, their electric field contributions are included in integrating outward to the next shell, and so on until the entire space is filled with shells of known discrete simulation charges. This defines the charge-density distribution over the space surrounding the coronating wire.

The applied voltage necessary to sustain i_c is equal to the onset voltage augmented by the potential contribution of each shell plus the potential of the images calculated at the wire surface. The I-V characteristics of corona computed by the method proposed by Horenstein (1980) agreed satisfactorily with those measured by experiment.

Sometimes the space charge, such as a cloud of ions developed by an avalanche, is of known magnitude traveling in an interelectrode space. The ion cloud may be approximated by point or ring charges (Zeitoun et al., 1976). As the space charges are known in magnitude, it is only necessary to calculate the charges simulating the electrodes. The potentials of contour points selected on electrode surface are expressed as

$$[P][q] + [P_s][q_s] = [V] \qquad (2.10)$$

where $[p_s]$ is the potential-coefficient matrix of the $[q_s]$ space charges and $[V]$ is the applied voltage.

The simulation charges $[q]$ of the electrodes are simply expressed as

$$[q] = [P]^{-1} \left\{ [V] - [P_s][q_s] \right\} \qquad (2.69)$$

The potential at any point in space is then calculated as

$$\Phi = \sum_{j=1}^{j=n} P_j q_j + \sum_{j=1}^{j=n_s} P_{sj} q_{sj} \qquad (2.70)$$

where n and n_s are, respectively, the number of simulation charges of the electrode and the number of discrete space charges. The electric field is calculated similarly.

The concept above was used for simulating electron avalanches growing in the ionization zone around stressed electrodes, such as in point-plane gaps. This was the basis for calculating the development of corna Trichel pulses in negative corona and burst pulses and onset-streamer pulses in positive corona (Ibrahim and Singer, 1982, 1984). Very recently, the concept was used to calculate the corona pulses from positive surges propagating along overhead transmission lines (Abdel-Salam and Stanek, 1987a).

2.6.4 Charge-Simulation Technique Combined with the Method of Residues

The charge-simulation technique was extended to the calculation of the electric field vector $\overline{E}$ directly from Poisson's equation. The space charge was represented by discrete values at the nodes of a grid that was superposed on the interrelectrode space. This method is more accurate in the calculation of $\overline{E}$ than would be possible from a FE solution of Poisson's equation for space potential Φ, where $\overline{E}$ is obtained by numerical differentiation of Φ. The updated estimates of the space-charge values are obtained from the continuity equations for current density using a weighted residual method (Oin et al., 1986). These charge densities are utilized for the calculation of $\overline{E}$ in the next iteration cycle.

The region of interest being defined by an artificial surface is subdivided into small triangular elements. The applied voltages are defined at ground and wire surfaces in wire-plane gaps. In the presence of corona, the electric field E consists of two components, the applied electrostatic component $\overline{E}_s$ and the component $\overline{E}_{sc}$ due to the space charge. The component $\overline{E}_s$ is computed by the CS technique as explained in Section 2.4.3. The component $\overline{E}_{sc}$ can be easily computed from the point form of Gauss's law.

To complete the iteration cycle, the space charges ρ at the grid nodes are computed knowing $\overline{E}$. For this purpose, the residual error R_{es} of the current continuity at each node is determined as

$$R_{es} = \nabla \cdot k\rho E \qquad (2.71)$$

Electric Fields

Setting the integral of R_{es} over the region of interest to zero leads to a set of algebraic equations that can be solved for nodal values of ρ. A weighting function for each element could be included in the integration of R_{es}.

The values of the electric field at the wire surface in each iteration cycle are compared with the onset value and the deviation is used to increase or decrease ρ appropriately to speed up the convergence. The CS technique combined with this method of weighted residuals has been used successfully in solving monopolar and bipolar dc ionized fields (Oin et al., 1986).

2.7 ELECTRIC STRESS CONTROL AND OPTIMIZATION

Field stresses are controlled in much high-voltage equipment, such as cable terminations, high-voltage bushings, and potential transformers. Measures have been suggested and employed with the aim of optimizing the stress throughout every part of the insulation of high-voltage equipment in order to achieve the most economical designs.

2.7.1 Electric Stress Control

For example, the stress-controlled capacitor bushing will be discussed briefly to shed some light on the problems involved. A capacitor bushing is used to bring a high-voltage conductor through the ground case, for example, a tank of a high-voltage transformer without excessive electric fields between the conductor and the edge of the hole in the tank. This is achieved in practice by the introduction of floating screens separating dielectric layers that are interleaved at equal distances between the conductor and the grounded case, thus creating a uniform radial field. The bushing of a uniform dielectric should have a hyperbolic surface profile.

2.7.2 Electric Stress Optimization

High-voltage insulators usually serve as supports and spacers of HV electrodes with respect to earthed frames as in gas-insulated systems or the ground plane in air (see Chapters 9 and 10). Flashover takes place along the insulator surface if the potential field is high enough to sustain a discharge. Therefore, the tangential field should be kept below the critical level everywhere along the insulator surface while keeping the overall length of the insulator at the minimum for a given voltage rating, to minimize the cost of material and installation.

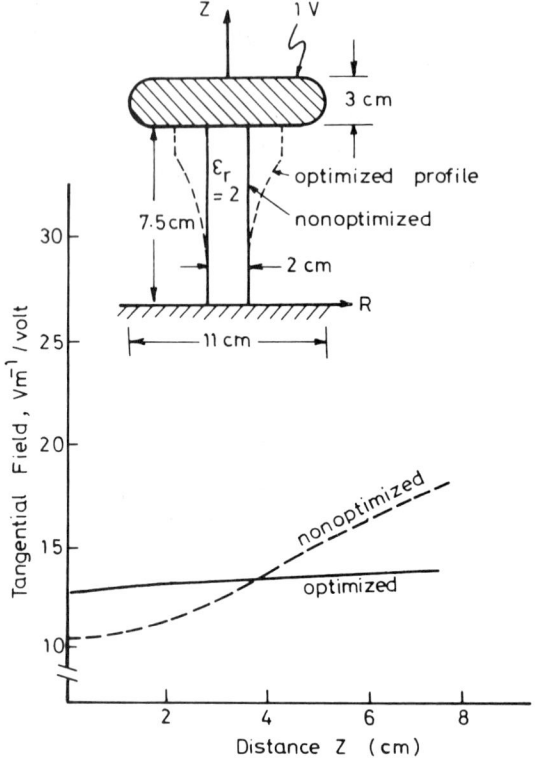

Fig. 2.20 Tangential-field distribution along a profile for optimized and nonoptimized insulators. [From Abdel-Salam and Stanek (1987b).]

Optimization of an HV insulator design was achieved by correcting its profile in elemental steps. At any point on the insulator surface the tangential field is derived from the potential difference between two test points close to the point under consideration. If the tangential field exceeds the set level, either the potential difference is decreased or the distance between the test points is increased (Gronwald, 1983). Recently, a method developed to calculate the tangential field component was applied to mathematical expressions of the profile to be corrected. The method was based on a modified charge-simulation technique whereby the magnitudes of the simulation charges were determined by minimizing a novel error function formulated over the HV electrode and insulator surface (Fig. 2.20) (Abdel-Salam and Stanek, 1987b).

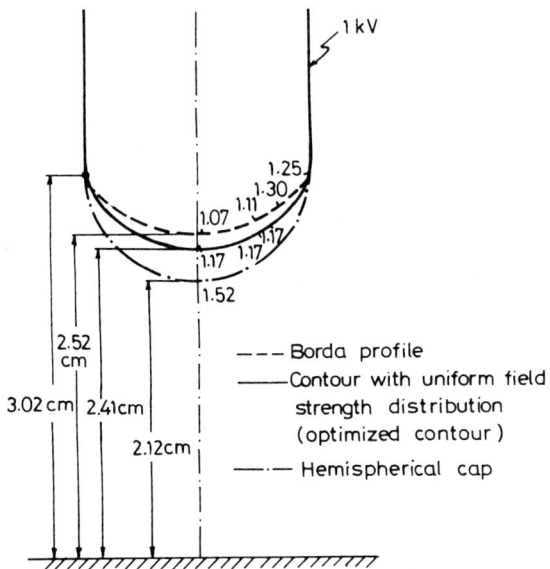

Fig. 2.21 Optimization of a rod-plane gap. Borda profile: field values = 1.07–1.25 kV/cm; optimized contour; field values = 1.17 kV/cm; hemispherical cap: field values = 1.25–1.52 kV/cm.

Shaping of high-voltage electrodes within insulator systems is aimed at bringing the surface field stress down to the minimum possible, thus raising the withstand voltage of a given gap (e.g., a limited space inside a high-voltage laboratory). Such electrode optimization was based on an interactive process of charge simulation in which the contour points are shifted after each computation of field values (Fig. 2.21). Other optimization techniques for electrodes were based on "optimized" simulation charge obtained by minimizing an objective function (Moller and Yousef, 1984).

REFERENCES

Abdel-Salam, M. (1987). *Journal of Physics D: Applied Physics,* 20:629.
Abdel-Salam, M. and El-Mohandes, Th. M. (1987). *Proc. Int. Symposium on High Voltage Engineering,* Braunschweig, West Germany, Paper 33.11.

Abdel-Salam, M. and El-Mohandes, Th. M. (1989). *IEEE Trans.*, IA-25, September/October.
Abdel-Salam, M. and Ibrahim, A. (1977). *IEEE-PS paper A-77-131-6*.
Abdel-Salam, M. and Khalifa, M. (1974). *Acta Physica*, 36:201.
Abdel-Salam, M. and Shamloul, D. (1988). *Proc. IEEE-IAS Annual Meeting*, Pittsburgh, Pa., p. 1677.
Abdel-Salam, M. and Stanek, E. K. (1987a). *IEEE Trans.*, IA-23:481
Abdel-Salam, M. and Stanek, E. K. (1987b). *IEEE Trans.*, EI-22:47.
Abdel-Salam, M., Farghally, M., and Abdel-Sattar, S. (1982). *IEEE Trans.*, PAS-101:4079.
Abdel-Salam, M., Farghally, M., and Abdel-Sattar, S. (1983). *IEEE Trans.*, EI-18:110.
Abdel-Salam, M., Farghally, M., Abdel-Sattar, S., and Shamloul, D. (1984). *Proc. 4th Int. Symposium on Gaseous Dielectrics*, Knoxville, Tenn., p. 492.
Anis, H. and Abd-Allah, M. (1989). *Proc. Middle East Power System Conference*, Cairo, January, pp. 116–121.
Cendes, Z. J. and Hamann, J. R. (1981). *IEEE Trans.*, PAS-100:1806.
Chen, H. S. and Pearmain, A. J. (1983). *Proc. 4th Int. Symposium on High Voltage Engineering*, Athens, Paper 11.07.
Comber, M. and Zaffanella, L. (1974). *IEEE Trans.*, PAS-93:81.
Daffe, J. and Olsen, R. G. (1979). *IEEE Trans.*, PAS-98:1609.
Davis, J. L. and Hoburg, J. F. (1986). *J. Electrost.*, 18:1–22.
El-Arabaty, A., Abdel-Salam, M., and Mansour, E. (1977). *IEEE-PES paper A-77-236-3*.
Foo, P. Y. and King, S. Y. (1976). *Proc. IEE*, 123:702.
Gronwald, H. (1983). *Proc. 4th Int. Symposium on High Voltage Engineering*, Athens, Paper 11.01.
Hornstein, M. N. (1980). *Proc. IEEE-IAS Annual Meeting*, Philadelphia, p. 1081.
Ibrahim, A. A. and Singer, H. (1982). IEE Conference Publication, *Gas Discharges & Their Application*, Institution of Electrical Engineers, Stevenage, Herts, England, p. 128.
Ibrahim, A. A. and Singer, H. (1984). *Proc. 4th Int. Symposium on Gaseous Dielectrics*, Knoxville, Tenn., p. 106.
James, M. L., Smith, G. M., and Wolford, J. C. (1985). *Applied Numerical Methods for Digital Computation*, Harper & Row, Publishers, Inc., New York.
Jennings, A. (1977). *Matrix Computation for Engineers and Scientists*, John Wiley & Sons, Inc., New York.
Khalifa, M. and Abdel-Salam, M. (1973). *Proc. IEE*, 120:1574.

Khalifa, M. and Abdel-Salam, M. (1974a). *IEEE Trans.*, *PAS-93*: 1699.
Khalifa, M. and Abdel-Salam, M. (1974b). *IEEE Trans.*, *PAS-93*: 720.
Khalifa, M., Abdel-Salam, M., Aly, F., and Abou-Seada, M. (1975). *IEEE—PES paper A-75-563-7*.
Khan, M. J. and Alexander, P. H. (1982). *IEEE Trans.*, *EI-17*: 325.
Moller, K. and Yousef, F. (1984). *ETZ-Archiv*, 6:143.
Mukherjee, P. K. and Roy, C. K. (1983). *Proc. 4th Int. Symposium on High Voltage Engineering*, Athens, Paper 12.09.
Oin, B. L., Sheng, J. N., Yan, Z., and Gela, G. (1986). *IEEE papers 86T and D-513-6*.
Okubo, H. and Metz, D. (1978). *Archiv fuer Electrotechnik*, 60: 27.
Okubo, H., Ikeda, M., Honda, M., and Tanari, T. (1982). *IEEE Trans.*, *PAS-101*:4039.
Ryan, H. M., Ali, S. M. G., and Powell, C. W. (1983). *Proc. 4th Int. Symposium on High Voltage Engineering*, Athens, Paper 12.12.
Scott, M., Mattingley, J., and Ryan, H. (1974). *IEEE Trans.*, *EI-9*:18.
Shen, L. C. and Kong, J. A. (1983). *Applied Electromagnetism*, Brooks/Cole Publishing Co., Pacific Grove, Calif.
Shephard, M. S. and Gallagher, R. H. (1979). *Finite Element Grid Optimization*, American Society of Mechanical Engineers, New York.
Silvester, P. P. and Ferrari, R. L. (1983). *Finite Elements for Electrical Engineers*, Cambridge University Press, Cambridge.
Singer, H. (1981). *ETZ-Archiv*, 3:265.
Singer, H. (1984). *Archiv fuer Electrotechnik*, 67:309.
Singer, H., Steinbigler, H., and Weiss, P. (1974). *IEEE Trans.*, *PAS-93*:1660.
Steinbigler, H. (1979). *Proc. 3rd Int. Symposium on High Voltage Engineering*, Milan, Paper 11.11.
Takuma, T., Ikeda, T., and Kawamoto, T. (1981a). *IEEE Trans.*, *PAS-100*:2802.
Takuma, T., Kawamoto, T., and Fujinami, H. (1981b). *IEEE Trans.*, *PAS-100*:4665.
Tan, K. X. and Steinbigler, H. (1985). *COMPEL*, 4:209.
Timascheff, A. (1975). *IEEE Trans., PAS-94*:104.
Zeitoun, A., Abdel-Salam, M., and El-Ragheb, M. (1976). *IEEE—PES paper A-76-418-4*.
Zienkiewicz, O. C. (1977). *The Finite Element Method*, McGraw-Hill Book Company (U.K.), Ltd., Maidenhead, Berkshire, England.

3
Ionization and Deionization Processes in Gases

M. ABDEL-SALAM *Assiut University, Assiut, Egypt*

3.1 INTRODUCTION

In the process of partial or complete breakdown of a gas gap there are several mechanisms of electron and ion generation or annihilation in operation either singly or in combination. In the present chapter we summarize briefly the more significant mechanisms for ionization and deionization in a gas discharge. The processes considered include:

1. Ionization by cosmic rays, x-rays, and nuclear radiation
2. Ion generation by electron impact, photoionization, interaction with metastable atoms, thermal ionization, and electron detachment
3. Deionization by recombination and by diffusion

Also, in becoming attached to a neutral particle, an electron produces a negative ion. Before discussing these processes it would be best first to describe the behavior of gaseous dielectrics in an electric field. This calls for a review of some basic principles of the kinetic theory of gases as they pertain to gaseous ionization and deionization processes.

3.2 KINETIC THEORY OF GASES

Many of the properties of gases can be well represented by kinetic theory, which treats the gas molecules as tiny balls with no need to specify their exact size, shape, or internal construction. These properties include gas pressure, temperature, distribution of molecular speeds, and energies.

3.2.1 Kinetic Interpretation of Gas Pressure

Let us consider a gas composed of moving molecules, each with mass m_0. The cumulative effect of many molecules striking against a wall leads to a more-or-less steady force (pressure) on the wall, depending on the number of molecules hitting the wall per second. Consider the ith molecule of the gas in a cubic container; it has a velocity component v_{ix} in the X direction (Fig. 3.1). Assume that collision with the wall reverses the velocity component v_{ix}, so that the time Δt_i between two successive collisions of the ith molecule with the right-hand wall is

$$\Delta t_i = \frac{2D}{v_{ix}}$$

where D is the container-side length.

The impulse exerted on the molecule during a collision is equal to $2m_0 v_{ix}$. If an average force F_{ix} is exerted by the molecule on the wall during time Δt_i, then with m_0 being the mass of the gas molecule,

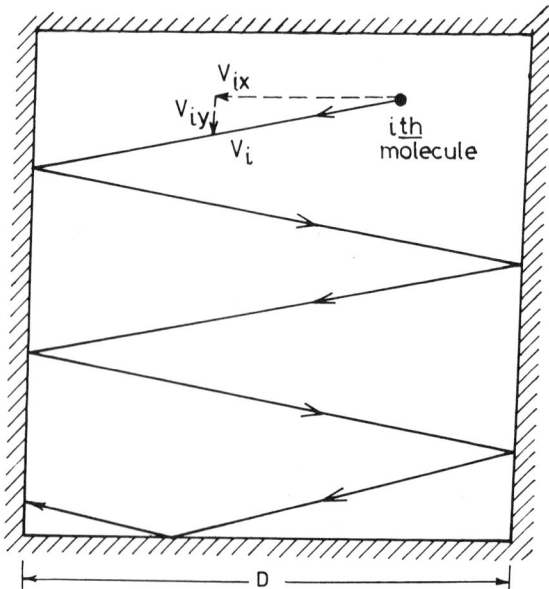

Fig. 3.1 Motion of a gas molecule in a cubic container.

Ionization and Deionization Processes in Gases

$$F_{ix} = \frac{2m_0 v_{ix}}{\Delta t_i} = \frac{m_0 v_{ix}^2}{D}$$

The total force F_x on the wall due to ND^3 molecules in the container will be the sum of the average force due to each. N is the number of molecules per unit volume.

$$F_x = \sum_{i=1}^{i=ND^3} F_{ix} = \frac{m_0}{D} ND^3 (\overline{v_x^2})$$

where $\overline{v_x^2}$ is the mean square of the x-component velocities of the molecules. Since the number of molecules is very large and there is no preferred direction in the container,

$$\overline{v_x^2} = \overline{v_y^2} = \overline{v_z^2} = \frac{\overline{v^2}}{3}$$

where $\overline{v^2}$ is the average of the squared velocities of the molecules. The gas pressure p then becomes

$$p = \frac{F_x}{D^2} = \frac{1}{3} m_0 N \overline{v^2} = \frac{2}{3} N \left(\frac{1}{2} m_0 \overline{v^2} \right) \tag{3.1}$$

3.2.2 Kinetic Interpretation of Gas Temperature

If a mass m of a gas of molecular weight M is confined to a volume G, the general gas law applies for the pressure p:

$$pG = \frac{m}{M} RT \tag{3.2}$$

where T is the absolute gas temperature and R is the universal gas constant. Combining equations (3.1) and (3.2) and realizing that $m = NGm_0$, we have,

$$T = \frac{1}{3} \frac{M}{R} \overline{v^2} = \frac{1}{3} \frac{N_0}{R} m_0 \overline{v^2} = \frac{2}{3k} \left(\frac{1}{2} m_0 \overline{v^2} \right) \tag{3.3}$$

where the mass of molecule m_0 is the molecular weight M divided by Avogadro's number N_0, and k is Boltzmann's constant. Thus the absolute temperature of the gas is a measure of the square velocity of the molecules within it. Thus the temperature T is directly proportional to the average translational kinetic energy of the molecules in a gas at that temperature.

3.2.3 Distribution of Molecular Speeds

Not all molecules in a gas travel with equal speed. The speed variation among the molecules is usually expressed by plotting the number N_v of molecules having speeds within a unit interval ($\Delta v = 1$ m/s) centered about the speed v (Fig. 3.2). The total number N_t of molecules is represented by the area under the curve, that is,

$$N_t = \int_0^\infty N_v \, dv$$

The speed probability distribution function $P_r(v)$ is the percentage of gas molecules having speeds within the range $v \pm dv/2$. The expression for $P_r(v)$ is

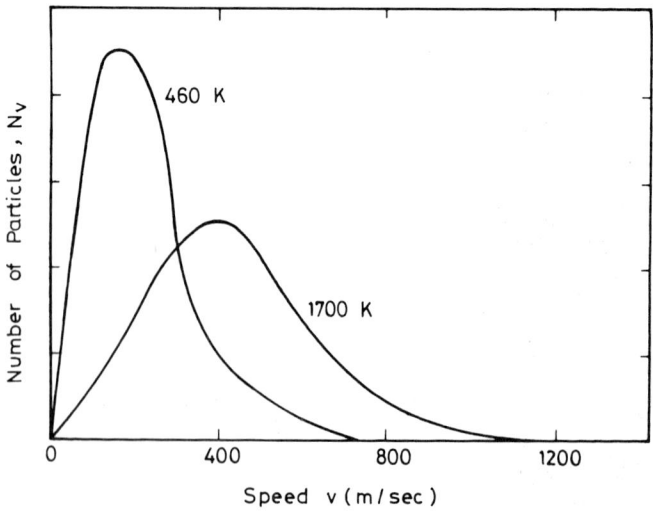

Fig. 3.2 Number of molecules N_v versus speed v at 460 and 1700 K. The vertical axis is linear in arbitrary units.

Ionization and Deionization Processes in Gases 69

$$P_r(v) = \frac{4}{\pi}\left(\frac{m_0}{2kT}\right)^{3/2} v^2 \exp\left(\frac{-m_0 v^2}{2kT}\right) \qquad (3.4)$$

Note that $P_r(v) \rightarrow 0$ only as $v \rightarrow 0$ and as $v \rightarrow \infty$. Therefore, all speeds are possible for the molecules.

3.2.4 Maxwell-Boltzmann Distribution Function

Very often interest is directed to how the gas molecules are distributed in regard to their energies. Since $P_r(v)\,dv$ is the fraction of molecules having speeds in the range dv centered on v, it is convenient to represent the equivalent kinetic energy distribution as $P_r(U)\,dU$ as

$$U = \frac{1}{2} m_0 v^2 \qquad \text{and} \qquad dU = m_0 v\, dv$$

Therefore, the fraction of molecules with translational kinetic energies in the range dU centered on U is given by

$$P_r(U)\, dU = \frac{2}{\sqrt{\pi}\,(kT)^{3/2}} \sqrt{U}\, e^{-U/kT}\, dU \qquad (3.5)$$

This distribution is shown in Figure 3.3, where the most probable energy occurs for $U = (1/2)kT$. The average value $\overline{U}$ of the kinetic energy is given by

$$\overline{U} = \int_0^\infty U P_r(U)\, dU = \frac{3}{2} kT \qquad (3.6)$$

Since $\overline{U} = (1/2)m_0 \overline{v^2}$, then $\overline{v^2} = 3kT/m_0$, as deduced before [Equation (3.3)].

The question is: What would happen if the gas is to exist in a region where its potential energy varies from place to place? The Maxwell-Boltzmann distribution law tells us how the number of molecules N changes with the total energy V. The number of molecules N_1 that will be found in state 1, say, is related to the number N_2 in state 2 by the relation

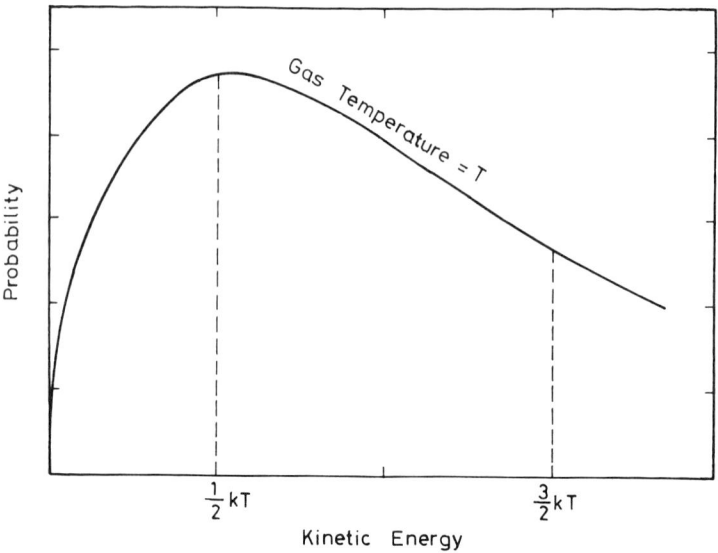

Fig. 3.3 Probability $P_r(U)$ versus kinetic energy U. The vertical axis is linear.

$$\frac{N_1}{N_2} = \exp\left(\frac{V_2 - V_1}{kT}\right) \tag{3.7}$$

which is known as the Maxwell-Boltzmann distribution.

3.3 BEHAVIOR OF A GASEOUS DIELECTRIC IN AN ELECTRIC FIELD

The atoms of a gas can absorb, transport, and release energy. The energy would be acquired from or given up to other atoms of the gas. The internal energy of an atom is associated with the energy levels of its electrons. Whenever an electron drops or rises from one energy level to another, a quantum of energy is released or absorbed as a photon. These photons have definite magnitudes corresponding to the specific energy levels the electrons can acquire in the atoms (McDaniel, 1964).

The electrons in the atom have a tendency to revert to their lowest possible energy levels. When all the electrons occupy their permitted energy levels, the atom has a "normal state." When

Ionization and Deionization Processes in Gases 71

some electrons are at higher energy levels but without leaving the atom, it is in an "excited state" lasting for a very short duration (10^{-8} s). The energy required is known as the "excitation energy" W_e. The atom has to release this energy to return to its normal state. The radiated photon has a wavelength λ:

$$\lambda = \frac{ch}{W_e} \tag{3.8}$$

where c is the velocity of light and h is Planck's constant.

Some atoms have certain energy levels to which an internal electron may be raised but from which it cannot revert easily to its normal energy level. This state of the atom is known as a "metastable state." It ordinarily gives up its energy to other atoms in the gas, to the electrodes, or to the walls of the container. The duration of a metastable state is on the order of 0.1 s.

If an electron acquires enough energy, it gets separated from its atom by ionization. What remains of the atom after ionization is a positive ion. The electron and the ion resulting from the ionization move independently. Under the action of an applied field, they become accelerated and acquire kinetic energies. It may be noted that the ion possesses both ionization energy and kinetic energy. It will release the ionization energy only when an electron recombines with it to form a normal atom. Negative ions, formed by electrons attached to neutral atoms or molecules, have much less internal energy than do positive ions. Negative ions have a larger probability of occurrence if the gas is electronegative (i.e., its molecules have a strong affinity for electron attachment).

Electrons moving in a gas under the action of an electric field are bound to make numerous collisions with the gas molecules. The electrons are accelerated and their energies rise between collisions by amounts depending on their free paths and the applied electric field intensity. The electron velocity component along the field direction taken over several free paths per unit field intensity is termed the electron mobility.

Similarly, the positive ions receive energy from the field. When these positive ions collide with gas molecules, they lose a considerable part of their energy (about 50% on the average), as the ion and gas molecule have closely comparable masses. Therefore, their energies are lower than those of electrons in similar situations. Thus the positive ions are less capable of ionizing the gas molecules. An ion needs to have a kinetic energy twice the ionization energy in order to ionize the atom it collides with, whereas an electron only needs a kinetic energy equal to the ionization energy. In general, ions have very large masses and small speeds. They are therefore responsible for producing space-charge clouds in the interelectrode region.

Photons have neither mass nor charge. They are thus independent of the applied field. As they give up their energy to atoms or molecules, they vanish. Photons bombarding a cathode give up their energies to help release electrons from the cathode surface. Excited atoms and metastable atoms are not influenced by the electric field because they are neutral.

3.4 ION-GENERATION PROCESSES

3.4.1 Ionization by Electron Impact

Ionization by electron impact is probably the most important ionization process for a gas discharge. Its effectiveness depends on the electron energy. When the electron collides with an atom or molecule, kinetic energy is exchanged. If no excitation or ionization results, the collision is called "elastic." If it is inelastic, the gas atom or molecule becomes excited or ionized by the energy acquired from the incident electron. This means that a portion of the kinetic energy of the electron prior to impact has been converted to potential energy of the atom or molecule. An excited atom or molecule may become ionized by a subsequent collision with another slow-moving electron. This process becomes significant only when densities of electrons are high. Very fast electrons are also poor ionizers, as the period of interaction between the electron and the atom becomes too short, and therefore the amount of energy transferred from the electron to the atom decreases significantly. For every gas there exists an optimal electron energy that gives a significant ionization probability.

Ionization Cross Section

In some models, the atoms, molecules, and electrons are assumed to be rigid solid-material balls with the aim of deriving very useful expressions, such as the concept of collision cross section and free path. Thus, when the distance between an electron and an atom is equal to the sum ($r_e + r_a$) of their radii, a collision must have taken place (Nasser, 1971). The collision cross section

$$\sigma = \pi(r_e + r_a)^2 \tag{3.9}$$

The values of the ionization cross sections for various gases resulting from collision with electrons of energies between 10 and 1000 eV are given in Figure 3.4. If a gas has N molecules per

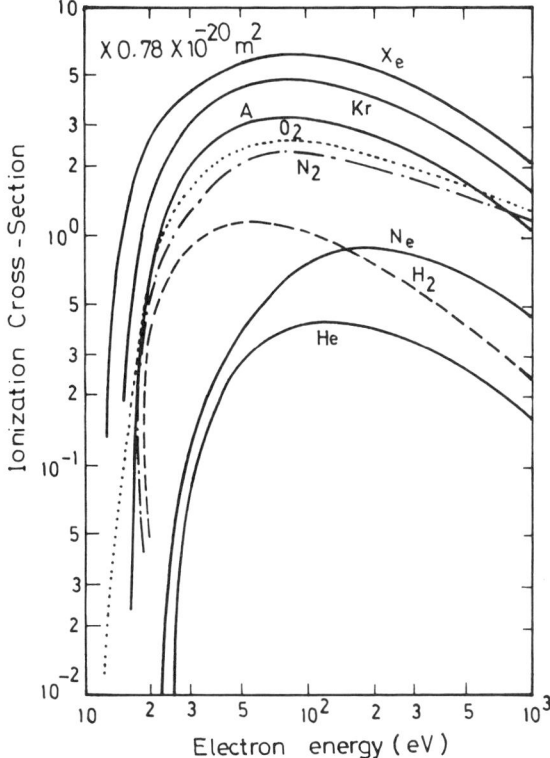

Fig. 3.4   Ionization cross sections for various gases due to electron impact. [From Rapp and Englander-Golden (1965).]

unit volume, the number of collisions per unit length of the electron path is equal to σN and the mean free path $\bar{\lambda}$ is

$$\bar{\lambda} = \frac{1}{\sigma N} = \frac{1}{\pi N (r_e + r_a)^2} \qquad (3.10)$$

The mean free path is a statistical average and indicates only the average distance between successive collisions. It varies from one gas to another (Table 3.1). Increased gas pressure p or decreased gas temperature proportionately increases its density and hence N. Therefore, the mean free path $\bar{\lambda}$ is inversely proportional

Table 3.1 Values of the Mean Free Path $\bar{\lambda}$ of Some Gas Molecules[a]

Gas	H_2	He	Ne	N_2	O_2	A	CO_2	Kr	Xe
$\bar{\lambda}$ (10^{-8} m)	11.77	18.62	13.22	6.28	6.79	6.66	4.19	5.12	3.76

[a] T = 288 K and p = 101.3 kPa.
Source: McDaniel (1964).

to p at constant temperature. The probability f_x that an electron will have a collision while traveling a distance x can be given by a relationship of the form

$$f_x = \exp\left(\frac{-x}{\bar{\lambda}}\right) \qquad (3.11)$$

where x is the distance traveled without having a collision.

The free-path distribution function can be derived by considering a group of n_0 molecules just after a collision. When these molecules are moving in a given direction, n of them will reach a distance x without collision. Note that n is related to n_0 through the probability f_x:

$$n = n_0 f_x = n_0 \exp\left(\frac{-x}{\bar{\lambda}}\right) \qquad (3.12)$$

Equation (3.12) gives the distribution of free paths where the ratio n/n_0 decays exponentially with distance x.

Ionization Coefficient

When an electron travels a distance equal to its free path λ_e in the direction of the field E, it gains an energy = $eE\lambda_e$. For the electron to ionize, its gain in energy should be at least equal to the ionization potential V_i of the gas:

$$e\lambda_e E \geqslant eV_i \qquad \text{or} \qquad \lambda_e \geqslant \frac{V_i}{E} \qquad (3.13)$$

Table 3.2 lists values of V_i for the molecules of some gases and atoms of some vapors.

The ionization coefficient α, known as "Townsend's first ionization coefficient," is the number of ionizing collisions per unit length

Table 3.2 Excitation and Ionization Potentials of Some Gases and Metal Vapors

Material	First excitation potential (eV)	Ionization potential (eV)	
		Single ionization	Double ionization
A	11.56 11.49 (metastable) 11.66 (metastable)	15.8	27.5
He	20.9 19.8 (metastable)	24.6	54.1
Ne	16.58 16.53 (metastable) 16.62 (metastable)	21.6	40.9
Xe	8.39 8.28 (metastable) 9.4 (metastable)	12.1	21.2
H_2	11.2	15.4	
N_2	6.1	15.5	
O_2		12.2	
F_2		17.8	
SF_6		19.3	
H_2O		12.6	
CO_2		14.4	
CO		14.1	
Cl_2		13.2	
S		10.3	
K		4.3	
Cu		7.7	
Cs		3.9	

Source: Meek and Graggs (1953).

of the path. It is equal to the number of free paths ($= 1/\bar{\lambda}_e$) times the probability of a free path being more than the ionizing length λ_{ie} (Kuffel and Abdullah, 1970). It can be expressed as

$$\alpha = \frac{1}{\bar{\lambda}_e} \exp\left(\frac{-\lambda_{ie}}{\bar{\lambda}_e}\right) = \frac{1}{\bar{\lambda}_e} \exp\left(\frac{-V_i}{\bar{\lambda}_e E}\right) \qquad (3.14)$$

or

$$\frac{\alpha}{p} = \Gamma \exp\left(\frac{-AV_i p}{E}\right) = F\left(\frac{E}{p}\right) \qquad (3.15)$$

as found experimentally. Typical values for the constant Γ of equation (3.15) are 11253.7, 9003.0, and 3826.3 m^{-1} for air, N_2 and H_2, respectively. Typical values of α/p of the technically interesting gases (air, N_2, and SF_6) versus E/p at standard temperature and pressure (STP) are given in Figure 3.5.

3.4.2 Photoionization

It should be pointed out that photoionization does not occur simply due to radiation emitted from the gas itself when excited atoms return to their ground state but also takes place as a result of external radiations, such as cosmic rays, x-rays, or nuclear radiations. Ionization by cosmic rays is a continuous process that produces ions and electrons everywhere since these rays are capable of penetrating most conventional walls. The fact that electrons and ions are always available in air has great significance and consequences in many applications. Without free electrons, it would not be easy to produce a spark, for example, or to ignite the combustible mixture in internal combustion engines. Measurements have shown that the intensity of cosmic rays increases sharply with altitude to reach 20 times the sea-level value at about 16 km, and then decreases gradually with further height. Its rate of ionization also increases steadily from a very small value at sea level to about 80×10^6 ion pairs per cubic meter per second at an altitude of 10 km (Nasser, 1971). It must be kept in mind that the insulation of high-voltage systems at high altitudes is subject to reduced air density and to increased ionization by cosmic rays.

X-rays and γ-rays ionize gases because of their high energies. Their energies range from a few electron volts to 100 MeV or more. Because of their highly energetic photons, their ionizing processes are bound to differ from those of radiations with relatively lower energies such as ultraviolet as follows:

Ionization and Deionization Processes in Gases

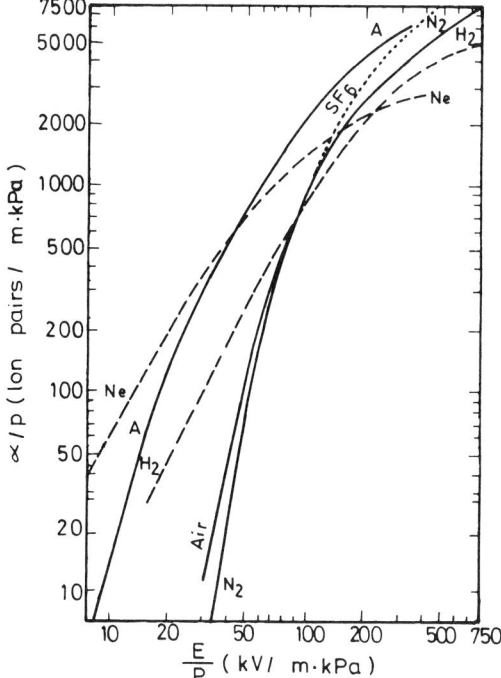

Fig. 3.5 α/p as a function of E/p for different gases. [From Brown (1966), and Bahalla and Craggs (1962).]

1. When a gas atom absorbs an x-ray photon, a loosely bound valence electron may be liberated using part of the quantum energy. The excess energy of the photon, which is usually very high, will be converted to kinetic energy for the electrons. This electron may well ionize more atoms by collision, forming an avalanche-like ionization.
2. A high-energy x-ray photon may knock out an electron from an inner shell of the atom. The high energy absorbed may be adequate to give the electron enough momentum to leave the atom. In the ionized atom, the missing electron is substituted by another from the next-higher shell. This process is accompanied by the release of energy in the form of the emission of another x-ray of lower energy than the primary x-ray quantum. This new x-ray photon can ionize additional atoms, just as in the preceding case.

3. The atom does not completely absorb the incident x-ray quantum and the so-called Compton effect occurs with scattering of x-rays with longer wavelengths.

If A represents a neutral atom in the gas, e^- an electron, KE the kinetic energy of the electron, and A* the excited atom of the gas, the reaction describing the excitation of the atom is

$$A + e^- + KE \longrightarrow A^* + e^- \qquad (3.16)$$

On recovering from the excited state in about 10^{-8} s, the atom radiates a photon that may in turn ionize another molecule whose ionization potential (eV_i) is equal to or less than the photon energy. The photoionization process may be described by the reaction

$$A + h\nu \longrightarrow A^+ + e^- \qquad (3.17)$$

where $h\nu$ is the photon energy and e^- is the ejected photoelectron.

The excess of the quantum $h\nu$ over eV_i may be imparted to the released electron as kinetic energy. If the photon energy is less than eV_i, it may still be absorbed by the atom, which gets excited only to a higher-energy level (i.e., a photoexcitation).

A photon emitted from an atom returning to its ground state may be absorbed by another atom of the same gas, exciting it. The same thing may continue to happen within a gas until the photon is lost to the boundaries. Such a state of affairs is known as "resonance" radiation and is encountered frequently in ionized gases.

Ionization by a beam of photons can be analyzed in the same way as that produced by a beam of electrons, and the concept of free path may be extended to photons. Thus the photon mean free path $\bar{\lambda}_{ph}$ is expressed as

$$\bar{\lambda}_{ph} = \frac{1}{\sigma N} = \frac{1}{\pi r_a^2 N} \qquad (3.18)$$

and the photon number n_{ph} decreases with the distance x traveled, that is,

$$n_{ph} = n_{pho} e^{-x/\bar{\lambda}_{ph}} \qquad (3.19)$$

where n_{pho} is the photon number at their origin.

Being radiation, photons get reduced in intensity by dI as they travel a distance dx at x from their origin. dI is proportional to dx and I at the distance x. Then

Ionization and Deionization Processes in Gases

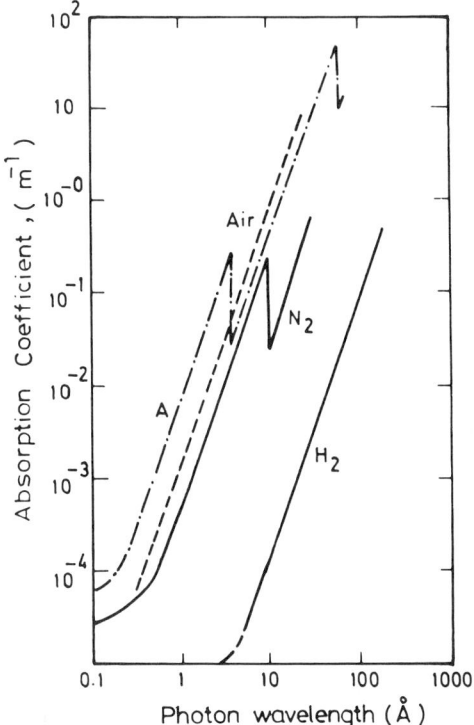

Fig. 3.6 Absorption coefficient for x-rays in different gases as a function of wavelength at a temperature of 0°C and pressure of 133 Pa.

$$-dI = \mu I \, dx$$

where μ is the absorption coefficient of the gas. Integrating, we get

$$I = I_0 e^{-\mu x} \tag{3.20}$$

where I_0 is the intensity of photons at their origin.

Comparing equations (3.19) and (3.20), we get

$$\mu = \frac{1}{\bar{\lambda}_{ph}} \tag{3.21}$$

Figure 3.6 shows μ as a function of the photon wavelength for different gases.

3.4.3 Ionization by Interaction with Metastables

For the atoms of inert gases and of some elements in group II of the periodic table, the lifetime for their metastable states extends to seconds. These states are represented by A^m and associated with energies V^m.

If V^m of an atom A^m exceeds the ionization V_i of another atom B, then on collision, ionization may result according to the reaction

$$A^m + B \longrightarrow A + B^+ + e^- \tag{3.22}$$

For $V^m < V_i$, the reaction would produce only excitation of the atom B.

Another possibility for ionization by metastables is when $2V^m$ for A^m is greater than V_i for A; the reaction may proceed as follows:

$$2A^m + A \longrightarrow A_2^* + A$$

$$A_2^* \longrightarrow A + A^+ + e^- \tag{3.23}$$

where A_2^* is a diatomic molecule in the excited state. Sometimes, an atom in the metastable state may be ionized by absorbing another photon before returning to its ground state. This process is known as step ionization.

$$A^m + h\upsilon \longrightarrow A^+ + e^- \tag{3.24}$$

As mentioned before, ionization by metastable interactions comes into operation long after excitation. It has been shown by some research work that these reactions are responsible for the long time lags of breakdown observed in certain gases (Hartmann and Gallimberti, 1975).

3.4.4 Ionization by Nuclear Particles

Nuclear raditions include α- and β-particles, as well as photons of γ-rays. The energies of α-particles are on the order of MeV. Beta-particles are fast electrons having energies on the same order of magnitude. Both types of particles make collisions with the gas atoms and cause ionization by collision. γ-rays were discussed in Section 3.4.2.

The ionization effects of particles, heavier than electrons, can be considered by applying the principles of elastic and inelastic collisions. The most important particles are the proton and the α-particle. When they are injected into a gas, they undergo collision with the gas atoms, thus losing their kinetic energy to the gas atoms. Because of their heavy masses their direction of travel is not affected much by the collisions. They proceed in almost straight lines, leaving ionized atoms along their wakes until they finally lose all their kinetic energies and come to relative rest (Fikry et al., 1978). It is interesting to note that as in the case of β-rays, the number of ionizations produced by the particle is much greater toward the end of its path when it has slowed down appreciably, thus having a longer interaction with the gas molecules.

The distance traveled by the particle until it is stopped or "absorbed" is a well-defined range that depends on the particle's energy and mass and on the density of the absorbing medium (Table 3.3). The particle range in gases is, of course, much greater than in solids. The range of the heavy particles (α-particles and protons) increases very rapidly with energy (Table 3.3). For equal energies, the proton has a greater range because of its lower mass.

3.4.5 Thermal Ionization

If a gas is heated to a sufficiently high temperature, most of its molecules get dissociated into atoms (Fig. 3.7). Also, ionization of its atoms and molecules may result. Following are the possibilities for thermal ionization:

1. Ionization by collision of the gas atoms with each other. Because of the high temperature, the velocities and kinetic energies are very high and can cause ionization.
2. Photoionization resulting from the thermal emission by the hot gas.

Table 3.3 Absorption Ranges of α-Particles, Protons, and β-Particles in Air at STP

Energy (MeV)	Range (m)		
	α-particle	Proton	β-particle
1	0.005	0.023	3.14
5	0.035	0.34	30.00
10	0.107	1.17	41.00

Source: Nasser (1971).

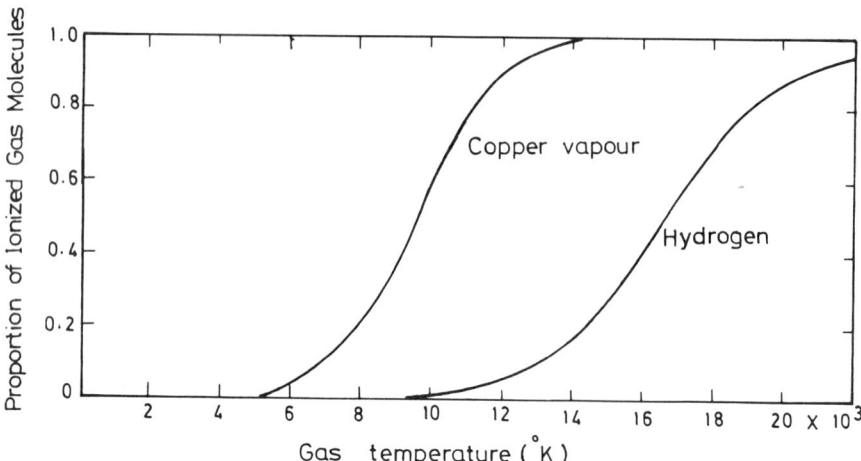

Fig. 3.7 Proportion of ionized gas molecules as a function of gas temperature.

3. Ionization by collision with high-energy electrons produced by the preceding two processes. In fact, the ionization coefficient α increases with temperature (Abdel-Salam, 1976).

Thermal ionization is the chief source of ionization in flames and high-pressure arcs (see Chapter 6). The process is described by the equation

$$A + W_i \rightleftarrows A^+ + e^- \tag{3.25}$$

where A represents a neutral atom, A^+ a singly ionized atom, e^- the electron removed from the atom, and W_i the ionization energy.

3.4.6 Electron Detachment

Under certain conditions in high electric fields, electrons may get detached from negative ions. This process, however, requires a large concentration of negative ions. Loeb (1965) carried out experiments on electron detachment in oxygen and found that it occurs for E/p = 68 V/m·Pa. In compressed electronegative gases such as SF_6, detachment has been considered as one of the main processes involved in the breakdown phenomena over a wide range of gas pressures (Maller and Naidu, 1981; Abdel-Salam, Radwan, Ali, 1978). Gas

Ionization and Deionization Processes in Gases 83

discharges under impulse voltages may be initiated by electrons getting detached from negative ions available in the gas (Allen et al., 1981; Sommerville et al., 1984; Abdel-Salam and Turkey, 1988). Detachment may be ascribed to the following events:

1. Photodetachment by absorption of radiation by a negative ion:

$$A^- + h\nu \rightleftharpoons A + e^- \qquad (3.26)$$

2. Collision of ions with a fast atom:

$$A^- + A \rightleftharpoons 2A + e^- \qquad (3.27)$$

3. Collision and subsequent association with neutral atoms; that is, associative detachment leading to the loss of the excessive electron:

$$A^- + B \rightleftharpoons AB + e^- \qquad (3.28)$$

4. Collision of molecular negative ions with fast excited atoms, leading to a vibrational excited state of the negative ion and subsequent loss of the excessive electron:

$$A_2^- + B^* \rightleftharpoons A_2^* + B + e^- \qquad (3.29)$$

5. Recombination of negative and positive atomic ions to form a diatomic molecule:

$$A^- + B^+ \rightleftharpoons AB^+ + e^- \qquad (3.30)$$

This is a very likely process at pressures above a few hundred pascal.

3.5 DEIONIZATION PROCESSES

3.5.1 Deionization by Recombination

Whenever positively and negatively charged particles are present, there is a chance for recombination to take place. The potential energy and the relative kinetic energy of the recombining electron-ion or ion-ion pair is released as a quantum of radiation:

$$A^+ + B^- \longrightarrow AB + h\nu \qquad (3.31)$$

In this expression B^- may be an electron or a negative ion.

The rate of recombination is directly proportional to the concentrations of both positive ions and negative ions or electrons. If n_+ and n_- are the number of positive and negative ions per unit volume, then

$$\frac{dn_+}{dt} = \frac{dn_-}{dt} = -\zeta \, n_+ n_- \tag{3.32}$$

where ζ is a constant known as the recombination coefficient. In general, $n_+ = n_- = n$, and we have

$$\frac{dn}{dt} = -\zeta n^2 \tag{3.33}$$

If recombination is allowed to take place from $t = 0$ to $t = t$ and the density of charged particles at $t = 0$ is n_0, by integrating between these limits we get

$$n = \frac{n_0}{1 + n_0 \zeta t} \tag{3.34}$$

which indicates that the density of particles decreases hyperbolically with time.

The densities n_+ and n_- no doubt increase with the gas pressure. The variation of ζ with pressure is shown in Figure 3.8,

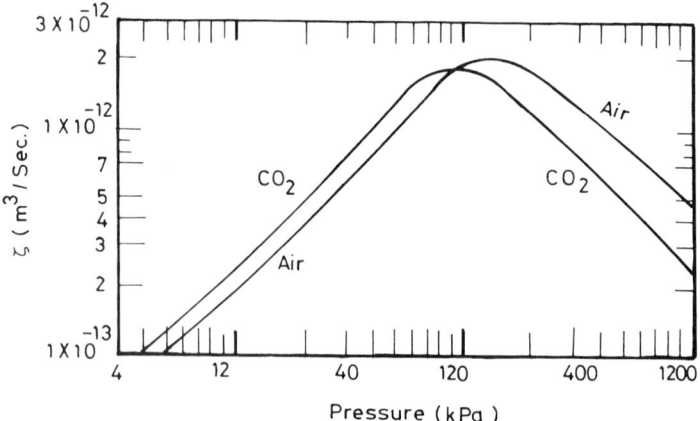

Fig. 3.8 Effect of gas pressure on recombination coefficients (ζ) for air and CO_2.

where the recombination is particularly important at high pressure. On the other hand, ζ does not show a significant increase with temperature.

From experiments in gases in which negative ions can form very easily, such as oxygen, it was found that the recombination of positive ions with negative ions takes place much faster than does the recombination of a positive ion with a free electron in a gas where negative ions cannot form, such as in the rare gases (Nasser, 1971). This is true because the two species of ions have almost equal masses and average velocities and the time they remain in close vicinity is therefore long.

3.5.2 Electron Attachment

When an electron gets attached to a neutral molecule or atom, it produces a negative ion. Negative ions play an important role in the breakdown of the technically interesting gases such as O_2, N_2, SF_6, and air. Even their presence as a small impurity can change the behavior of an ionized gas drastically. Negative-ion formation is very useful in quenching arcs in power circuit breakers in high-voltage generators and accelerators.

In electronegative gases, a general criterion for negative-ion stability can be inferred from consideration of the neutral atom, which is stable because it possesses the lowest energy level of all the possible states. Hence, for a negative ion to exist and be stable for some time, its total energy must be lower than that of the atom in its ground state. The difference in total energy between the negative ion and the unexcited neutral atom is termed the "electron affinity" of the atom. This energy is released as a photon or as kinetic energy upon attachment.

The different modes of negative-ion formation by attachment are:

Radiative attachment: takes place when the excess energy is released as a quantum upon attachment. Such a process is reversible and can be expressed as

$$e^- + A \underset{\text{detachment}}{\overset{\text{attachment}}{\rightleftarrows}} A^- + h\nu \qquad (3.35)$$

Attachment through three-body collision: takes place when the excess energy released upon electron attachment is acquired by a third particle during the collision process as kinetic energy (U). This can be expressed as

$$e^- + A + B \longrightarrow A^- + (B + U) \tag{3.36}$$

Dissociative attachment: takes place in molecular gases when the excess energy is not radiated but is used to separate the two atoms into a neutral particle and an atomic negative ion. Such a process is reversible and is expressed as

$$e^- + XY \rightleftharpoons (XY)^{-*} \rightleftharpoons X^- + Y \tag{3.37}$$

The suppression of electrons from an ionized gas by attachment is expressed by the attachment coefficient η. This coefficient is defined by analogy with the ionization coefficient α as the number of attachments produced in the path of a single electron traveling a unit distance in the direction of the field. Typical values of η/p versus E/p (field-to-pressure ratio) at STP are given in Figure 3.9 for SF_6 and air (Maller and Naidu, 1981).

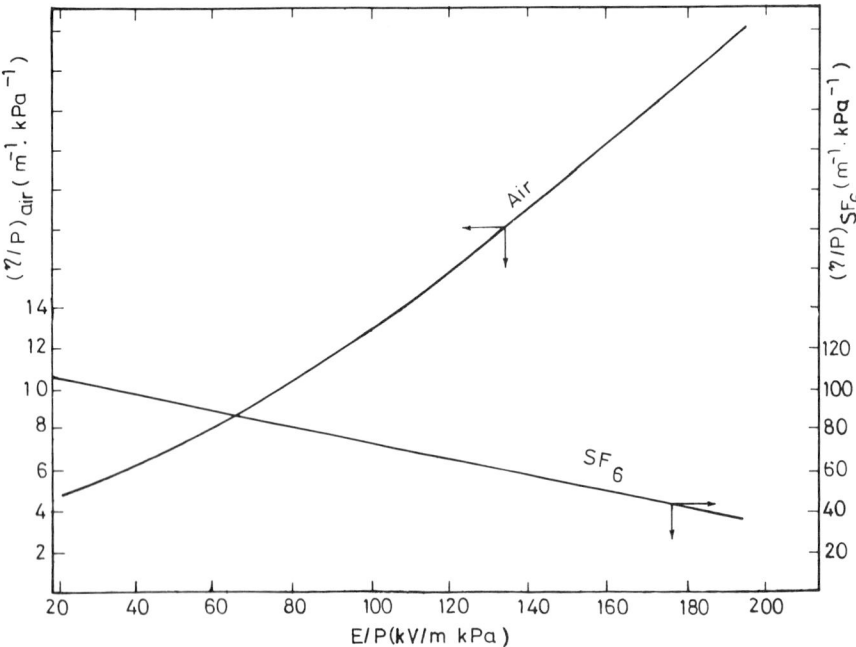

Fig. 3.9 η/p as a function of E/o for air and SF_6. [From Maller and Naidu (1981) and Bahalla and Craggs (1962).]

3.5.3 Deionization by Diffusion

During the extinction of arcs in circuit breakers, electron and ion diffusion is one of the most active deionizing agents. The faster the arc of the circuit breakers can be deionized, the higher their arc-interrupting capacity. In electrical discharges, whenever there is a nonuniform concentration of ions there will be diffusion of ions from regions of higher to regions of lower concentration. This diffusion will cause a deionizing effect in the regions of the higher concentration. The flow of ions along a concentration gradient constitutes a drift velocity similar to that under electric fields. The rate of ion flow is

$$J = -D \, \nabla n \qquad (3.38)$$

where $D = \bar{v}\bar{\lambda}/3$ m^2/s is the diffusion coefficient.

As the concentration gradient varies with time, the rate of change of ion density is

$$\frac{\partial n}{\partial t} = D \, \nabla^2 n = -\nabla \cdot \bar{J} \qquad (3.39)$$

which is known as the continuity equation. Its solution would give the concentration of ions at any time and at any point in space. For the case of diffusion from a cylindrical concentration, the average square of displacement will be given by (Nasser, 1971)

$$(\bar{r})^2 = 4Dt \qquad (3.40)$$

This equation is of greatest interest when we are dealing with discharges in cylindrical channels comprising electrons and positive ions. In the same gas, D is three orders of magnitude higher for electrons than for ions, owing to the greater average velocities $\bar{v}$.

3.6 GAS-DISCHARGE PARAMETERS AS INFLUENCED BY ATMOSPHERIC CONDITIONS

3.6.1 Ionization Coefficient, α

Curve fitting of the data measured in air results in accurate expressions of α/p (m^{-1} kPa^{-1}) as a function of E/p (V·m^{-1} kPa^{-1}) (Abdel-Salam and Abdellah, 1978). In SF$_6$, α/p is related to E/p (Boyd and Crichton, 1971);

$$\frac{\alpha}{p} = 12{,}937.5 \left(\frac{E}{p}\right) - 1230 \qquad (3.41)$$

In a gas mixture, the value of the ionization coefficient α_m for a mixture of two components is equal to the sum of the two individual coefficients (α_1 and α_2) multiplied by their respective partial pressure ratios:

$$\alpha_m = \frac{P_1}{p}\alpha_1 + \frac{P_2}{p}\alpha_2 \tag{3.42}$$

Humid air could be considered as a mixture of dry air and water vapor, and thus the effect of humidity on α could be evaluated (Abdel-Salam, 1985). The dependence of α on gas temperature T is obtained from equation (3.15) simply as $\alpha/p = \alpha T/p_0 T_0$ and $E/p = ET/p_0 T_0$, where T_0 and p_0 are the normal values (i.e., $T_0 = 293$ K and $p_0 = 101.3$ kPa) (Abdel-Salam, 1976).

3.6.2 Attachment Coefficient η

Measured values of η in air (Brown, 1966) made it possible to express η/p (m$^{-1}\cdot$kPa^{-1}) as a function of E/p (V$\cdot$m$^{-1}\cdot$kPa^{-1}) as

$$\frac{\eta}{p} = 9.738 - 304.31\left(\frac{E}{p}\right) + 3674.31\left(\frac{E}{p}\right)^2 \tag{3.43}$$

In SF$_6$, η/p is related to E/p (Boyd and Crichton, 1971) as

$$\frac{\eta}{p} = 1170.4 - 2250\left(\frac{E}{p}\right) \tag{3.44}$$

The effects of temperature and of gas mixtures on the attachment coefficient η_m are the same as those on α_m.

3.6.3 Photon Absorption Coefficient μ

The photon absorption coefficient μ (m^{-1}) of any gas at pressure p (kPa) was expressed (Khalifa et al., 1975) as

$$\mu(p) = \frac{\mu_0 p}{101.3} \tag{3.45}$$

where μ_0 is the value of μ at STP. It equals 330 for air, 250 for nitrogen, and 600 for SF$_6$ (Alexandrov, 1972). The absorption coefficient $\mu_m(p)$ for a gas mixture can be evaluated using a relation similar to (3.42). Its dependence on temperature is similar to that of α.

3.6.4 Electron Drift Velocity v_e

Measured values of v_e (Naidu and Prasad, 1972) made it possible to express v_e (m/s) as a function of E/p (V·m^{-1}·kPa^{-1}) in air by Badaloni and Gallimberti (1972) and in nitrogen and SF_6 by Abdel-Salam and Abdellah (1978):

$$(v_e)_{air} = \begin{cases} 113.7 \times 10^4 \left(\dfrac{E}{p}\right)^{0.715} & \text{for } \dfrac{E}{p} \leq 75{,}000 \text{ V·m}^{-1}\text{·kPa}^{-1} \\ 93.69 \times 10^4 \left(\dfrac{E}{p}\right)^{0.62} & \text{for } \dfrac{E}{p} > 75{,}000 \text{ V·m}^{-1}\text{·kPa}^{-1} \end{cases}$$

(3.46)

$$(v_e)_{N_2} = \left[2.8 + 2205\left(\dfrac{E}{p}\right)\right] \times 10^6 \quad (3.47)$$

$$(v_e)_{SF_6} = \begin{cases} 1312.5 \times 10^5 \left(\dfrac{E}{p}\right) & \text{for } 25{,}000 \leq \dfrac{E}{p} < 127{,}500 \text{ V·m}^{-1}\text{·kPa}^{-1} \\ 1380 \times 10^5 \left(\dfrac{E}{p}\right) & \text{for } 127{,}500 \leq \dfrac{E}{p} < 150{,}000 \text{ V·m}^{-1}\text{·kPa}^{-1} \end{cases}$$

(3.48)

In a gas mixture, the electron drift velocity $(v_e)_m$ is estimated, using the electron mobility in the mixture, as $(v_e)_m = (k_e)_m E$. The mobility $(k_e)_m$ is expressed in terms of the components $(k_e)_1$ and $(k_e)_2$ and the partial pressure ratios as

$$\dfrac{1}{(k_e)_m} = \dfrac{1}{(k_e)_1}\dfrac{p}{p_1} + \dfrac{1}{(k_e)_2}\dfrac{p}{p_2} \quad (3.49)$$

Equation (3.49) was used by Abdel-Salam (1985) to determine $(k_e)_m$ in humid air. The dependence of the electron velocity v_e on the gas temperature can be determined from equations (3.46) to (3.49) simply by replacing E/p by $ET/p_0 T_0$.

3.6.5 Electron Diffusion Coefficient D_e

A fundamental relation was developed by Loeb (1965) correlating the mobility k_e to the diffusion coefficient D_e:

$$\dfrac{k_e}{D_e} = \dfrac{e}{kT_e} = \dfrac{e}{z_m[(1/2)m_0 \overline{v}^2]} \quad (3.50)$$

where kT_e is the energy of thermal agitation at the effective absolute temperature of the gas, k is Boltzmann's constant, and z_m is the ratio between the energy of thermal agitation and the mean thermal energy of the gas molecules. z_m was expressed (Badaloni and Gallimberti, 1972) as

$$z_m = \begin{cases} 1870\left(\dfrac{E}{p}\right)^{0.71} & \text{for } \dfrac{E}{p} \leq 2250 \text{ V}\cdot\text{m}^{-1}\cdot\text{kPa}^{-1} \\ 238\left(\dfrac{E}{p}\right)^{0.49} & \text{for } \dfrac{E}{p} > 2250 \text{ V}\cdot\text{m}^{-1}\cdot\text{kPa}^{-1} \end{cases} \quad (3.51)$$

The effect of pressure, temperature, and humidity on z_m are similar to their effects on α, η, and μ.

REFERENCES

Abdel-Salam, M. (1976). *Journal of Physics D: Applied Physics*, 9: L-148.
Abdel-Salam, M. (1985). *IEEE Trans.*, *IA-21*:35.
Abdel-Salam, M. and Abdellah, M. (1978). *IEEE Trans.*, *IA-14*: 516.
Abdel-Salam, M. and Turkey, A. (1988). *IEEE Trans.*, *IA-24*: 1031.
Abdel-Salam, M., Radwan, R., and Ali, Kh. (1978). *IEEE—PES paper A-78-601-7*.
Alexandrov, G. N. (1972). *Proc. IEE Gas Discharges Conference*, London, pp. 398–401.
Allen, N. L., Berger, G., Dring, D., and Hahn, N. (1981). *Proc. IEE, 128A*:565.
Badaloni, S. and Gallimberti, I. (1972). *Basic Data of Air Discharges*, Università di Padova, Padova, Italy.
Bahalla, M. S. and Craggs, J. D. (1962). *Proceedings of the Physical Society*, London, 80:151.
Boyd, H. A. and Crichton, G. C. (1971). *Proc. IEE, 118*:1872.
Brown, S. C. (1966). *Basic Data of Plasma Physics*, The MIT Press, Cambridge, Mass.
Fikry, L., Abdel-Salam, M., Zeitoun, A., and Goher, M. (1978). *IEEE Trans.*, *IA-14*:510.
Hartmann, G. and Gallimberti, I. (1975). *Journal of Physics D: Applied Physics*, 8:670.
Khalifa, M., El-Debeiky, S., and Abdel-Salam, M. (1975). *Proc. Int. High Voltage Symposium*, Zurich, Switzerland, pp. 343–347.

Kuffel, E. and Abdullah, M. (1970). *High Voltage Engineering*, Pergamon Press, Ltd., Oxford.

Loeb, L. B. (1965). *Electrical Coronas: Their Basic Physical Mechanism*, University of California Press, Berkeley, Calif.

Maller, V. A. and Naidu, M. S. (1981). *Advances in HV Insulation and Arc Interruption in SF_6 and Vacuum*, Pergamon Press, Ltd., Oxford.

McDaniel, E. W. (1964). *Collision Phenomena in Ionized Gases*, John Wiley & Sons, Inc., New York.

Meek, J. M. and Craggs, J. D. (1953). *Electrical Breakdown of Gases*, Clarendon Press, Oxford.

Naidu, M. S. and Prasad, P. (1972). *Journal of Physics D: Applied Physics*, 5:1090.

Nasser, E. (1971). *Fundamentals of Gaseous Ionization and Plasma Electronics*, John Wiley & Sons, Inc., New York.

Rapp, D. and Englander-Golden, P. (1965). *Journal of Chemical Physics*, 43:1464–1470.

Sommerville, I. C., Farish, O., and Tedford, D. J. (1984). *Proc. 46h Int. Symposium on Gaseous Dielectrics*, Knoxville, Tenn., pp. 137–144.

4
Electrical Breakdown of Gases

M. ABDEL-SALAM   *Assiut University, Assiut, Egypt*

4.1   INTRODUCTION

Air has been the insulating ambient most commonly used in electrical installations. Among its greatest assets, in addition to its abundance, is its self-restoring capability after breakdown. Liquid and solid insulants in use often contain gas voids that are liable to breakdown. Therefore, the subject of electrical breakdown of gases is indispensable for designers and operators of electrical equipment.

An electrical discharge in a gas gap can be either a partial breakdown (corona) over the limited part of the gap where the electrical stress is highest or a complete breakdown. The complete breakdown of an entire gap initially takes the form of a spark requiring a high voltage, and through it a relatively small current flows. Depending on the source and the gas-gap conditions, the spark may be either extinguished or be replaced by a highly conductive conducting arc. Depending on the circumstances, the arc may be maintained, as in arc furnaces, or extinguished, as in circuit breakers. In this chapter we present a brief discussion of the breakdown mechanisms in uniform and nonuniform fields, under dc, ac, and impulse voltages, of air and SF_6 at various pressures and temperatures.

4.2   PRE-BREAKDOWN PHENOMENA IN GASES

4.2.1   Generation of Electron Avalanches

Ionization by electron impact is probably the most important process in the breakdown of gases. However, as will be shown later, this process alone is not sufficient to produce breakdown. Consider a uniform field between electrodes immersed in a gas. Electrons may, for example, originate from the cathode by ultraviolet radiation, or from the gas volume by ionization of neutral molecules, as noted in Chapter 3. They could also be produced at a later stage by photons from the discharge itself. If an electric field E is applied across a gap, the electrons present will be accelerated toward the anode, gaining energy with distance of free travel.

As mentioned in Chapter 3, if the electron acquires enough energy, it can ionize a gas molecule by collision. Leaving a positive ion behind, the new electron, together with the primary electron, proceed along the field and repeat the process. At a distance x from the cathode, the number of electrons will thus have increased to n_x. A further increase dn_x is

$$dn_x = \alpha n_x dx \qquad (4.1)$$

where α is Townsend's first ionization coefficient. In a uniform field (i.e., constant α) with the initial number of electrons n_0 emitted from the cathode, their number at x will be

$$n_x = n_0 \exp(\alpha x) \qquad (4.2)$$

It is well to emphasize at this stage the statistical fluctuating nature of the impact ionization process and the fact that α is only an average value for the number of ionizations per unit length of electron drift along the field. It can be shown (Raether, 1964) that the size of an electron avalanche from a single starting electron follows the exponential distribution

$$p_r(n) = \frac{1}{\bar{n}} \exp\left(\frac{-n}{\bar{n}}\right) \qquad (4.3)$$

where $P_r(n)$ is the probability of occurrence of an avalanche with size n, $\bar{n}$ is the average size, $\bar{n} = e^{\alpha d}$, and d is the gap length.

The head of the avalanche is built up of electrons while its long tail is populated by positive ions (Fig. 4.1). The track is wedge-shaped, due to the diffusion of the drifting electron swarm. Because of the difference in drift velocities of electrons (v_e) and ions (v_+), the latter are virtually stationary during the time required for the electrons to reach the anode. The transit time τ_e for the electrons of the avalanche to cross the gap is of the order of nanoseconds.

Figure 4.2 shows the components of the current pulse produced by an avalanche started by one electron leaving the cathode. Its

Electrical Breakdown of Gases

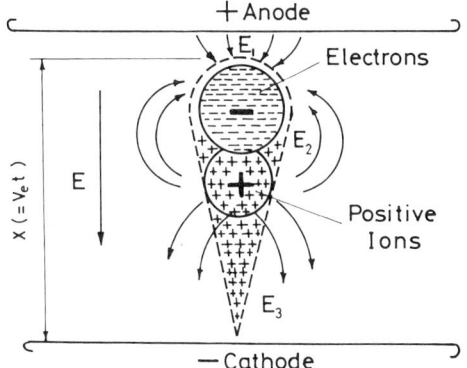

Fig. 4.1 Distribution of charge carriers in an avalanche and their contribution to the applied uniform electric field. $E_1 > E$; $E_2 < E$; $E_3 > E$.

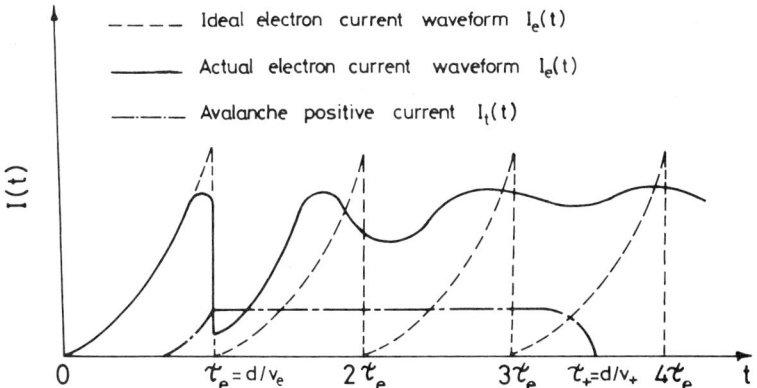

Fig. 4.2 Current due to electrons and positive ions of an avalanche. Electron current waveform when secondary electrons are produced by photons received by the cathode.

initial part is carried by electrons moving rapidly toward the anode, while the latter is carried by positive ions drifting slowly toward the cathode.

4.2.2 Avalanches with Successors

The "primary avalanche" process as described above is completed when the ions have entered the cathode. If, however, the amplification of the avalanche ($e^{\alpha d}$) is increased, the probability of additional electrons being liberated in the gap by other (secondary) mechanisms is increased, and these electrons can initiate new avalanches. The secondary ionizing agents—positive ions, excited atoms, photons, metastables (Section 3.4)—are presented quantitatively by a coefficient defined as the average number of secondary electrons produced at the cathode corresponding to one ionizing collision in the gap. γ is called Townsend's second ionization coefficient. It is a function of E/p in the same manner as α but has a much smaller magnitude (compare Figs. 4.3 and 3.5) (Naidu and Kamaraju, 1982).

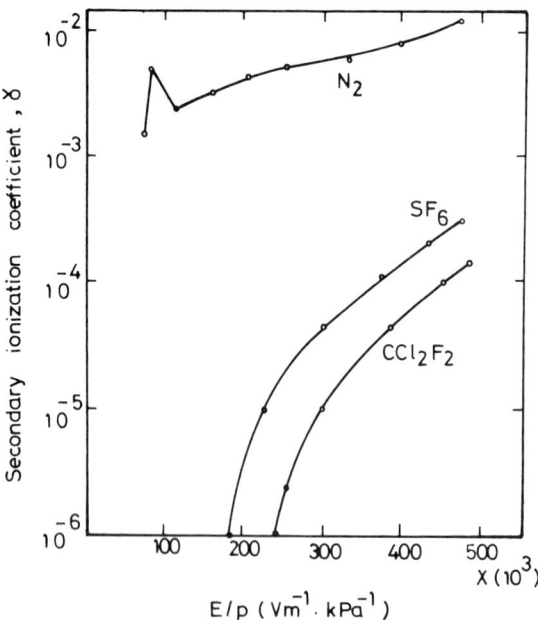

Fig. 4.3 Secondary ionization coefficient (γ) as a function of (E/p) for different gases.

Electrical Breakdown of Gases 97

The number of secondary electrons produced at the cathode during the life of the first avalanche is given by

$$\beta = \gamma(e^{\alpha d} - 1)n_0 \qquad (4.4)$$

These new electrons start the second generation of avalanches. The generation interval after which the succeeding avalanche starts depends on the secondary process. It is common in air discharges that the secondary process is photoelectric and the new avalanches start after almost $\tau_e = d/v_e$ (Fig. 4.2). The dashed line shows the idealized case, which assumes that all of the radiation producing the photo effect at the cathode is produced at the moment the electrons enter the anode. In fact, however, the radiation intensity is proportional to $e^{\alpha v_e t}$, and secondary electrons are liberated from the cathode during the transit time of the avalanche electrons, which results in the solid line of Figure 4.2.

Such generations of rapidly succeeding avalanches can produce a space charge of the slow positive ions in the gap. These space charges enhance the electric field somewhere in the interelectrode space, with a subsequent rapid current growth leading to breakdown.

4.3 BREAKDOWN IN STEADY UNIFORM FIELDS

Two typical gas breakdown mechanisms have been known: the Townsend mechanism and the streamer (or channel) mechanism. For several decades there has been controversy as to which of these mechanisms goverened spark breakdown. It is now widely accepted that both mechanisms operate, each under its own most favorable conditions. The avalanche process described above is basic for both mechanisms of breakdown.

4.3.1 Townsend Mechanism

Townsend observed that the current through a uniform-field air gap at first increased proportionately with the applied voltage in the region $(0-V_1)$ and remained nearly constant at a plateau value I_{01} (Fig. 4.4). I_{01} corresponds to the photoelectric current produced at the cathode by external irradiation. At voltages higher than V_2, the current rises above I_{01} at a rate that increases rapidly with the voltage until a spark results. If the illumination level at the cathode is increased, the plateau I_0 rises proportionately, but the voltage V_s at which sparking occurs remains unaltered, provided that there is no space-charge distortion for the electric field

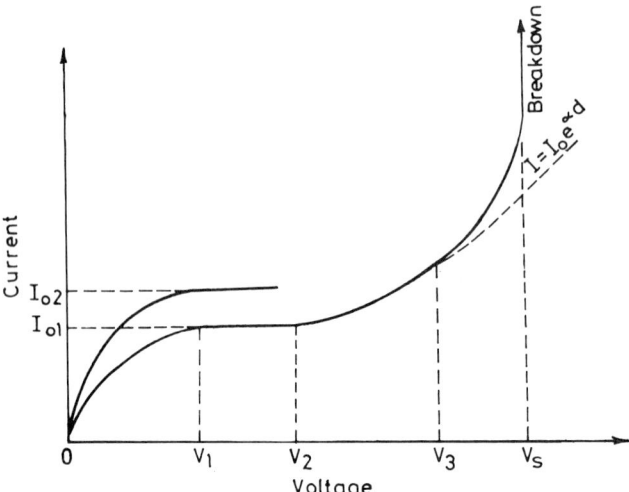

Fig. 4.4 Average current growth to breakdown as a function of the applied voltage.

between the electrodes. The increase of current in the region V_2–V_3 is ascribed to ionization by electron impact, while the secondary (γ) process accounts for the sharper increase of current in the region V_3–V_s and eventual spark breakdown of the gap.

Current Growth Equations

It has been proved in Section 4.2.1 that each electron leaving the cathode produces on the average ($e^{\alpha d} - 1$) new electrons (and positive ions) while traversing the distance d. Having n_0 electrons produced at the cathode by the external source of radiation, let

n_0' = number of secondary electrons produced at the cathode

n_0'' = total number of electrons leaving the cathode

so that:

$$n_0'' = n_0 + n_0' \qquad (4.5)$$

Electrical Breakdown of Gases 99

Each electron leaving the cathode produces (on the average) $(e^{\alpha d} - 1)$ collisions in the gap. Therefore, the number of ionizing collisions in the gap equals $n_0''(e^{\alpha d} - 1)$. By definition,

$$n_0' = \gamma \, n_0''(e^{\alpha d} - 1)$$

thus

$$n_0'' = \frac{n_0}{1 - \gamma(e^{\alpha d} - 1)} \tag{4.6}$$

The number of electrons arriving at the anode is

$$n_d = n_0'' e^{\alpha d}$$

so that

$$n_d = \frac{n_0 e^{\alpha d}}{1 - \gamma(e^{\alpha d} - 1)}$$

In the steady state, the circuit current I will be given by

$$I = \frac{I_0 e^{\alpha d}}{1 - \gamma(e^{\alpha d} - 1)} \tag{4.7}$$

4.3.2 Townsend's Criterion for Spark Breakdown

Equation (4.7) describes the growth of the average current in the gap before spark breakdown occurs. At low field strengths V/d, $e^{\alpha d} \to 1$, so that $I = I_0 e^{\alpha d}$ in the region V_2-V_3. As V increases, $e^{\alpha d}$ and $\gamma e^{\alpha d}$ increase until $\gamma e^{\alpha d}$ approaches unity, and I approaches infinity [Equation (4.7)]. In this case the current will be limited only by the resistance of the power supply and the conducting gas. This condition is defined as "breakdown" and its "Townsend criterion" is thus

$$\gamma(e^{\alpha d} - 1) = 1 \tag{4.8}$$

Normally, $e^{\alpha d} \gg 1$, and the expression becomes simply

$$\gamma e^{\alpha d} = 1$$

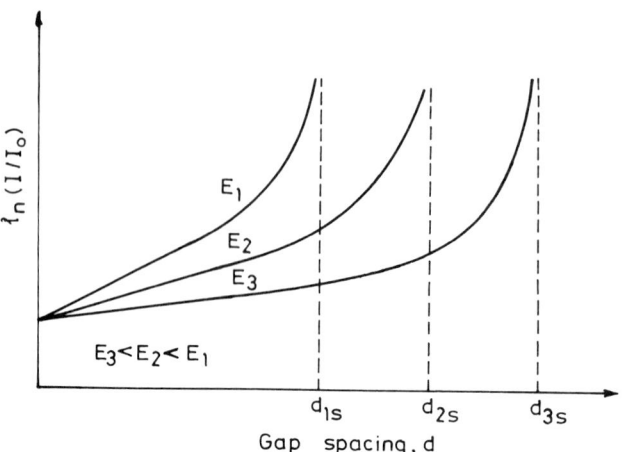

Fig. 4.5 Townsend-type $\ln(I/I_0)$ versus d plot.

Thus, for a given gap d, breakdown occurs when γ and α acquire the corresponding critical values. Both coefficients are functions of the electric field strength E. For a given photoelectric current I_0 and applied field E_1, say, the circuit current I varies with the gap d according to equation (4.7) and as shown in Figure 4.5. A maximum length d_{1s} is reached when the criterion (4.8) is fulfilled (i.e., spark breakdown occurs and the current I rises sharply).

At the lower values of d, $\gamma e^{\alpha d} \ll 1$ in equation (4.7), so that $I \sim I_0 e^{\alpha d}$, and the plot of $\ln(I/I_0)$ versus d is linear, with slope α. As d increases, $\gamma e^{\alpha d}$ increases and upcurving occurs until when $\gamma e^{\alpha d} \sim 1$, I approaches infinity and sparking occurs at $d = d_s$. The spark breakdown voltage corresponding to each value of E could easily be calculated (Abdel-Salam and Stanek, 1988).

4.3.3 Paschen's Law

Remembering from Chapter 3 that coefficients α and γ are both functions of the gas pressure p and the electric field $E = V/d$, thus $\alpha = pF_1(E/p)$, $\gamma = F_2(E/p)$, and equation (4.8) could be rewritten as

$$\left(F_2 \frac{V}{pd}\right)\left[\exp\left(pdF_1 \frac{V}{pd}\right) - 1\right] = 1 \tag{4.8a}$$

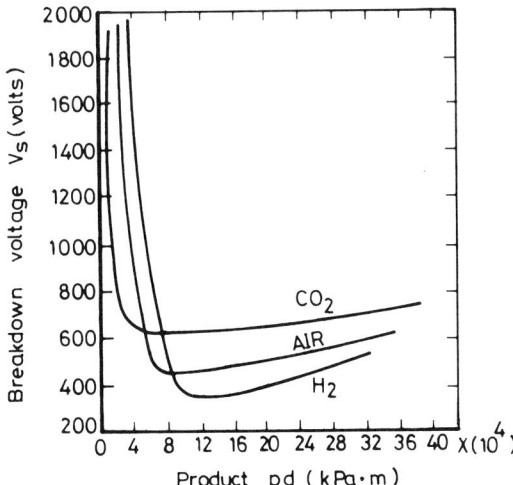

Fig. 4.6 Breakdown voltage V_S versus the product pd for different gases. (Courtesy of F. Llewellyn-Jones and Oxford University Press, 1957.)

Equation (4.8a) gives the breakdown voltage V implicitly in terms of the product of gas pressure p and electrode separation d. The breakdown voltage V_s is the same for a given value of the product pd:

$$V_s = F(pd) \tag{4.9}$$

which is the well-known Paschen's law.

The relation between V_S and pd is plotted in Figure 4.6. To explain the shape of the curve, consider a gap of fixed spacing. As the pressure decreases from a point to the right of the minimum, the gas density decreases and the electron free path increases. Consequently, an electron makes fewer collisions with gas molecules as it travels toward the anode. Since each collision entails some loss of energy, it follows that a lower electric field would still furnish the electrons with kinetic energies sufficient for the ionizing collisions.

When the minimum of the curve is reached, the density is low and there are relatively few collisions. It is necessary now to take into account the fact that an electron does not necessarily ionize a molecule on colliding with it, even if the energy of the electron exceeds the ionization energy. The electron has a finite chance of ionizing, which depends on its energy (Chapter 3). If the density and hence the number of collisions decreases, breakdown can occur only if the chance of ionizing is increased by an increase in voltage, which is shown to the left of the minimum. The minimum breakdown voltages and the corresponding pd values are given in Table 4.1 (Naidu and Kamaraju, 1982).

The discussion so far has ignored temperature variations. A more general statement of Paschen's law is therefore $V = F(\rho d)$, where ρ is the gas density, which takes into account the effect of temperature (Abdel-Salam, 1976). The validity of Paschen's law has been confirmed experimentally up to 1100°C (Alston, 1968). Further increase in temperature ultimately results in failure of Paschen's law because of significant thermal ionization above 2000 K, as shown in Chapter 3.

Table 4.1 Minimum Breakdown Voltages for Various Gases

Gas	V_S (min) (V)	pd at V_S (min) (Pa·m)
Air	327	0.754
Argon	137	1.197
H_2	273	1.530
Helium	156	5.320
CO_2	420	0.678
N_2	251	0.891
N_2O	418	0.665
O_2	450	0.931
SO_2	457	0.439
H_2S	414	0.798

Source: Naidu and Kamaraju (1982).

Empirical relations were suggested by many workers to express the breakdown voltage of uniform-field air gaps at atmospheric pressure (Alston, 1968).

$$V_s = 2440d + 61 \sqrt{d} \quad \text{kV} \tag{4.10}$$

where d is the gap length in meters.

4.3.4 Streamer Mechanism

The very short time lags of spark breakdown, measured when high overvoltages were applied to uniform field gaps, are not consistent with the Townsend mechanism, which is based on the generation of a series of successive avalanches. Also, it was difficult to envisage how the Townsend mechanism would apply for long gaps where the sparks were observed to branch and to have irregular character of growth. As a consequence, and following the experimental results of Raether (1964), the streamer theory of the spark was proposed by Loeb (1955) and Meek for the positive streamer, and independently by Raether for the negative streamer. According to both versions, spark discharge develops directly from a single avalanche, of which the space charge transforms it into a plasma streamer. Thus the conductivity grows rapidly, and breakdown occurs through its channel.

The principal features of both versions are the photoionization of gas molecules in the space ahead of the streamer and the local enhancement of the electric field by the space charge at its tip. The space charge produces a distortion of the field in the gap, as evident from Figure 4.1.

Version by Loeb and Meek

The positive streamer from the anode toward the cathode in uniform-field gaps is explained as follows: When the avalanche has crossed the gap, the electrons are swept into the anode, the positive ions remaining in a cone-shaped volume extend across the gap (Fig. 4.7a). A highly localized space-charge field is produced near the anode, but elsewhere in the gap the ion density is low. However, in the gas surrounding the avalanche, photoelectrons are produced by photons emitted from the densely ionized gas constituting the avalanche stem. These photoelectrons initiate auxiliary avalanches, which are directed by both the space-charge field and the externally applied field. Evidently, the greatest multiplication in these auxiliary avalanches will occur along the axis of the main avalanche, where the space-charge field supplements the applied field. Positive ions left behind by these avalanches effectively

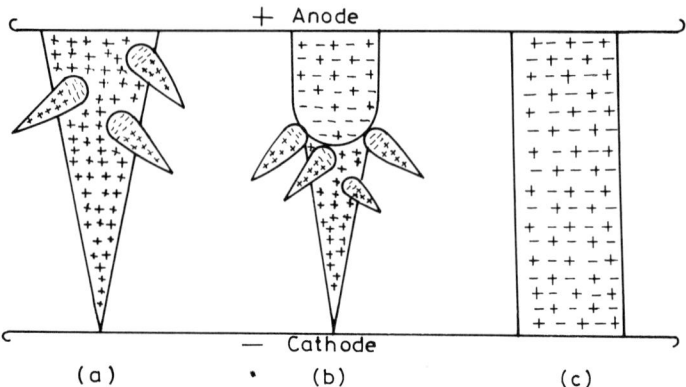

Fig. 4.7 Cathode-directed streamer as envisaged by Meek and Loeb: (a) first avalanche crossed the gap, (b) streamer extending from the anode, (c) streamer crossed the gap.

branch, lengthen, and intensify the space charge of the main avalanche toward the cathode. The process thus develops a self-propagating streamer, which effectively extends the anode toward the cathode (Fig. 4.7b). Ultimately, a conducting filament of highly ionized gas bridges the whole gap between the electrodes (Fig. 4.7c).

The transition from an electron avalanche into a streamer is considered to occur when the radial field E_r produced by the positive ions at the head of the avalanche is of the order of the externally applied field E, that is,

$$E_r = \kappa E \qquad (4.11)$$

where $\kappa \sim 1$ and

$$E_r = \frac{5.27 \times 10^{-7} \alpha \exp(\alpha x)}{(13,500 x/p)^{1/2}} \quad \text{V/m}$$

with α in m^{-1}, x in meters, and p is the gas pressure in pascals. Equation (4.11) expresses the criterion of breakdown by Loeb and Meek. The breakdown voltage is that at which the avalanche grows to the limit that its field E_r becomes equal to E.

Raether's Version

A slightly different criterion was proposed independently by Raether for negative anode-directed streamers. He postulated that streamers would develop when the initial avalanche produces a sufficient number of electrons ($e^{\alpha x}$), such that its space-charge field E_r is comparable to the applied field E. The total field thus enhanced would promote secondary anode-directed avalanches ahead of the initial one, forming a negative streamer. These secondary avalanches are initiated by photoelectrons in the space ahead of the streamer (Fig. 4.8).

Experiments for breakdown in air proved the validity of the Townsend mechanism in uniform fields with values of pd at least up to about 15 kPa·m (Alston, 1968). At higher pd values, the streamer mechanism plays the dominant role in explaining the breakdown phenomena.

The streamer mechanism has been widely used to explain breakdown in nonuniform fields, as in point–plane or point–point gaps. Similar but more practical arrangements are encountered in cases of imperfections on high-voltage conductors such as a nick of a strand or from airborne substances such as insects, dust or leaf particles, bird droppings, and other nonmetallic materials. Details of breakdown in nonuniform field gaps are discussed in Sections 4.6 and 4.7.

In long gaps, if the voltage gradient at the stressed electrode exceeds the corona onset level, the ionization activity in the gap

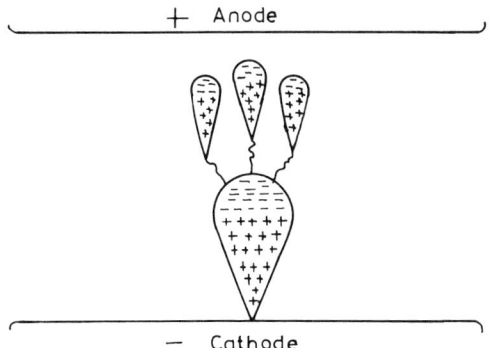

Fig. 4.8 Anode-directed streamer as envisaged by Raether.

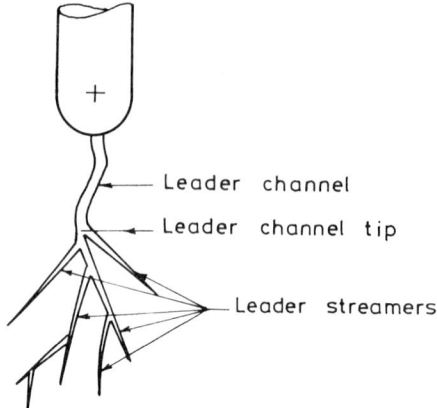

Fig. 4.9 Leader channel and leader streamers.

increases. As a result, a highly ionized and luminous filamentary channel, called the leader channel, develops at the electrode and propagates toward the other electrode. At the tip of the leader channel, filamentary branches called leader streamers exist where most of the ionization activity takes place (Fig. 4.9). The electrons produced due to ionization in the leader streamers feed through the leader channel into the stressed electrode. Depending on the value of the instantaneous voltage gradient at the stressed electrode and the leader channel length, the leader streamer either stops after having crossed a part of the gap, or reaches the plane electrode, causing a return ionizing wave to develop there. In the latter case, the ionizing wave advances toward the leader channel tip and leads to the final jump. At this stage, the leader channel tip advances very rapidly, bridging the entire gap and causing a complete breakdown.

4.4 BREAKDOWN IN ELECTRONEGATIVE GASES AND THEIR MIXTURES

The last three decades have witnessed intense research into electronegative gases such as SF_6 (sulfur hexafluoride) and CCl_2F_2 (Arcton-12). For such gases the electron attachment coefficient η is significant and their dielectric strength is considerably higher than that of air. Those studies have led to the development of modern gas-insulated systems (GIS). Thus a high insulation strength could be achieved without severely raising the gas static pressure (see Chapter 10).

Townsend's equation can be modified to include the effects of both ionization and attachment processes. With the attachment coefficient being quite significant in such gases, Townsend's criterion for breakdown equation (4.8) gets modified to

$$\gamma \frac{\alpha}{\alpha - \eta} [e^{(\alpha-\eta)d} - 1] = 1 \qquad (4.12)$$

For $\eta > \alpha$, and for large gaps, this criterion is approximated by

$$\alpha = \frac{\eta}{1 + \gamma} \qquad (4.13)$$

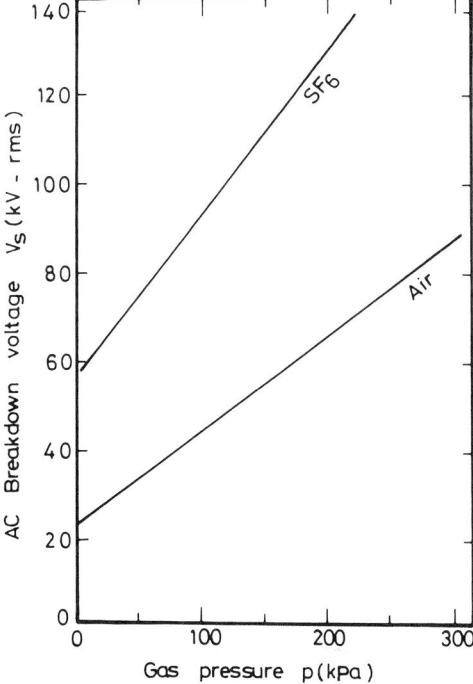

Fig. 4.10 Breakdown voltages of air and SF_6 for 0.75-m-diameter uniform-field electrodes having 2.1-cm gap spacing as a function of pressure.

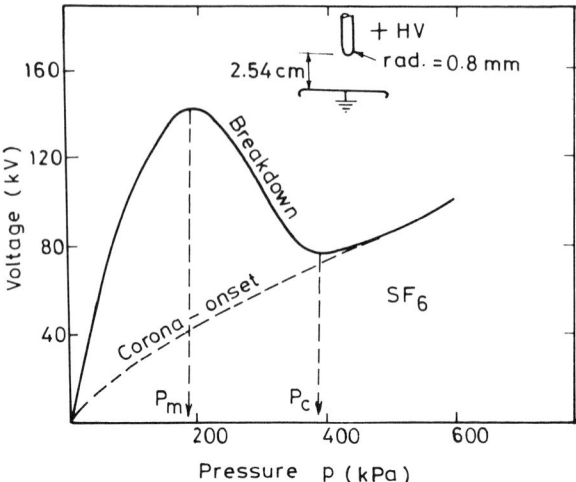

Fig. 4.11 Positive dc breakdown and corona-onset voltages in SF_6 as functions of pressure.

This is a condition that depends only on E/p, and sets a limit for E/p below which no discharges should be possible, whatever the value of d. As γ is usually much smaller than unity ($\gamma \sim 10^{-4}$), equation (4.13) can safely reduce to $\alpha = \eta$. The limiting values of E/p below which no discharges are possible are 86.3 V/m·Pa for SF_6 and 89.3 V/m·Pa for CCl_2F_2 compared to 23.6 V/m·Pa for air. With an increase in gas electronegativity (i.e., with an increase in η), the limiting value of E/p for breakdown increases. This is why the breakdown voltage measured in electronegative gases such as SF_6 is substantially higher than that measured in air (Fig. 4.10). It is widely used as an insulating as well as arc-quenching medium in high-voltage equipment such as circuit breakers and metal-clad switchgear (GIS).

Breakdown voltages in gases normally show anomalies in their increase with gas pressure. In SF_6, the breakdown voltage of a positive point-plane gap increases with gas pressure, reaching a maximum value at a pressure P_m (Fig. 4.11). Above P_m, the breakdown voltage drops with further increase of pressure until a critical value P_c is reached. At pressures above P_c, breakdown occurs without any preceding corona and the breakdown voltage resumes a slow increase with pressure (Abdel-Salam et al., 1978). The high breakdown voltage at P_m is attributed to the so-called "corona stabilization (i.e., a field modification arising from space-charge effects). The pressure P_m decreases with an increase in gap spacing. The decrease in breakdown voltage at pressures above P_m is attributed to the enhanced corona-streamer propagation across the gap due to the effectiveness of photoionization at the higher gas densities. Above P_c, field emission from the cathode becomes the significant factor for breakdown to occur unpreceded by corona (Khalifa et al., 1977). The dielectric strength of highly electronegative gases is drastically reduced by the presence of foreign conducting particles, as discussed in Chapter 10.

4.5 EFFECTS OF GAS PARAMETERS

The breakdown voltage of a given gap depends on the gas parameters (α, η, μ, . . .), which in turn are functions of the electric field and of such factors as the gas pressure, temperature, and humidity (Section 3.6). Therefore, the breakdown voltage of a given gas gap decreases at higher temperatures and increases with the static pressure. The combined factor is the relative air density (Abdel-Salam and Stanek, 1988). It decreases only slightly at higher humidities.

In many gases and at pressures around the atmospheric level, little difference has been observed between the breakdown voltages obtained with electrodes of different materials, provided that the surface remained clean and smooth. At high pressures, however, surface cleanliness and smoothness considerably improve the breakdown characteristics (Chapter 10). Point discharges that exist at the irregularities of a rough surface cause a considerable pre-breakdown current and result in a decrease in positive breakdown voltages. Field emission from irregularities on metallic cathodes contributes significantly to the pre-breakdown current (Khalifa et al., 1977). Therefore, the breakdown versus pressure characteristics show saturation (i.e., Paschen's law ceases to apply beyond a certain pressure). With successive sparking, microscopic protrusions on the electrode surfaces tend to get burned off and breakdown voltage rises to a higher plateau.

4.6 BREAKDOWN IN NONUNIFORM DC FIELDS

Most practical gas gaps have nonuniform fields. Examples include wire-to-plane gaps and coaxial cylinders, where the applied field and the first Townsend coefficient α vary across the gap. Electron multiplication is thus governed by the integral of α over the path. Townsend's criterion for the spark takes the form

$$\int_0^d \alpha \, dx = \ln\left(1 + \frac{1}{\gamma}\right) \tag{4.14}$$

where d is the gap length.

Meek's criterion for the spark is obtained from equation (4.11) after modification as

$$\alpha(x) \exp\left(\int_0^x \alpha \, dx\right) = G(x,p) \tag{4.15}$$

where $G(x,p)$ is avaluated from uniform-field data at the same gas pressure p.

In a nonuniform field gap, if the maximum field occurring at the highly stressed electrode is less than five times the average field, the discharge phenomena would be similar to those in a uniform field. In more divergent fields, however, a different phenomenon sets in (Waters, 1978). At certain voltages below breakdown level, ionization may be maintained locally at the highly stressed electrode (Chapter 5). No criterion has yet been well established for the advance of streamer in nonuniform fields, although computations and explanations for streamer growth have been reported (Abdel-Salam et al. Hashem, 1976).

Once corona starts, the electric field becomes distorted by space charge, and here the dependence of the breakdown voltage on the electrode configuration is much more complex than the dependence of the corona onset voltage. In sphere-plane and point-to-plane gaps, if the stressed electrode is positive, the space charge acts as an extension of the electrode, but if the sphere is negative, the space charge acts as a screen that decreases the field in its vicinity, and thus tends to raise the breakdown voltage (Fig. 4.12). Because of these polarity effects, characteristics obtained with the pointed electrode positive are usually more important for practical applications. Figure 4.13 illustrates schematically the dependence of the voltage on the gap length and sphere diameter. There are three main regions:

Region I. At small gaps, depending on the sphere diameter, the field is almost uniform, and the breakdown voltage depends mainly on the gap length.

Region II. For moderate gap lengths the field shows significant nonuniformity. Therefore, the breakdown voltage increases with

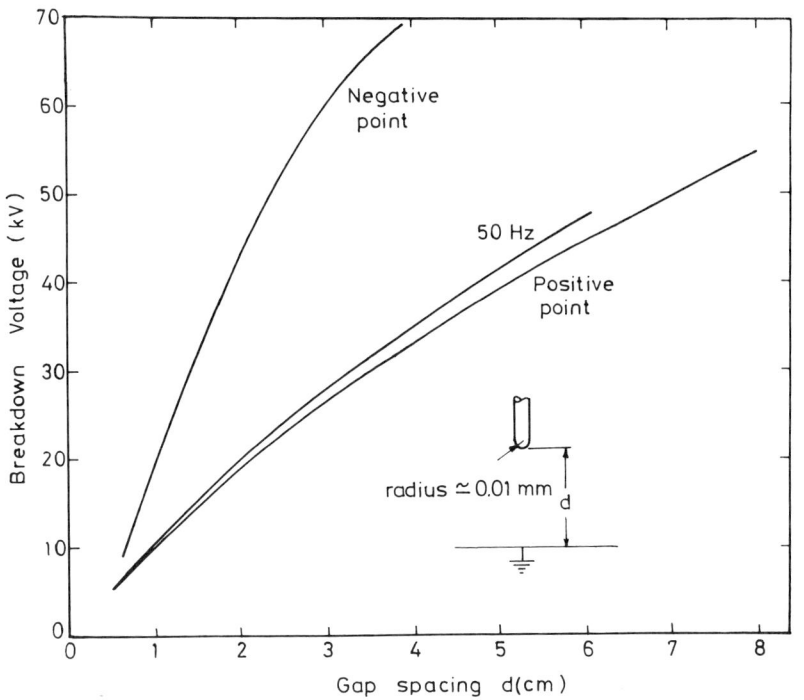

Fig. 4.12 Breakdown voltage of a point-to-plane gap in atmospheric air with dc voltage of both polarities. Breakdown voltage with 50-Hz voltage is included for comparison.

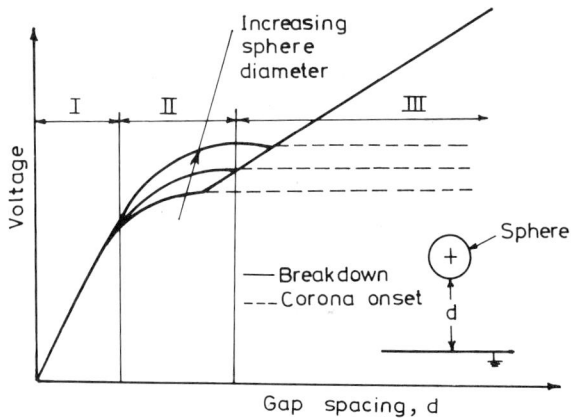

Fig. 4.13 Breakdown and corona-onset voltages versus the gap spacing for sphere of different diameters.

the sphere diameter as well as with the gap length. The effect of the sphere diameter on the field magnitude becomes more significant as the gap increases.

Region III. For gaps exceeding twice the sphere diameter, breakdown is preceded by corona. The maximum field, and therefore the corona onset voltage, are influenced by the sphere diameter, whereas the breakdown voltage depends mainly on the gap length.

4.7 BREAKDOWN IN NONUNIFORM AC FIELDS

The breakdown process gets completed in an interval on the order of 10^{-6} to 10^{-8} s. This represents an extremely small fraction of half a cycle of the power frequency. Therefore, the mechanism of breakdown is essentially the same as under dc. The only difference is that the ions in the gas will be subjected to a slowly alternating field. If the applied ac voltage magnitude is such that at the voltage peak the discharge onset conditions are reached, electron avalanches will be produced in the same way as under dc. The space charges produced would have ample time to leave the gap before the field reverses polarity. The maximum gap spacing (L_{max}) in which this is possible is the distance the ions move under such conditions. In uniform fields, this distance is about 1 m in atmospheric air under power frequency. Spacings encountered in high-voltage power transmission systems exceed L_{max} and the ions are constrained to the vicinity of the conductors; recombination occurs among the outgoing and the returning ions (Abdel-Salam et al., 1984).

4.8 BREAKDOWN UNDER IMPULSE VOLTAGES

Impulse overvoltages arise in power systems due to lightning or switching surges. They represent a principal factor in the design of equipment insulation (Chapter 14).

Therefore, it is important to appreciate the fact that the breakdown mechanism under impulse voltages is different from cases of steady dc or low-frequency ac. Under impulse, a time lag is observed between the instant the applied voltage is sufficient to cause breakdown and the actual event of breakdown, if it occurs. The two basic relevant phenomena are the appearance of electrons for initiating the avalanches and their ensuing temporal growth.

In the case of steady or slowly varying fields, there is usually no difficulty in finding an initiatory electron from natural sources (e.g., cosmic rays or detachment from negative ions) (Allen et al.,

1981). However, under an impulse voltage of short duration ($\sim 10^{-6}$ s), depending on the gap volume, natural resources may not be sufficient to provide the initiating electron at the appropriate site in time for the breakdown to occur. The probability of breakdown increases from zero to 100% over a suitable voltage range. The time t_s that elapses between the application of a voltage greater than V_s, the gap's static breakdown voltage, and the appearance of a suitably placed initiatory (seed) electron is called the statistical time lag t_s of the gap, because of its statistical nature (Berger, 1973). After such a seed electron appears, the subsequent time t_f required for the breakdown of the gap to materialize is known as the formative time lag. The sum ($t_f + t_s$) is the total time lag, or the time to breakdown (TBD).

For example, in a positive point-to-plane gap, if the seed electron is too close to the point, it develops an avalanche of insufficient size for the streamers to develop. Also, streamers cannot form if the seed electron is too far from the anode, where the attachment coefficient exceeds the ionization coefficient. Thus a critical volume is defined (Fig. 4.14a), which at the static breakdown voltage V_s should theoretically be reduced to a point on the axis of the electrode system. The critical volume grows axially and laterally with the increase in applied voltage (Fig. 4.14a) (Abdel-Salam and Turkey, 1988). The probability that a negative ion appears in the critical volume to give birth to an electron expresses the distribution of the statistical time lag t_s shown in Figure 4.14b).

For breakdown to occur, the applied impulse voltage V must be greater than V_s. Breakdown on the front of an applied voltage wave is shown in Figure 4.15. The overvoltage $V - V_s = \Delta V$ clearly depends on ($t_s + t_f$) and on the rate of rise of the applied voltage.

The differences among the volt-time characteristics for different air-gap geometries, for internal and external insulation of equipment, and for overvoltage protective devices provide the basis of insulation coordination. This important subject is discussed in Chapter 15.

4.8.1 Statistical Time Lags

It is interesting to examine some of the factors controlling t_s. If θ is the rate at which electrons are produced in the gap by external irradiation, P_{r1} the probability of an electron appearing in the critical volume where it can lead to a spark, and P_{r2} the probability that such an electron will lead to a spark, then the average statistical time lag t_s equals $1/(\theta P_{r1} P_{r2})$. Also, if a gap has survived breakdown for a period t, then the probability that it will break down in the next time interval dt is $\theta P_{r1} P_{r2}$ dt. This will be

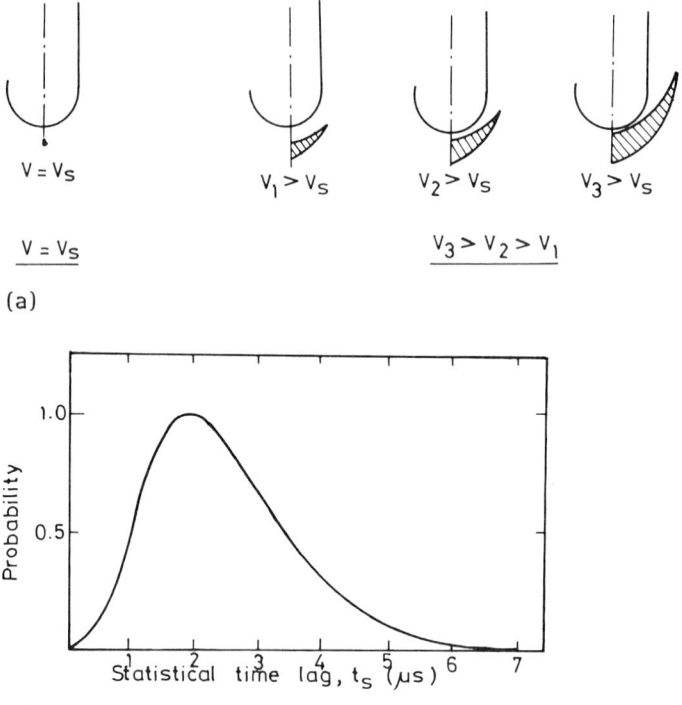

Fig. 4.14 Initiation of positive corona under surge voltages: (a) variation of critical volume with applied voltage, (b) distribution of time lag (t_s) for the initiation of positive corona. [From Goldman and Goldman (1978).]

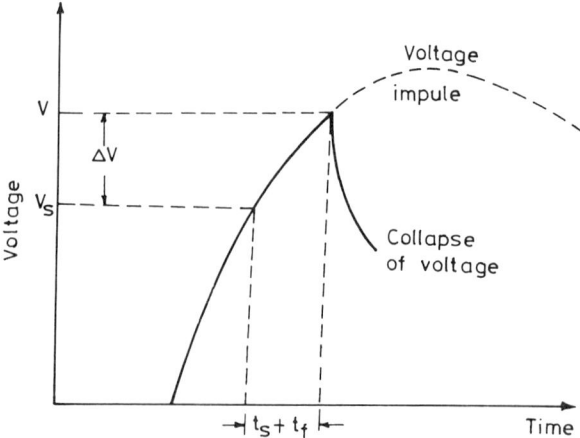

Fig. 4.15 Breakdown on the front of the applied impulse voltage wave.

Electrical Breakdown of Gases 115

independent of t if $\theta P_{r1} P_{r2}$ is independent of t. The probability of breakdown with a time lag t is $\exp(-t/t_s)$.

4.8.2 Formative Time Lags

Once an initiatory electron is made available in the gas gap, it will under the applied voltage start an electron avalanche with the subsequent processes, such as secondary avalanches, streamers and leaders, eventually culminating in the sparkover of the gap. The formative time lag t_f consumed in these processes was measured experimentally by Kohrman and others (Nasser, 1971). For overvoltages of less than 1% for air gaps relatively long, time lags t_f ($\geqslant 20$ μs) were observed, which led to the belief that at such small overvoltages a Townsend mechanism may be at work. At higher overvoltages, t_f was of the order of 1 μs, which supported the streamer mechanism.

4.8.3 Breakdown of Long Gaps Under Switching Impulses

In highly non-uniform-field gaps, the space charge resulting from pre-breakdown corona causes a marked distortion of the applied electrostatic field. The establishment of such space charge takes a finite time; consequently, the breakdown characteristics will be affected by the rate at which the voltage rises. It has been observed that with wavefronts of 50 to 300 μs, positive breakdown voltages of non-uniform-field gaps were minimum (Fig. 4.16). With the shorter wavefronts, corona would develop and produce a large space charge in the gap, which delays the advancement of the leader stroke and leads to an increase in the breakdown voltage. With much longer fronts, the space charge of the corona gradually fills the zone near the stressed electrode and reduces the voltage gradient there, again raising the breakdown voltage. From Figure 4.16 it is noticed that the critical front duration corresponding to the minimum breakdown voltage increases with gap length. Harada and his colleagues (1973) have given an empirical formula for the critical front duration:

$$T_c = 40 + 35 d \quad \mu s \qquad (4.16)$$

where the gap length d is in meters.

Because of the time and expense involved in high-voltage tests on large structures, any method of extrapolating known results is economically attractive and important for the design of EHV systems. Given the breakdown voltage of a rod-plane gap, that for any other gap length could be estimated by

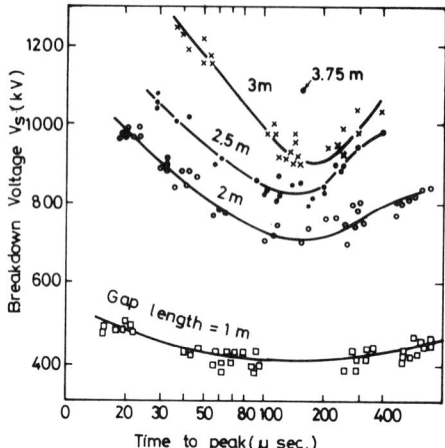

Fig. 4.16 Dependence of the breakdown voltage of a rod-gap on the voltage front duration for various gap lengths. [From Bazelyan et al (1961).]

$$V_{s+} = 500 K_g d^{0.5} \quad \text{kV} \tag{4.17}$$

where K_g is the gap factor for the configuration. Paris and his coworkers (1973) deduced the gap factors for various electrode configurations, ranging from 1.15 for conductor-plane gaps, to 1.3 for rod-rod gaps, and up to 1.9 for large conductor-to-rod gaps. The gap factor method has also been applied to lightning impulses (Paris et al., 1973).

Semiempirical models have been proposed by researchers to calculate the breakdown voltage of long air gaps of up to 50 m (e.g., Alexandrov and Podporkyn, 1979; Waters, 1978). Very recently, Rizk (1989) developed a mathematical model for calculating the continuous leader inception and breakdown voltages of long air gaps under positive switching impulses with critical time-to-crest. The model is based on assuming the presence of the following factors:

1. Axial propagation of the leader.
2. Constant charge injection during propagation (q_1 = 45 µC/m).
3. Constant velocity of leader propagation (v = 1.5 cm/µs).
4. Resemblance between the leader and the electrical arc with a conductance that varies expotentially with the lifetime of the leader (time constant τ = 50 µs). Subsequently, the voltage

Electrical Breakdown of Gases

gradient within the leader varies from an initial value E_i (= 400 kV/m) to an ultimate value E_∞ (= 50 kV/m).

5. Constant voltage gradient through the leader corona streamer (E_s = 400 kV/m).

The agreement between the values predicted by the model, including the height of the final jump, leader voltage drop, and 50% breakdown voltage, and those measured experimentally is excellent in light of the aforementioned assumptions and the unique values of q_1, v, E_i, E_∞, and E_s. The model dealt with rod, sphere-, and conductor-plane gaps.

For rod-plane gaps, Rizk (1989) developed an expression for the breakdown voltage V:

$$V = \frac{1556 + 50d}{1 + 3.89/d} + 78 \qquad \text{kV for } d > 4 \text{ m} \qquad (4.18)$$

which predicts breakdown voltages that agree reasonably well with findings of the Renardier group (Paris et al., 1973).

4.9 HIGH-FREQUENCY BREAKDOWN

In uniform field gaps, breakdown starts with an avalanche process, as described above for the cases of dc and low-frequency ac. As the frequency f of the applied field increases to reach very high levels, the discharge behavior starts to differ. To explain this point, let the field in the gap be expressed as $(V/d) \sin 2\pi ft$, V being the applied peak voltage and d the gap length. Then the maximum distance L_{max} that any positive ion can travel during one half-cycle is

$$L_{max} = \int_0^{\frac{1}{2f}} k_+ \frac{V}{d} \sin 2\pi ft \, dt = \frac{k_+ V}{\pi f d} \qquad (4.19)$$

k_+ being the positive ion mobility. For $d > L_{max}$, most positive ions will not be able to reach the cathode before the applied voltage reverses sign. In other words, for a given d the critical frequency f_c at which all positive ions can just be cleared from the gap during one half-cycle is

$$f_c = \frac{k_+}{\pi d^2} V \qquad (4.20)$$

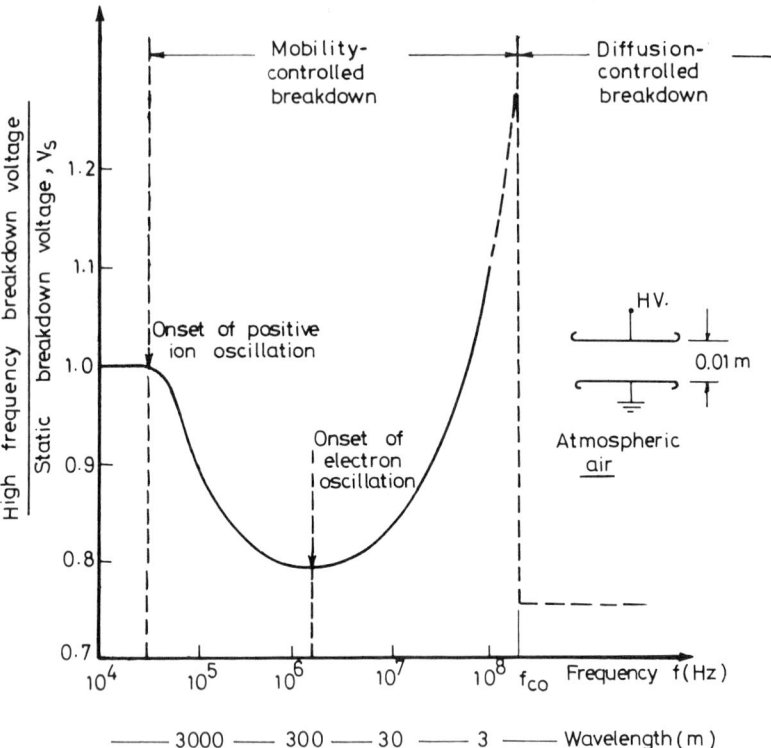

Fig. 4.17 Ratio of high-frequency breakdown voltage to static breakdown voltage as a function of frequency for a uniform air gap. [From Gänger (1953).]

At frequencies $f > f_c$ the cloud of positive-ion space charge will oscillate between the electrodes while new avalanches grow and add to its density and size until instability and breakdown occur. This accumulating space charge will no doubt distort the field in the gap. Therefore, breakdown occurs at lower field strengths than under dc for power-frequency ac (Fig. 4.17). At frequencies $f < f_c$ the breakdown conditions are quite similar to those of static fields.

The foregoing analysis of ion motion can be extended to the electrons oscillating between the electrodes. In analogy with f_c for ions, there will be a critical frequency above which electrons would have no time to reach the opposite electrode. They will oscillate in the gap and collide with the gas molecules. When the

Electrical Breakdown of Gases   119

field is adequately high, they will produce more and more electrons until breakdown is completed with no participation from the electrodes (MacDonald, 1966). This critical frequency, f_{ce}, depends on the electron mobility k_e, the electrode spacing d, and the magnitude of the applied voltage V.

$$f_{ce} = \frac{k_e V}{\pi d^2} \qquad (4.21)$$

Since electron drift velocities are two orders of magnitude higher than those of positive ions under the same conditions, the magnitude of f_{ce} is about two orders of magnitude higher than f_c (Fig. 4.17) (Nasser, 1971).

The shape of the curve in Figure 4.17 can be explained as follows. The decrease in breakdown voltage in the lower part of the frequency range above f_c is caused by the distortion of the electric field in the gap by the positive-ion space charges accumulating in the gap. At the higher frequencies, however, the electrons in the gap oscillate at increased frequencies and some of them would fail to reach the anode during the half cycle in which they were created. They would thus remain in the gap and partly neutralize the positive-ion space charge and the distortion in the field—hence the parabolic part of the curve.

The breakdown voltage even exceeds its static level at the higher frequencies where the electrodes do not contribute to the ionization process as they do under dc or power-frequency ac. At a certain frequency f_{co} the breakdown mechanism is controlled by diffusion; with the electrodes playing no role, the breakdown voltage drops sharply.

When $f < f_{co}$ electrons are lost by virtue of their mobility to the electrodes and the breakdown mechanism is called the mobility-controlled mechanism. When $f > f_{co}$, the electrons are lost by diffusion and the breakdown mechanism is known as the diffusion-controlled breakdown mechanism. At extremely high frequencies, electrons and ions oscillate in the gap, giving rise to higher currents whose phase relationship to the applied voltage is controlled by the rate of electron-ion recombination relative to the frequency. The breakdown field strength can be determined by relating the field-dependent ionization rate to that of charge loss by diffusion (MacDonald, 1966).

All this discussion was confined to uniform-field gaps. Nonuniform-field gaps also show differences between their high- and low-frequency performances. Such nonuniform fields are experienced in cases of transmitter antennas and their insulation. For point-to-plane gaps the high-frequency breakdown voltages increase with gap length, the same as at low frequency, but are different in magnitude.

REFERENCES

Abdel-Salam, M. (1976). *Journal of Physics D: Applied Physics*, 9:L-149.
Abdel-Salam, M. and Stanek, E. K. (1988). *IEEE Trans.*, *IA-24*: 1025.
Abdel-Salam, M. and Turkey, A. (1988). *IEEE Trans.*, *IA-24*: 1031.
Abdel-Salam, M., Khalifa, M., and Hashem, A. (1976). *Proc. IEEE-IAS Annual Meeting*, Chicago, pp. 507-512.
Abdel-Salam, M., Radwan, R., and Ali, Kh. (1978). *IEEE-PES paper A-78-601-7*.
Abdel-Salam, M., Farghally, M., Abdel-Sattar, S., and Shamloul, D. (1984). *Proc. 4th Int. Symposium on Gaseous Dielectrics*, Knoxville, Tenn.
Alexandrov, G. N. and Podporkyn, G. V. (1979). *IEEE Trans.*, *PAS-98*:597.
Allen, N. L., Berger, G., Dring, D., and Hahn, R. (1981). *Proc. IEE*, *128*:565.
Alston, L. L. (1968). *High Voltage Technology*, Oxford University Press, Oxford.
Bazelyan, E. M., Brago, E. N., and Stekolnikov, I. S. (1961). *Soviet Physics-Doklady*, 52:101.
Berger, G. (1973). Ph.D. thesis, Université de Paris-Sud, Paris.
Gänger, B. (1953). *Der elektrische Durschlag von Gasen*, Springer-Verlag, Berlin.
Goldman, M. and Goldman, A. (1978). "Corona Discharges," in *Gaseous Electronics* (M. N. Hirsh and H. J. Oskam, eds.), Academic Press, Inc., New York, pp. 219-290.
Harada, T., Aihara, Y., and Aoshima, Y. (1973). *IEEE Trans.*, *PAS-92*:1085.
Khalifa, M., Abdel-Salam, M., Radwan, R., and Ali, Kh. (1977). *IEEE Trans.*, *PAS-96*:886.
Llewellyn-Jones, F. (1957). *Ionization and Breakdown of Gases*, Methuen & Company, Ltd., London.
Loeb, L. B. (1955). *Basic Processes of Gaseous Electronics*, University of California Press, Berkeley, Calif.
MacDonald, A. D. (1966). *Microwave Breakdown in Gases*, John Wiley & Sons, Inc., New York.
Naidu, M. S. and Kamaraju, V. (1982). *High Voltage Engineering*, Tata McGraw-Hill Publishing Co., Ltd., New Dehli.

Nasser, E. (1971). *Fundamentals of Gaseous Ionization and Plasma Electronics*, John Wiley & Sons, Inc., New York.
Paris, L., Tashchini, A., Schneider, K. H., and Weck, K. H. (1973). *Electra*, 29:29.
Raether, H. (1964). *Electron Avalanches and Breakdown in Gases*, Butterworth & Company (Publishers) Ltd., London.
Rizk, F. (1989), *IEEE Trans.*, PD-4:596–606.
Waters, R. T. (1978). "Spark Breakdown in Nonuniform Fields," in *Electrical Breakdown of Gases* (J. M. Meek and J. D. Craggs, eds.), John Wiley & Sons, Inc., New York, pp. 385–532.

5
The Corona Discharge

M. KHALIFA   *Cairo University, Giza, Egypt*

5.1 INTRODUCTION

"Corona" literally means the disk of light that appears around the sun. The term was borrowed by physicists and electrical engineers to describe generally the partial discharges that develop in zones of highly concentrated electric fields, such as at the surface of a pointed or cylindrical electrode opposite to and at some distance from another. This partial breakdown of air is quite distinct in nature and appearance from the complete breakdown of air gaps between electrodes. The same applies for other gases.

The corona is also distinct from the discharges that take place inside gas bubbles within solid and liquid insulation, although the underlying phenomena of gas discharges are the same. The corona discharge is accompanied by a number of observable effects, such as visible light, audible noise, electric current, energy loss, radio interference, mechanical vibrations, and chemical reactions. The chemical reactions that accompany corona in air produce the smell of ozone and nitrogen oxides.

Corona has long been a main concern for power transmission engineers because of the power loss it causes on the lines and the noise it causes in radio and TV reception. On the other hand, corona does have several beneficial applications, as in Van de Graaff generators, electrostatic precipitators, electrostatic printing, electrostatic deposition, ozone production, and ionization counting (Berg and Hauffe, 1972).

5.2 MECHANISM OF CORONA DISCHARGE

The discharge process depends on the polarity of the applied voltage. Therefore, it will be discussed first for each polarity under dc.

5.2.1 Positive Corona

At the onset level, and slightly above, there exists a small volume of space at the anode where the field strength is high enough for ionization by collision. When a free electron is driven by the field toward the anode, it produces an electron avalanche (Chapter 3). The cloud of positive ions produced at the avalanche head near the anode forms an eventual extension to the anode. Secondary generations of avalanches get directed to the anode and to these dense clouds of positive ions (Fig. 5.1). This mode of corona consists of what are called onset streamers. If conditions are favorable, the high field space at the anode may suit the formation of streamers extending tangentially onto the anode called "burst-pulse streamers" (Loeb, 1965).

At slightly higher voltages a cloud of negative ions may form (Fig. 5.1) near the anode surface such that the onset-type streamers become very numerous. They are short in length, overlap in space and time, and the discharge takes the form of a "glow" covering a significant part of the HV conductor surface (Fig. 5.2b). The corresponding current through the HV circuit becomes a quasi-steady current (Fig. 5.2). This is in contrast with current pulses corresponding to onset streamers (Fig. 5.2) (Giao and Jordan, 1967; Khalifa, 1979). The positive current pulse corresponds to a succession of generations of electron avalanches taking place in the ionization zone at the anode (Khalifa and Abdel-Salam, 1974a).

At still higher voltages the clouds of negative ions at the anode can no longer maintain their stability and get ruptured by violent pre-breakdown streamers, corresponding to irregular, high-amplitude current pulses (Figs. 5.1 and 5.2). If we continue to raise the voltage, breakdown eventually occurs across the air gap. Figure 5.3 presents the range for each type of corona discharge with positive dc voltage applied across a point-to-plane gap (Nasser, 1971).

5.2.2 Negative Corona

At the onset level and slightly higher, the corona at the cathode has a rapidly and steadily pulsating mode known as Trichel pulse corona. Each current pulse corresponds to one main electron avalanche occurring in the ionization zone (Fig. 5.4) (Khalifa and Abdel-Salam, 1974b; Zeitoun et al., 1976). In this case the ionization zone extends from the cathode surface outward and as far as the point where the field becomes too weak for the ionization by collision to compensate for the electron attachment. Beyond such a point, more and more of the avalanche electrons get attached

Corona Discharge

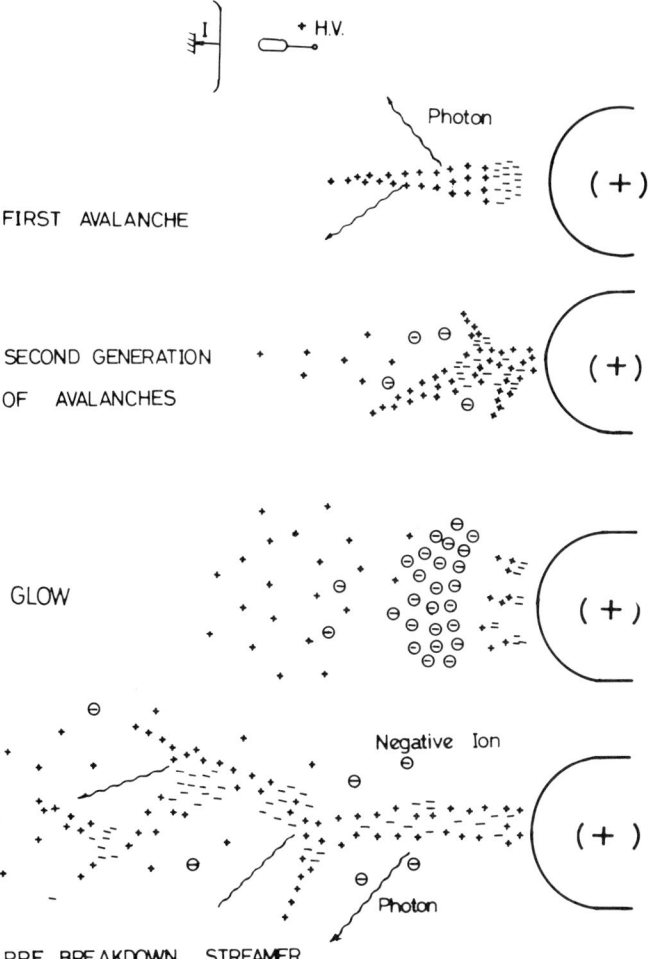

Fig. 5.1 Development of the first and subsequent generations of avalanches in positive corona discharges.

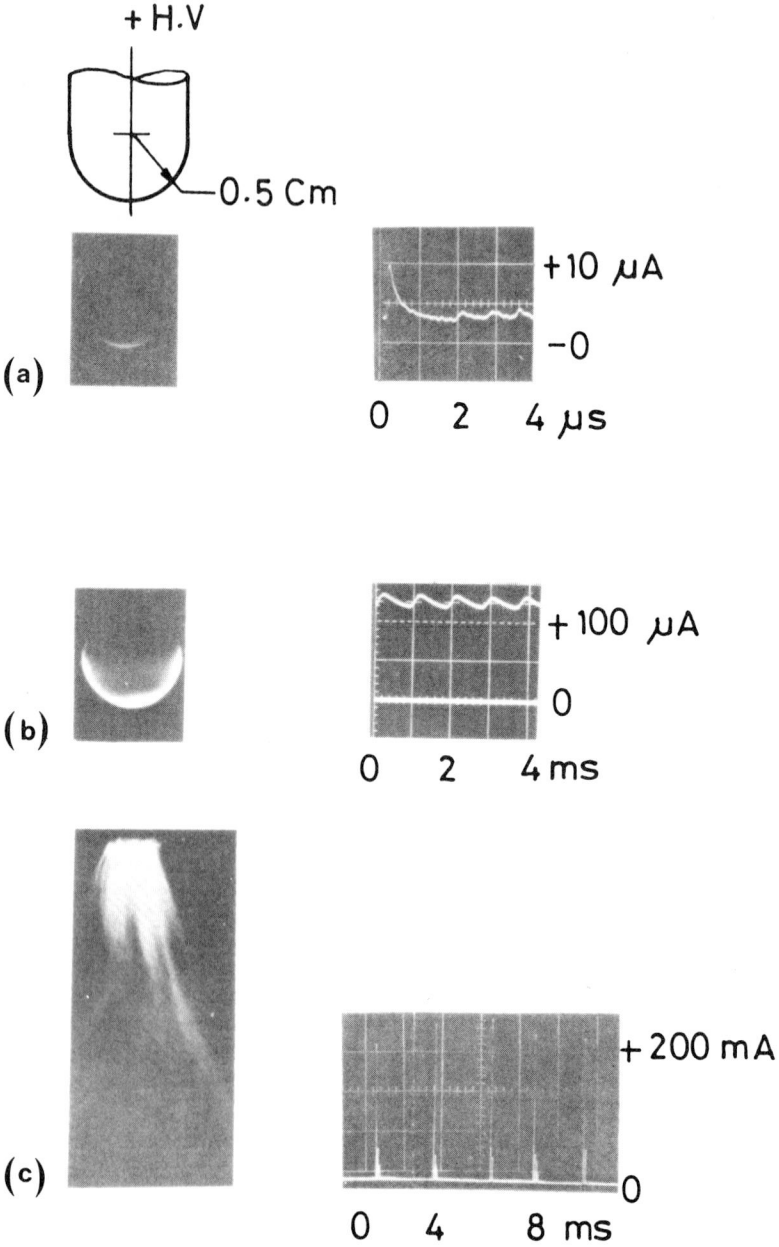

Fig. 5.2 Photographs and corresponding current oscillograms of different modes of positive corona. [From Giao and Jordan (1967).]

Corona Discharge 127

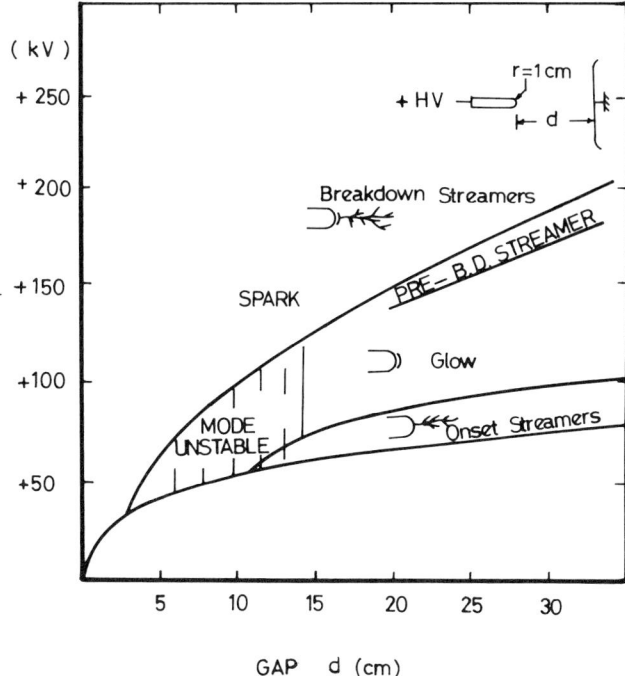

Fig. 5.3 Onset voltages of various positive corona modes and sparkover voltage as functions of point-to-plane gap spacing. (Courtesy of E. Nasser and Wiley-Interscience.)

to gas molecules and form negative ions which continue to drift very slowly away from the cathode. During the process of avalanche growth, some photons get radiated from the avalanche core in all directions (Fig. 5.4). The photoelectrons thus produced can start subsidiary avalanches that are directed from the cathode. The motion of the electrons and negative ions away from the cathode and that of positive ions toward it correspond to the corona current pulses flowing through the high-voltage circuit, as shown in Figure 5.5a, and could easily be computed.

With an increase in applied voltage, the Trichel pulses increase in a repetitive rate up to a critical level at which the negative corona gets into the steady "negative glow" mode (Figs. 5.4 and 5.5b). At still higher voltages pre-breakdown streamers appear, eventually causing a complete breakdown of the gap (Figs. 5.4 and 5.5c).

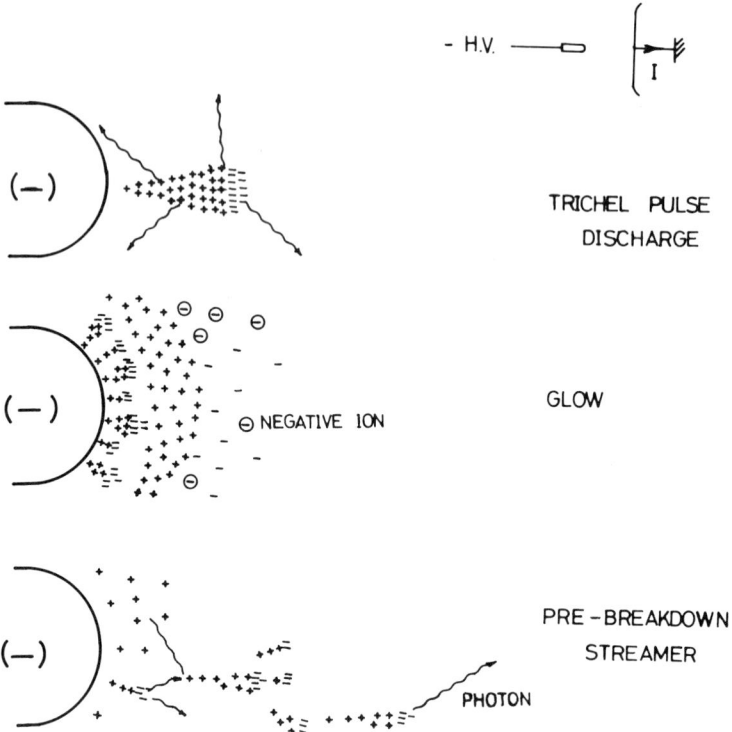

Fig. 5.4 Development of electron avalanches in negative corona discharges.

5.2.3 AC Corona

The basic difference between ac and dc coronas is the periodic change in direction of the applied field under ac, and its influences on the residual space charge left over from the discharge during preceding half-cycles (Fig. 5.6). Thus positive onset streamers and burst-pulse streamers may appear only over an extremely small range of voltage at onset, followed by a positive glow. Both negative Trichel pulses and negative glow can be observed in an ac corona. If the applied voltage has a suitable magnitude depending on the electrode geometry, both positive and negative glows and streamer coronas can be observed in each cycle.

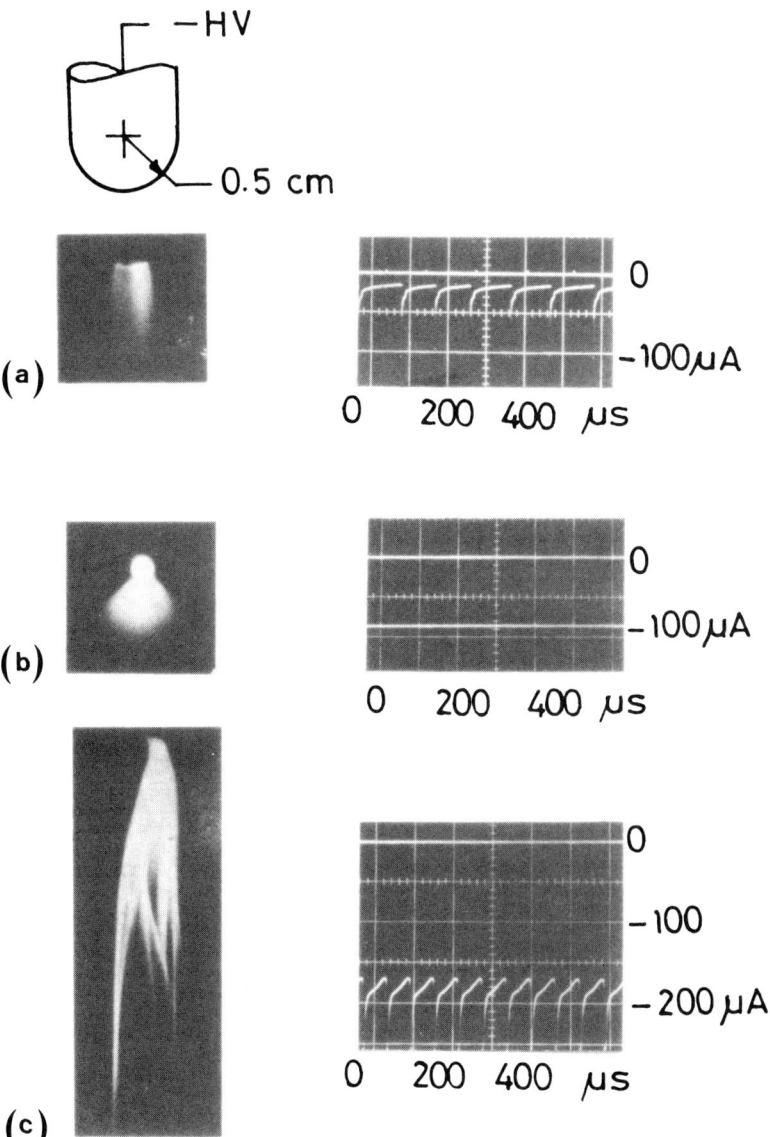

Fig. 5.5 Photographs and corresponding current oscillograms of various modes of negative corona. [From Giao and Jordan (1967).]

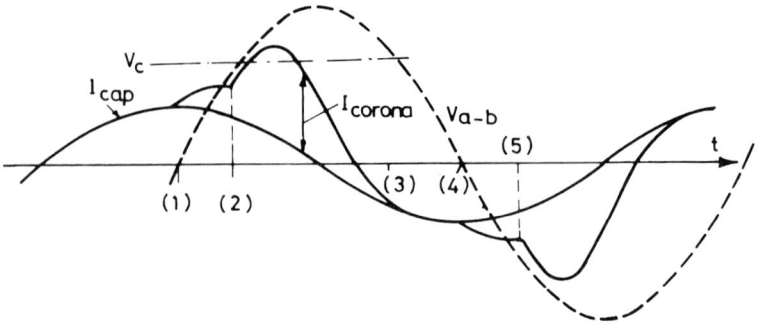

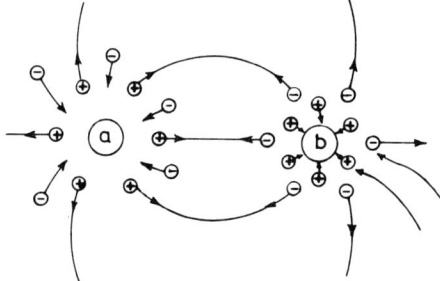

Period (1)−(2)

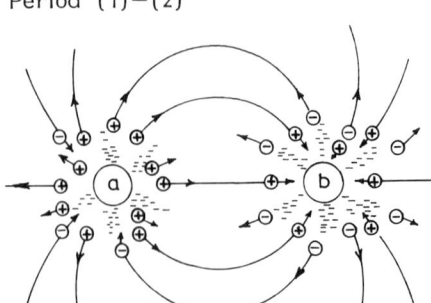

Period (2)−(3)

Fig. 5.6   AC corona current for a symmetrical gap between two parallel conductors a and b. Note that when the voltage V_{a-b} is below the corona-onset level V_c, the corona current corresponds to the motion and recombination of residual ions in the gap between the high-voltage conductors during the periods shown in the oscillogram.

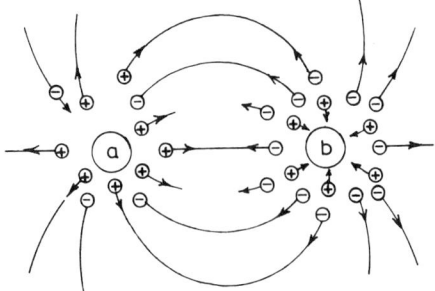

Period (3)–(4)

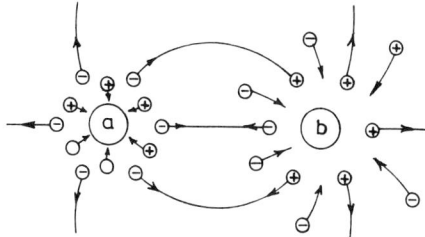

Period (4)–(5)

Fig. 5.6 (Continued)

5.2.4 Impulse Corona

Under impulse voltages, the corona starts in an air gap almost clear of any space charge. Therefore, electron avalanches and streamers extend over significant distances. The onset streamers produced and their branches can easily form traces on photographic films in contact with the anode or cathode, in what is known as Lichtenberg figures (Fig. 5.7). Such a figure cannot be produced by dc because of the choking effect of accumulating clouds of space charges. Under ac a very large number of traces get superimposed on each other.

On long HV transmission lines, the current corresponding to corona discharges under the traveling extra-high-voltage surges has the beneficial effect of reducing the surge peak and front steepness, as shown by experiment and computation (Chapter 14). Thus their stresses on the power system insulation are relieved.

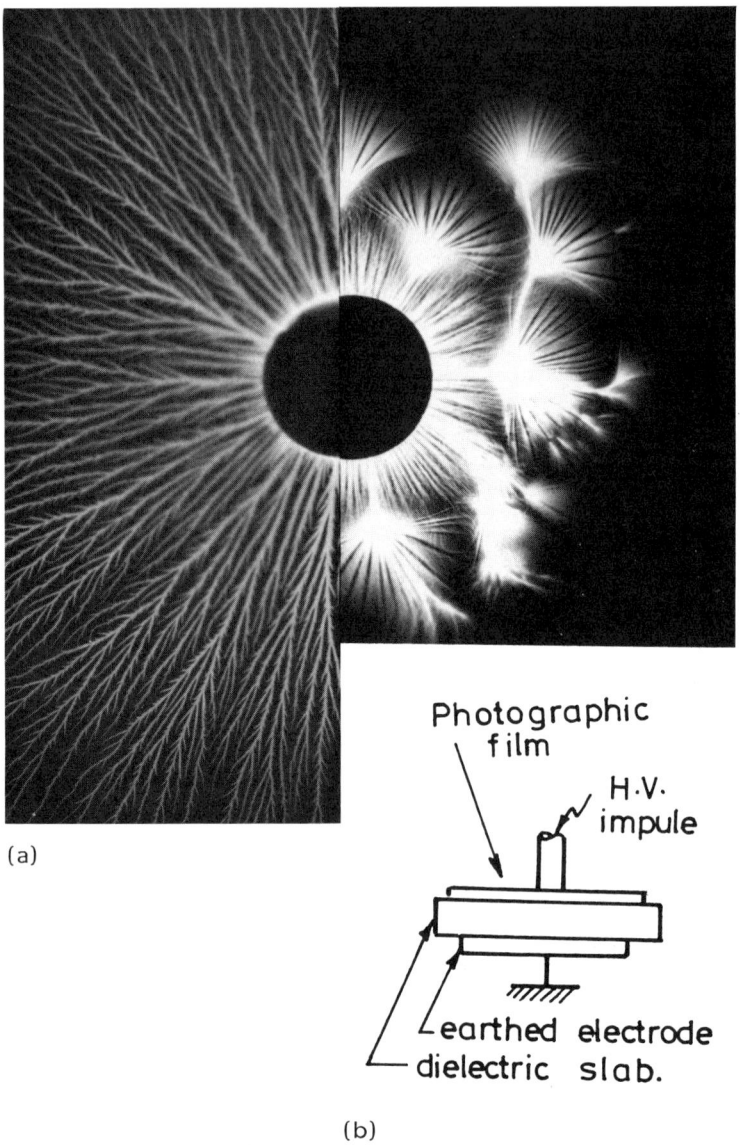

Fig. 5.7 Photographic traces of corona discharge under impulse voltage (Lichtenberg figures): (a) positive polarity, (b) negative polarity.

5.3 THE CORONA ONSET LEVEL

It has been well established by experiment and by computation that corona discharge starts at the surfaces of HV electrodes and conductors when their surface voltage gradients reach a critical value E_0 (Abdel-Salam and Khalifa, 1977). The magnitude of E_0 depends on the voltage polarity and on the pressure and temperature of the ambient gas. Humidity has a minor effect. The air pressure P (kPa) and temperature Θ (°C) are usually combined into one factor δ, the relative air density, referred to STP. Thus

$$\delta = \frac{2.94P}{273 + \Theta} \tag{5.1}$$

with δ ranging between 0.9 and 1.1, experimentally measured E_0 values fit in the following relations:

For ac:

$$E_0 = 30\delta \qquad kV_{peak}/cm \tag{5.2}$$

For dc:

$$E_\pm = A_\pm \delta + B_\pm \sqrt{\frac{\delta}{r}} \qquad kV/cm \tag{5.3}$$

A_+ and A_- are in the respective ranges 31 to 39.8 and 29.4 to 40.3; B_+ and B_- are, correspondingly, 11.8 to 8.4 and 9.9 to 7.3.

Under ac, a slightly higher field strength E_V corresponds to corona being clearly visible on the conductor surface and could be expressed as $E_V = 30\delta(1 + 0.3/\sqrt{\delta r})$ kV_{peak}/cm, r being the conductor radius in centimeters.

5.3.1 The Corona Onset Voltage

This could easily be calculated using E_0 or $E_\pm$ once the conductor arrangement and dimensions are known. Methods of field calculations were discussed in Chapter 2. It should be realized that field calculations would normally be based on the assumption of perfectly clean smooth conductors; different from practical conditions. At any point of microroughness on a practical conductor surface the field would be highly concentrated and the critical field strength for corona onset would be reached there while the average field strength over the entire surface would be considerably lower. A corresponding surface factor should be taken into account while estimating the corona onset voltage V_0 for the conductor arrangement. This factor is usually taken as about 0.6 for new rough stranded conductors and about 0.85 for weathered conductors.

Example

A three-phase single-circuit overhead line has a bundle of two conductors per phase arranged as shown in Figure 5.8. It is evident that the surface field strength is higher at the middle phase than at the outer phases. The difference is normally about 7% for practical line dimensions. The field strength at the surface of the middle-phase conductors is in terms of the line voltage V:

$$E_s = \frac{V}{2r \sqrt{3} \ln (D/\sqrt{rd})} \qquad (5.4)$$

For a given conductor radius r, relative air density δ, and conductor surface factor, the line voltage corresponding to corona onset on the middle phase could be estimated. The effect of conductor bundling and of height above ground could also be calculated.

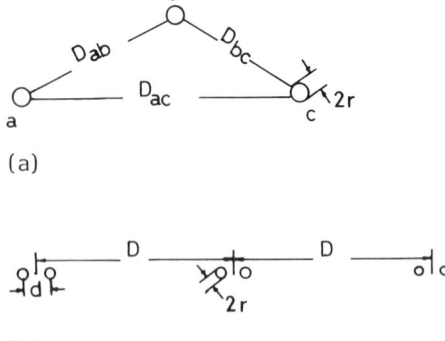

(a)

(b)

Fig. 5.8 Conductor arrangement for a three-phase transmission line with single conductors (a) and with bundle conductors at the same level (b).

Corona Discharge 135

5.3.2 Computation of DC Corona Onset Voltages

Onset voltages of dc corona could be estimated using the relation (5.3). They could also be computed according to an algorithm based on the ionization and deionization process acting in the corona discharge of either polarity (Khalifa and Abdel-Salam, 1974a).

Case of Positive Polarity

The first avalanche proceeds toward the anode and ends at its surface. It develops a cloud of positive ions, and photons are emitted from its core. If at least one photoelectron is produced during the lifetime of the first avalanche, the discharge would be self-sustaining and the applied voltage would equal the corona onset voltage. Also, at higher voltages the electron and ion populations of the successive avalanches and their motion could be computed and hence the positive corona current pulse (Khalifa and Abdel-Salam, 1974a).

Case of Negative Polarity

In this case electron avalanches start at the conductor surface and extend outward up to the distance where almost all of their electrons have formed negative ions by attachment to gas molecules. The numbers of electrons, ions, and photons could be computed. Photons were shown to be the main contributors to the emission of electrons from the cathode. The criterion for onset is that at least one photoelectron is emitted by the photons of the first avalanche, to keep the discharge self-sustaining. The computed onset voltages were in good agreement with experiment (Khalifa and Abdel-Salam, 1974b). Also, the current pulse shapes could be computed.

Corona Onset on Bipolar DC Lines

It is known that the onset voltage gradient for a positive corona is slighly lower than that for a negative corona. Therefore, the voltage level corresponding to positive corona onset can be calculated according to the method described above for monopolar corona. At the negative conductor, which has not yet reached its corona onset level, there are positive ions drifting toward it and coming from the positive corona discharge at the positive conductor. These ions enhance the field intensity at the negative conductor and cause a corona to start at a voltage slightly lower than the value calculated for the monopolar case. This slight difference could also be computed (Abdel-Salam and Khalifa, 1977).

5.3.3 Possible Corona in Compressed Air and SF_6

Because sulfur hexafluoride is an electronegative gas, it has a high affinity for electron attachment. This makes it more difficult for electron avalanches to grow. Thus corona and sparkover occur at voltages considerably higher than those in air. Above a certain critical gas pressure sparkover occurs across the gas gap without any preceding corona (Section 4.4). At such high pressures the coefficient of ionization by collision becomes lower than the coefficient of electron attachment: $\alpha < \eta$ (Chapter 3). Thus electrons produced at the cathode by photoemission and field emission would have to contribute more substantially to the discharge in order to maintain its stability. Taking them both into account has made possible the computation of breakdown voltages of gaps in compressed air and SF_6 (Khalifa et al., 1977; Abdel-Salam, 1978).

5.4 CORONA POWER LOSS

Empirical formulas have been suggested for evaluating corona losses on ac lines and on both monopolar and bipolar dc lines.

5.4.2 AC Lines

Empirical formulas were suggested early in this century by Peek and Peterson for estimating P_c, the fair-weather corona losses of overhead transmission lines (Begamudre, 1986). Because of several flaws, Peek's formula was superseded by that of Peterson, which takes the form

$$P_c = \frac{3.73K}{(D/r)^2} fV^2 \times 10^{-5} \qquad kW/conductor/km \qquad (5.5)$$

where f is the frequency, V the line voltage, and D and r the phase conductor separation and radius. K is a factor depending on the ratio of the operating voltage V to the corona onset line voltage V_0 (Fig. 5.9).

A much more recent and more scientific approach was the computer program developed by Abdel-Salam and his colleagues (1984) based on the physical phenomena of corona discharges (Shamloul, 1989).

In properly designed transmission lines, the corona loss in fair weather is usually insignificant. Typical values measured range from 0.3 to 1.7 kW/conductor/km for 500-kV lines and from 0.7 to 17 kW/conductor/km for 700-kV lines (Electrical Power Research Institute, 1979).

Corona Discharge 137

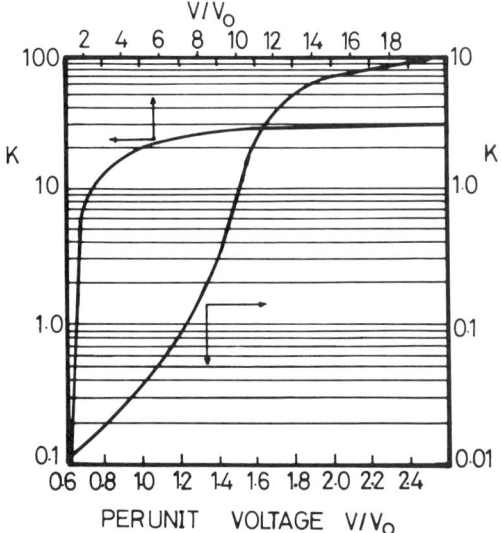

Fig. 5.9 Factor K to be used in Peterson's corona loss formula as a function of the per-unit operating voltage referred to the corona-onset voltage.

Effect of Conductor Bundling

For transmission lines with bundled conductors, Peterson's formula can be modified by including the capacitance geometric mean radius of the bundle instead of the single-conductor radius. Naturally, the separation between subconductors in the bundle has an effect on the amount of corona loss. There is an optimal separation between subconductors in the bundle that corresponds to minimum corona loss.

Effects of Weather

The principal weather parameters are air temperature, pressure, wind, humidity, rain, snow, and dust. The air temperature and pressure are included in the factor δ. No perceptible effect of wind could be noticed on the ac corona. The same goes for humidity unless it approaches 100%. With condensation, and much more seriously with rain, the corona losses increase 10-fold or more, depending on the rate of rainfall (Electrical Power Research Institute, 1979). The increase in corona loss due to rain and dust on conductors with large diameters is much greater than with smaller conductors, as the coronating points would be more numerous in the former case.

5.4.2 DC Lines

In unipolar dc lines, we have only ions of one polarity in the space between conductors, and between the conductors and ground. With bipolar lines, however, we have both positive and negative ions in the interconductor space and there is a high probability for ion recombination. This particular phenomenon accounts for the bipolar dc line corona loss being much higher than on monopolar lines, and even higher than on ac lines, as computed by Abdel-Salam and colleagues (1982).

In comparison with ac lines having equal per-unit effective voltages with respect to the corona onset levels, the fair-weather corona loss on a bipolar line was found to be about twice that on three-phase ac lines. On the other hand, increases in bipolar line corona loss with voltage and in foul weather are not as rapid as in the case of ac lines.

The effects of atmospheric humidity and wind could be measured and computed (Khalifa and Abdel-Salam, 1974b). The corona loss of monopolar lines could increase by about 20% if humidity increases from 60% to approaching 100%. A much more significant increase occurs if wind blows across the conductors of bipolar lines (Fig. 5.10).

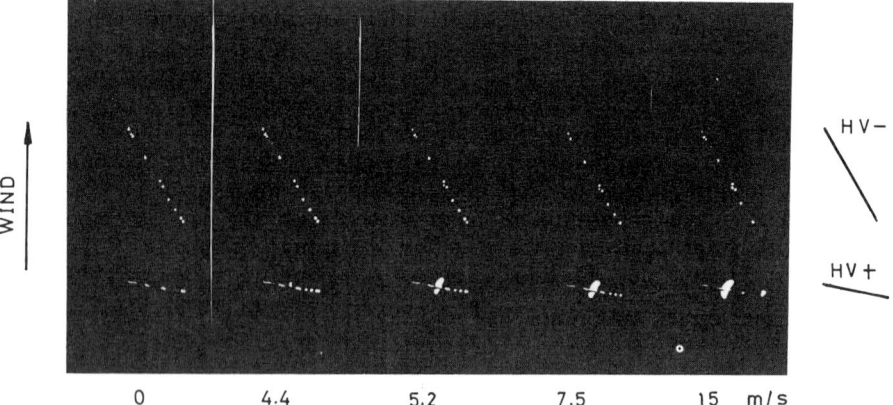

Fig. 5.10 Appearance of bipolar dc corona on two conductors, 6.5 mm in diameter and 45 cm apart, at ±63 kV, and at different wind speeds. The conductor arrangement is indicated.

Corona Discharge 139

5.5 CORONA NOISE

Corona noise includes interference with radio, television, and other wireless reception caused by corona. Also, audible noise is experienced near EHV lines and substations. Corona, undoubtedly interferes also with carrier signals transmitted along EHV lines.

The main source of corona radio noise is from positive streamers (Fig. 5.2), as their amplitudes are much higher than those of the negative Trichel pulses. As the pulses are random in amplitude, duration, and repetition rate, their noise is felt over a continuous spectrum. The noise level decreases at higher frequencies. This was shown by both measurement and computation.

While measuring the radio noise level and its lateral profile near bipolar HV dc lines, the highest level was recorded under the positive conductor while the noise contributed by the negative conductor was rather insignificant. For ac transmission lines the lateral decay of radio noise is less steep than for dc lines.

In the case of ac corona, positive streamer pulses would occur during the positive half-cycles and would also be the main source of noise. At voltages slightly above the onset level positive corona would tend to take the form of a steady glow, assisted by the negative ions produced during the preceding negative half-cycles. Therefore, HV dc lines are usually more noisy than ac lines at voltage gradients slightly above the onset levels, particularly in fair weather.

5.5.1 Effect of Line Conductor Size

Radio interference (RI) measurements under both ac and dc EHV lines have shown that the RI level rises with the voltage gradient (i.e., field strength E_{max} at the HV conductor surface) according to the relation

$$RI = C(E_{max})^n \quad dB \tag{5.6}$$

C being a constant. For dc lines, the exponent n has a value of 5 to 7 in fair weather and 1.5 to 3.5 in rain. For ac lines, on the other hand, the exponent n is about 7 to 8 in both fair and foul weather.

For the same conductor voltage gradient, in both ac and dc lines, the RI level was found to increase with the conductor radius according to the relation

$$RI = C_1 r^2 \quad dB \tag{5.7}$$

This relation was found to be independent of conductor bundling.

5.6 SOME INDUSTRIAL APPLICATIONS OF CORONA

Corona discharges have numerous applications in industry, as mentioned in Section 5.1. It may suffice here to explain in some detail three of these applications.

5.6.1 The Van de Graaff Generator

In such an apparatus, the ions of either polarity produced by corona at pointed electrodes are carried mechanically by a very highly insulating belt to the top of the apparatus, where they are conducted to a voluminous high-voltage terminal. As more and more charges accumulate, the generator voltage builds up as high as millions of volts. It is limited only by the dielectric strength of the terminal's insulation to ground. Compressed gas and/or grading rings are used to raise the output voltage (Fig. 5.11). The output current is limited to a few microamperes.

5.6.2 Electrostatic Precipitators

An electrostatic precipitator consists primarily of a group of thin, sometimes barbed wires energized at high-voltage dc, sometimes with superposed recurrent pulses. The intense corona discharges on these wires provide ions that get attached to particulates suspended in the flowing gas stream. Charged particles get diverted by the applied electric field, while the clean gas flows on (Fig. 5.12). Precipitator efficiency can reach or even exceed 99%, depending on several design parameters, including the dimensions and geometry of the gas duct, gas temperature and velocity, and average size and resistivity of the particulates, and the corona discharge intensity (Landham et al., 1987).

5.6.3 Photocopying Machines

Their operation is based on both corona discharge and photoconductivity (Seelentag, 1979). A photoconducting master plate or drum is initially charged by exposure to corona discharge uniformly over its surface. It then acts as a photographic film and captures a "charge image" of the original to be copied. The dark spots keep their charges, while the lighter ones lose theirs by photoconductivity. The image gets developed by exposing the master to a

Corona Discharge

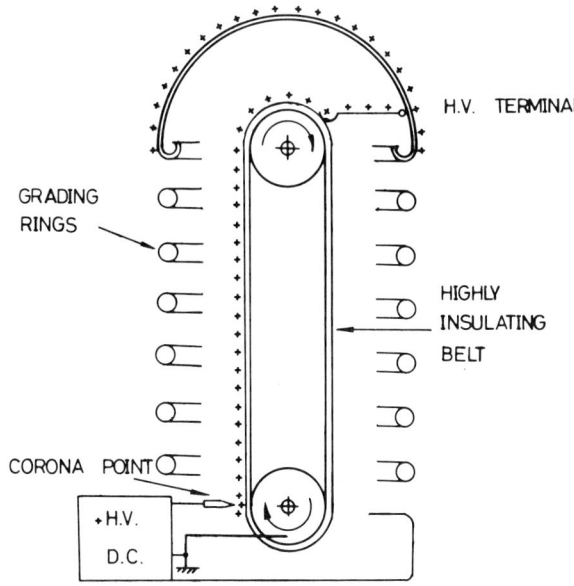

Fig. 5.11 Principle of operation of the Van de Graaff generator, shown producing positive high voltage.

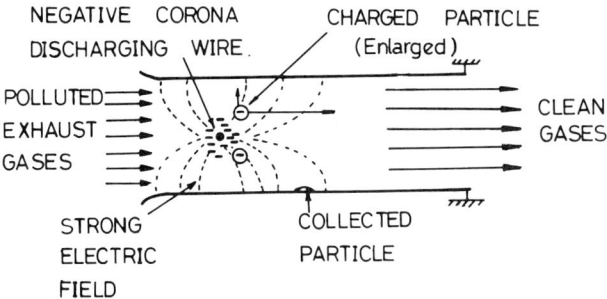

Fig. 5.12 Electrostatic precipitator's principle of operation.

toner and then transferring the image to a blank sheet of paper (Berg and Hauffe, 1972).

From the discussion above it is evident that corona discharges have several beneficial applications. On the other hand, they entail power losses on HV and EHV overhead transmission lines, particularly in foul weather. They also produce radio noise. Corona characteristics used to be estimated by empirical relations developed from experimental observations. However, since the 1970s, these characteristics could also be computed using the basic physical parameters of ionization and deionization.

REFERENCES

Abdel-Salam, M. (1978). *ETZ-Archiv, 99*(5):271-275.
Abdel-Salam, M. and Khalifa, M. (1977). *Proc. 13th Int. Conference on Phenomena in Ionized Gases*, Part I, Berlin, pp. 435-436.
Abdel-Salam, M., Farghally, M., and Abdel-Sattar, S. (1982). *IEEE paper 82WM-212-9.*
Abdel-Salam, M., Farghally, M., Abdel-Sattar, S., and Shamloul, D. (1984). *Proc. 4th Int. Symposium on Gaseous Dielectrics*, Knoxville, Tenn., pp. 492-497.
Begamudre, R. D. (1986). *Extra High Voltage AC Transmission Engineering*, John Wiley & Sons, Inc., New York.
Berg, W. and Hauffe, K. (1972). *Current Problems in Electrophotography*, Walter de Gruyter & Company, Mouton Publishers, West Berlin.
Electrical Power Research Institute (1979). *Transmission Line Reference Book, 345 kV and Above*, Project UHV, Electrical Power Research Institute, Palo Alto, Calif.
Giao, T. and Jordan, J. (1967). *IEEE publication 31-C-44*, pp. 5-15.
Khalifa, M. (1979). *Proc. 3rd Int. Symposium on High Voltage Engineering*, Milan, Paper 53-05.
Khalifa, M. and Abdel-Salam, M. (1974a). *IEEE Trans., PAS-93*: 720-726, 1693-1699.
Khalifa, M. and Abdel-Salam, M. (1974b). *Proc. 3rd IEE Int. Conference on Gas Discharges*, London, pp. 311-314.
Khalifa, M., Abdel-Salam, M., Radwan, R., and Ali, Kh. (1977). *IEEE Trans., PAS-96*:886-895.
Landham, E., Dubard, J., O'Brien, M., Lindsey, C., and Plulle, W. (1987). *Proc. IEEE Annual Meeting on Electrostatic Processes*, Atlanta, Ga., Paper 8-2.
Loeb, L. (1965). *Electrical Coronas—Their Basic Physical Mechanisms*, University of California Press, Berkeley, Calif.

Nasser, E. (1971). *Fundamentals of Gaseous Ionization and Plasma Electronics*, Wiley-Interscience, New York.

Seelentag, W. (1979). *Electrostatic Imaging, IEE Medical Electronics Monographs* 28–33 (W. Watson, ed.), Peter Peregrinus, Ltd., Stevenage, Herts, England.

Shamloul, D. (1989). Ph.D. thesis, Assiut University, Assiut, Egypt.

Zeitoun, A., Abdel-Salam, M., and El-Ragheb, M. (1976). *IEEE paper A-76-418-4*.

6
The Arc Discharge

M. KHALIFA   *Cairo University, Giza, Egypt*

6.1   INTRODUCTION

Arc discharge is encountered in the everyday use of power equipment. Whenever a circuit breaker or a load-break switch is opened while carrying a current, an arc strikes between its separating contacts. A persistent fault in a transformer, machine, or cable would eventually involve an arc. Therefore, information about the characteristics of arcs, the contact erosion they cause, and the factors conducive to their extinction is essential for the proper design, operation, and protection of such high-voltage equipment. Applications of the electric arc and its plasma include arc-discharge lamps, some furnaces, and processes in the manufacture of pure metals and some electronic devices.

6.2   ARCS IN CIRCUIT BREAKERS

When opening a switch or breaker, the contacts move apart and the contact area decreases rapidly until finally the contacts are physically separated. When the contact area decreases to a very small spot, the contact resistance increases considerably while the flowing current becomes highly concentrated. For a circular spot of contact with radius r, the contact resistance equals $\rho/2r$, ρ being the resistivity of the contact material. Thus, for a circular spot of radius 10 µm and a current of 10 A, the current density reaches about 3×10^6 A cm^{-2}. If the contacts are made of pure copper, the contact resistance can exceed 0.1 Ω. The corresponding power loss at such a microscopic spot would suffice for melting and even evaporating the hemisphere of metal at the contact spot within a period on the order of 1 µs.

 The metal vapor filling the space between the parting contact spots would thus furnish the conducting medium for the circuit

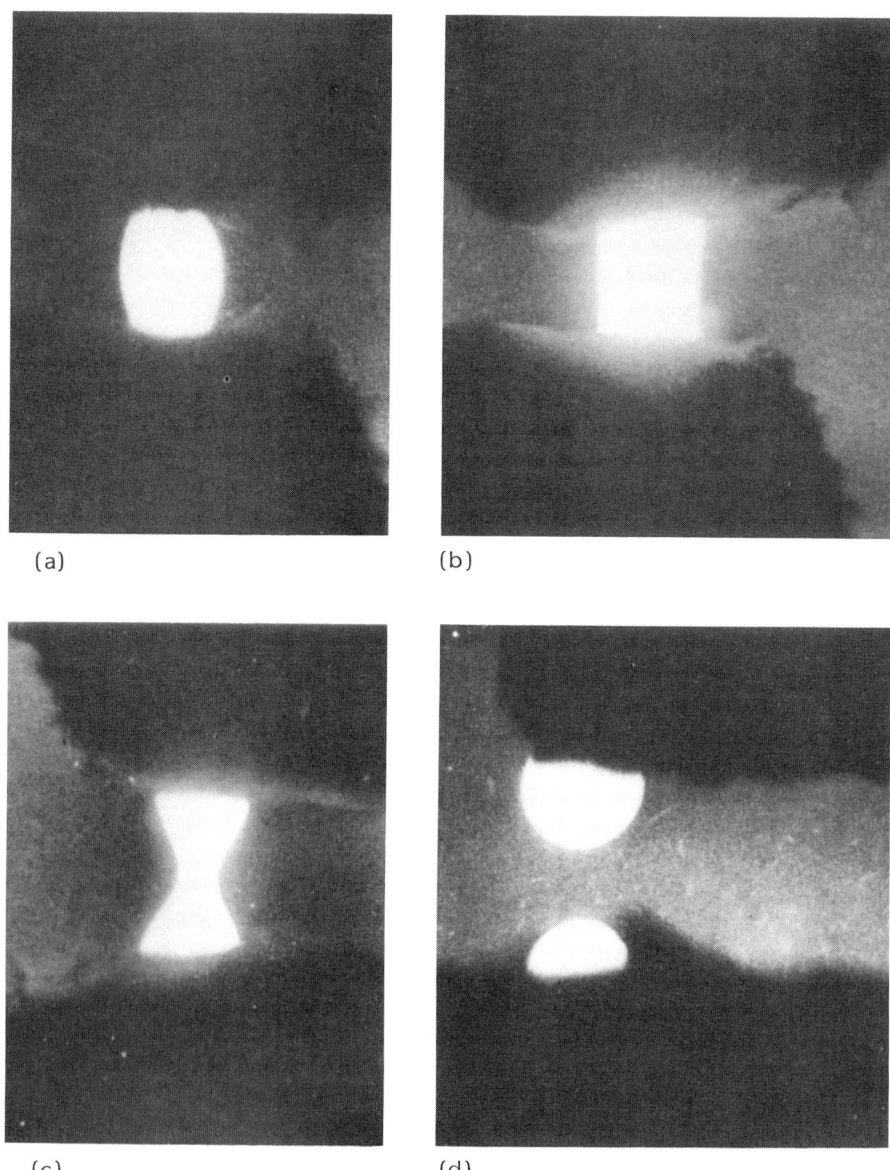

Fig. 6.1 Development of a molten metal bridge between separating iron contacts while carrying 40 A. Steps: a, b, c, d. Contact separation in step a is 0.5 mm. (Courtesy of F. Llewellyn-Jones and Oxford at Clarendon Press.)

Arc Discharge 147

current to flow in the form of an arc. If the electric circuit is such that no stable arc can exist, the molten metal bridging the microscopic gap between the contact spots would soon be broken as the spots part further (Fig. 6.1). Also, while closing the breaker or switch under voltage, a spark would occur because the contacts get very close. The very short arc that ensues would soon get extinguished.

6.3 REGIONS OF THE ARC

In air circuit breakers the arc burns mainly in an atmosphere of air; it burns mainly in hydrogen in oil circuit breakers; and it burns in composites of sulfur and fluorine in SF_6 circuit breakers. It burns mainly in an atmosphere of metal vapor in the case of vacuum circuit breakers. Under all these conditions the arc is composed of three principal regions: the cathode and anode regions and the arc column (Fig. 6.2), no matter what the total arc length is. Through all three regions the current is carried by electrons

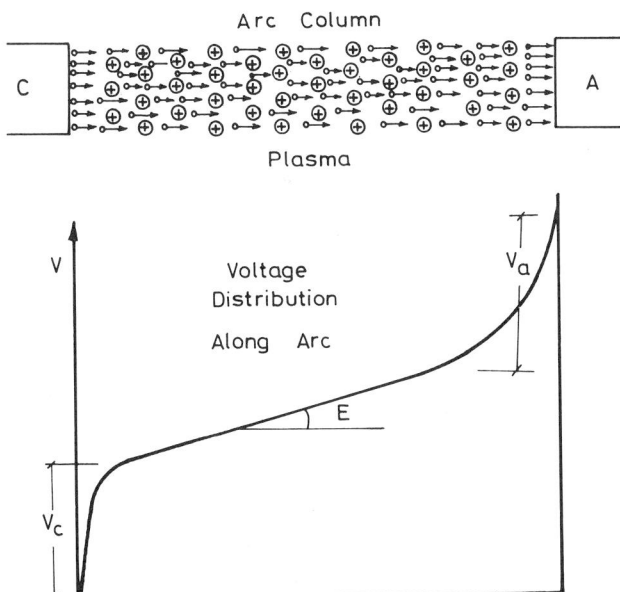

Fig. 6.2 Electrons and positive ions in the arc and its longitudinal voltage distribution.

and ions. In a steady arc, balance is struck between power input and losses.

6.3.1 Cathode Region

Electrons emitted from the cathode spot could be produced mainly by thermionic emission if the cathode is made of a high-melting-point metal (e.g., carbon, tungsten, or molybdenum). With cathodes of low melting point, electrons could be supplied by field emission from points of microroughness where the electric field would be highly concentrated. An additional important source of electrons at the cathode would be ionized metal vapor. This would occur in vacuum and other types of circuit breakers, and in mercury-vapor lamps.

The current is also carried partially by positive ions drifting slowly to the cathode from the plasma of the arc column. In the space between the surface of the cathode spot and the cloud of positive ions there is a high electric field (Fig. 6.2). Therefore, a significant cathode voltage drop builds up over the cathode region. Its magnitude and the width of the region depend on the arc current, the medium in which it is burning, and the cathode material.

6.3.2 Anode Region

The electrons from the arc plasma bombarding the anode spot and delivering all their energies keep the anode spot at a very high temperature. Positive ions are produced at the anode by thermal ionization of the gas and of any metal vapor near the anode spot. These ions drift slowly away from the anode into the arc plasma. Space-charge distribution near the anode produces a nonlinear field in the area and an anode voltage drop (Fig. 6.2).

Because of the different distributions of space charges at the cathode and anode, the cathode drop takes place across a considerably shorter distance than the anode drop (Fig. 6.2). The cathode region, however, would cover a sufficient number of free paths for the electrons leaving the cathode to reach the level of ionization by collision. They would thus liberate more electrons from the gas and metal vapor in the arc column.

6.3.3 Arc Column

At an arc's very high temperature, on the order of 10^4 K, the gas molecules are mostly dissociated. Many of their atoms and those of the metal vapor present would be ionized (Chapter 3). The degree of thermal ionization ζ of a gas depends on its temperature T, pressure p, and ionization potential V_i according to the well-known Saha relation:

$$\frac{\zeta^2}{1-\zeta^2} p = AT^{2.5} \exp \frac{-eV_i}{kT} \tag{6.1}$$

where e is the electronic charge and k is Boltzmann's constant. The magnitude of the constant A depends on the units used.

Inside the column the densities of positive and negative charge carriers are equal on the average and are comparable to that of neutral gas molecules. Therefore, the arc plasma exerts no electrostatic field. It has a significant electrical conductivity. Measurements have indicated that the conductivities of several gases (e.g., N_2, H_2, SF_6) are insignificant up to about 5000 K, rise steadily to about 30 S·cm at 10^4 K, and are about 80 S·cm at 2×10^4 K (Flurscheim, 1975). No doubt the extreme temperatures and high conductivities are confined to the core of the arc column. Both decrease sharply at some radius beyond which there is no current conduction to speak of. This effective radius of the arc column is a function of the arc current and the ambient gas and its pressure. The temperature distributions over the cross sections of 80-A arcs burning in O_2, N_2, and SF_6 are shown schematically in Figure 6.3. The interaction between the arc column and its surrounding ambient takes the form of diffusion of charge carriers and heat transfer.

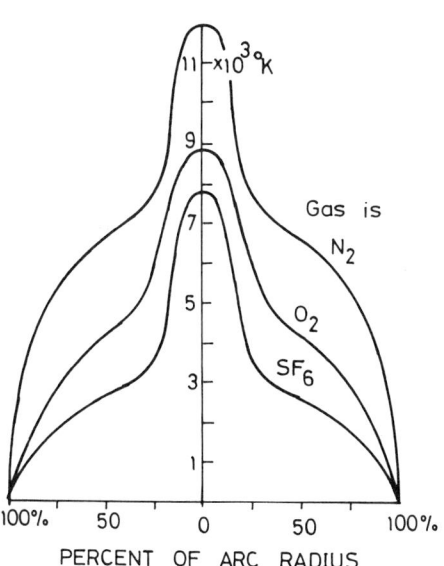

Fig. 6.3 Radial temperature distribution over the cross section of an arc as dependent on the ambient gas. [From Frind (1960).]

6.4 ENERGY BALANCE IN A STEADY ARC

If the arc carries a dc current of constant or slowly varying magnitude and is not subjected to turbulence in its environment, it can be described as a steady arc. Let us now look at the power balance for each region of this arc.

6.4.1 Cathode Region

The cathode spot receives both electrical and thermal power inputs, P_1 and P_2.

$$P_1 = aI_i(V_c + V_i) \tag{6.2}$$

where V_c is the cathode drop (Fig. 6.2), V_i the gas ionization potential, I_i the current component delivered by ions, and the factor a represents the fraction of the ions' energy delivered to the cathode spot and not to the arc plasma.

$$P_2 = G\beta(T_g - T_c) \tag{6.3}$$

where G is a geometrical factor, β the coefficient of heat transfer, and T_g and T_c are the temperature of the arc column near the cathode and that of the cathode spot, respectively. These two inputs are balanced by four components of power loss from the cathode spot: electron emission, heat conduction into the cathode volume, metal evaporation, and radiation from the cathode spot.

6.4.2 Anode Region

The input to the anode spot is the sum of the potential energy of the electrons falling through the anode drop, their kinetic energy delivered on impact, and the heat input to the anode spot. This input is consumed as heat conduction, radiation, and evaporation of metal from the anode spot. The sizes of the anode and cathode spots are functions of the arc current, the electrode material, and the ambient gas pressure.

6.4.3 Arc Column

Here the input is purely electrical. The power input per unit length of the arc column is $E(I_e + I_i)$, E being the voltage gradient along the arc, and I_e and I_i the components of the arc current

Arc Discharge

carried by electrons and ions. The current is mostly electronic since ion mobilities are about three orders of magnitude lower than that of electrons. The densities and mobilities of these charge carriers are functions of the arc column temperature.

The arc column loses some of its charge carriers by diffusion from its surface. It also loses heat by radiation and convection. The components of heat dissipation depend on the ambient. For example, in oil circuit breakers, energy is consumed in boiling some oil, dissociating some of its molecules producing hydrogen and hydrocarbon gases, expanding the gas bubble, and dissociating and ionizing some of the gas molecules.

Along the arc column, the power input maintains the ionization in the arc plasma, and compensates for the losses at the periphery and for any changes in the heat stored in the column. Because of the different ionization potentials and thermal conductivities and other thermal parameters, it was noted that for the same current in the range 10 to 1000 A, the arc column gradient E in hydrogen was about 10 times its value in oxygen when the arc was contained in a tube 2 cm in diameter. The current density varies along the axis of the arc column. This induces gas flow and static pressure gradients along the arc. This in turn could lead to ejection of molten metal and vapor from the electrode spots.

It could be shown that there is an interdependence among the arc current, column temperature, and radius. The arc always adjusts itself so that the system in equilibrium has minimum entropy. The arc column radius r and temperature T are such that the heat loss is minimal. Thus for a given arc current the column voltage gradient E is also a minimum consistent with the power balance mentioned above. Thus

$$\frac{dE}{dr} = 0 \quad \text{and} \quad \frac{dE}{dT} = 0$$

In other words, for a given arc current I, if the arc column cross section decreases, its conductance would decrease and the voltage gradient E would increase. If, on the other hand, the cross section increases, its heat loss to the surrounding would increase, its temperature would decrease, and so would its conductivity, and thus E would increase.

Experimental tests have shown that the arc column radius r is proportional to I^n, with n ranging between 0.25 and 0.6, depending on the type of cooling for the arc. In air-blast breakers it is volumetric cooling, whereas for a stationary arc, cooling is effected at its column surface.

6.5 STEADY-STATE ARC CHARACTERISTICS

As is evident from the previous discussion, the arc is by no means a simple element. The relation between the arc voltage and current depends on the arc length, electrode material, and ambient. The vacuum arc voltage was shown to depend principally on the cathode metal (Reece, 1975).

For arcs of fixed length in atmospheric air, the V-I characteristics follow inverse relations of the form shown in Figure 6.4. An approximate relation between the arc column voltage gradient E and its current I could be obtained as follows:

$$I = E\ [\pi r^2 (N_e k_e + N_i k_i)] \tag{6.4}$$

The electron and ion densities N_e and N_i and their respective mobilities k_e and k_i are functions of temperature. Thus equation (6.4) could be rewritten as

$$I = E(\pi r^2) F_1(T) \tag{6.4a}$$

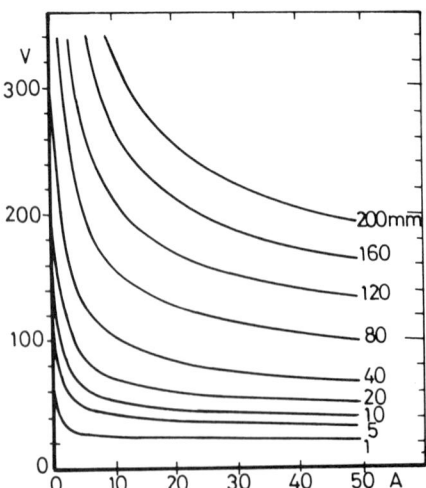

Fig. 6.4 DC voltage-current characteristics for arcs of different lengths burning in air between copper electrodes. [From Reider (1967).]

Arc Discharge

Under steady conditions, the input power is balanced by heat dissipation from the arc column surface, another function of temperature $F_2(T)$. Thus

$$E^2(\pi r^2)F_1(T) = 2\pi r F_2(T) \tag{6.5}$$

Eliminating r and differentiating, we get

$$E = CI^{-1/3} \tag{6.6}$$

where C is a constant.

Many experimental results on dc arcs in air fit relations between the total arc voltage V and current I of the form

$$V = a + b\ell + (c + d\ell)I^{-1} \tag{6.7}$$

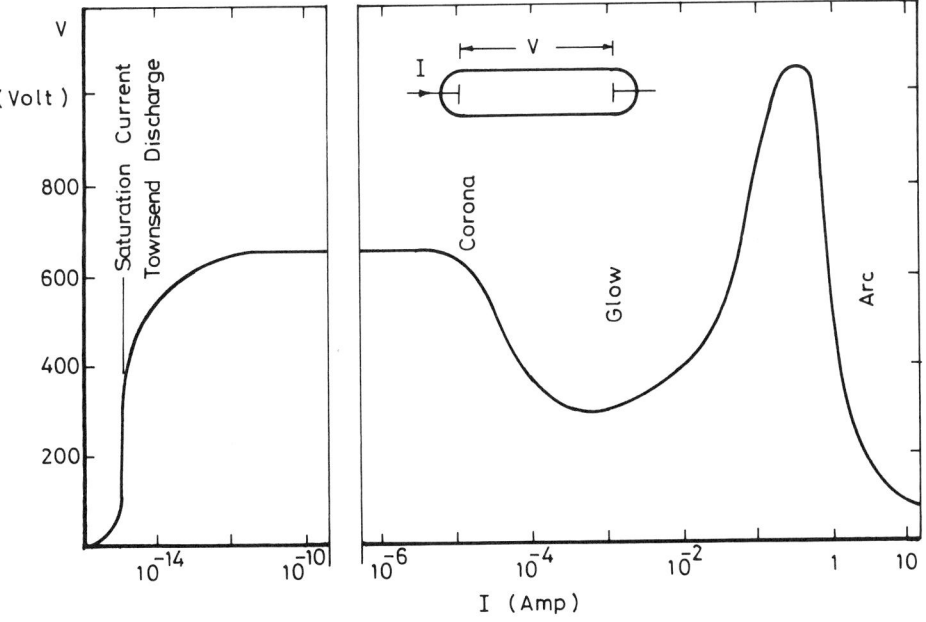

Fig. 6.5 Typical voltage-current characteristic for the various discharges in a gaseous gap. Neon is at a pressure of 130 Pa. The gap is 50 cm between disk electrodes 2 cm in diameter. (Courtesy of E. Nasser and Wiley-Interscience.)

For arc currents up to 20 A and lengths ℓ of about 5 cm, the constants of this relation have the following values: a = 17 V, b = 22 V/cm, c = 20 W, and d = 180 W/cm.

For the same current and length, the arc voltage increases with ambient pressure above atmospheric. In a rarefied atmosphere of about 10 Pa, however, and with a high enough source voltage, the arc gets unstable and turns into a glow discharge. We thus get the famous V-I characteristic shown in Figure 6.5. This is experienced in some gas-discharge lamps and tubes.

As the arc current is reduced to a very small value, a limit is reached when the arc input power is no longer sufficient to maintain the column temperature. The electron and ion densities in the arc plasma decrease so drastically that there is not enough conductivity for the current through the arc and it is extinguished.

6.6 MAGNETIC PHENOMENA IN ARCS

The arc is influenced by two magnetic fields, its own field and that of its feeding circuit. The circumferential field produced by the arc's current does exert a pressure to squeeze the arc column. The pressure could easily be calculated as being equal to $I^2/\pi r^2$. At the anode and cathode spots the arc radius is much smaller than anywhere along the column. Therefore, there is a high axial gradient of that pressure which can set up jets of plasma and metal vapor from these spots with velocities reaching 1 km/s (Barrault et al., 1972). Also, as current-carrying circuits under their own magnetic fields tend to increase their self-inductance, the arc will bow outward into a larger loop. Such phenomena are usually exploited in the design of dc circuit breakers for accelerating arc extinction (Chapter 11).

6.7 DYNAMIC ARC CHARACTERISTICS

Because of the thermal capacity of the arc column, a sudden rise in the magnitude of an otherwise steady arc current will have to be accompanied by an initial rise in the arc voltage so as to furnish the extra energy needed for building up the column ionization to the level corresponding to the new current magnitude. After some time, termed the thermal time constant Θ of the arc, the arc voltage would settle down to a steady value according to the V-I characteristic. Of course, the opposite will occur if the arc current suddenly drops. This can be visualized by thinking of the power balance for a unit length of the arc column. There, the input power

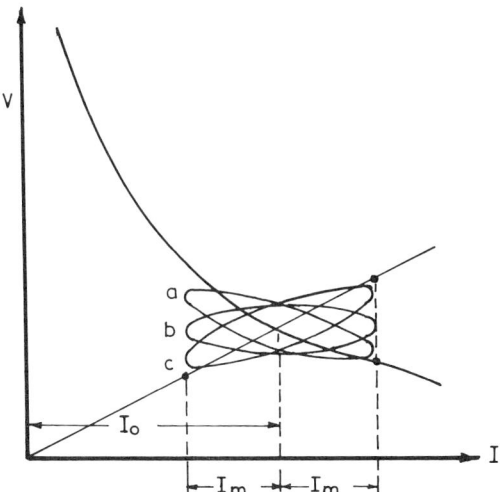

Fig. 6.6 Voltage-current characteristic of a dc arc with a superposed current modulation of amplitude I_m: (a) low frequency, (b) medium frequency, (c) high frequency.

$$P = EI = H + \frac{dQ}{dt} \qquad (6.8)$$

where H is the heat loss per second and Q is the heat stored in the thermal capacity.

If the current variation is an ac modulation superposed on the dc current magnitude I_0, the corresponding voltage variation will be such that the operating point will follow a loop (Fig. 6.6). The loop shape depends on the frequency f of the current modulation. At extremely low frequencies the variation will almost follow the static characteristic. On the other hand, at extremely high frequencies the arc behaves like a linear resistance. Its thermal capacity would prevent its temperature from varying at any significant fraction of such frequencies.

6.7.1 AC Arc Characteristics

Near the peak value of the arc current the voltage necessary to maintain it is relatively low (Fig. 6.7). As the current approaches zero, a higher and higher voltage is needed to maintain the arc. When the voltage across the arc is not high enough, it will be unstable and will get extinguished even before the zero point of the

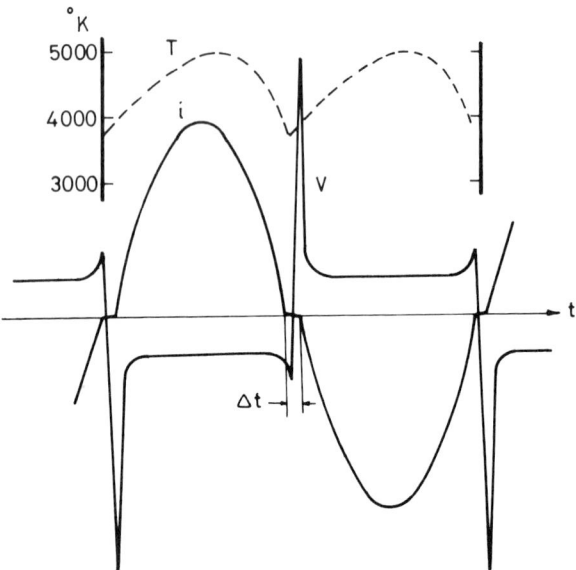

Fig. 6.7 Time variations of current, voltage, and temperature for a 50-Hz 10-A arc, 3 mm long, in air between copper electrodes.

sinusoid (Fig. 6.7). For the current to flow in the opposite direction in the following half-cycle, the arc gap has to break down again under a sufficiently high voltage. Therefore, the arc voltage at the beginning of the half-cycle is considerably higher than that at its end. There is also a period Δt of effectively zero current around the virtual zero point of the sinusoid (Fig. 6.7). The arc column temperature also varies during the ac cycle; the peak temperature lags behind the peak current because of the arc's thermal capacity.

6.8 THE ARC AS A CIRCUIT ELEMENT

Electric circuits subjected to analysis do sometimes contain arcs, as in faulty power systems or when arc furnaces are included. Accurate analysis requires a truly representative circuit to take the place of the arc. The arc is by no means a simple circuit element. This applies in both ac and dc circuits. According to dc arc characteristics such as the ones in Figure 6.4, the ratio V/I has a positive magnitude (the same as for a metal resistance), and both

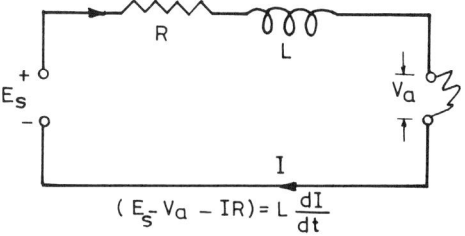

(a)

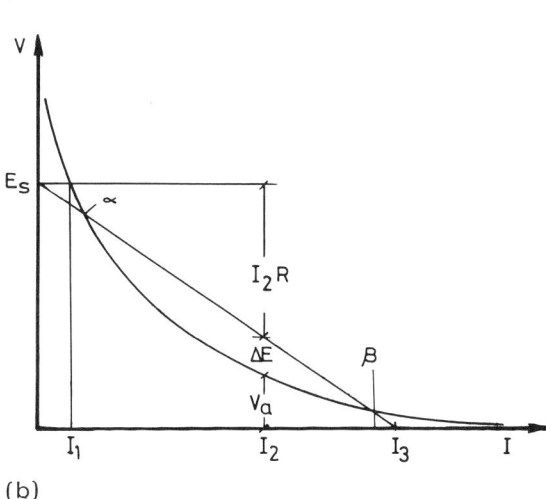

(b)

Fig. 6.8 Arc as an element in a dc circuit: (a) arc in the circuit, (b) voltages across the arc and the circuit (the arc is stable at the point β but not at α).

consume power. However, the arc equivalent resistance varies with the current and dV/dI is negative. Therefore, a stabilizing impedance must be included in series with the arc in both ac and dc circuits (Fig. 6.8). Looking at the arc's V-I characteristic, it can easily be proved that the stable operating conditions are represented by point β (Fig. 6.8b). If the current happens to swing below point α it will get extinguished.

For arcs of very small length, the voltage does not vary significantly over a wide current range, which means that the arc could be represented by a fixed back voltage opposing the source (Fig. 6.9). The equivalent circuit also comprises a resistance and

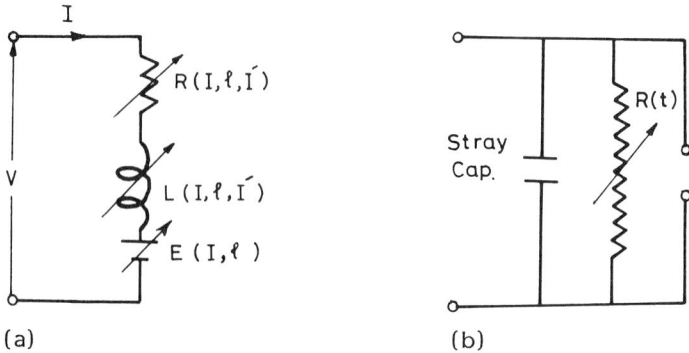

Fig. 6.9 Equivalent circuit of the arc: (a) cases of dc and ac during the burning period (R, L, and E are variables); (b) case of ac during the extinction period (Δt in Figure 6.7).

an inductance that are functions of the arc length ℓ, its current I, and the current's rate of change with time I'. The magnitudes of these parameters could be obtained from experimental tests. Such a circuit (Fig. 6.9a) would apply for both dc and ac circuits during periods when the arc is burning. During periods of effective current zero, Δt (Fig. 6.7), however, the arc can be represented by a high resistance that increases with time. The nonlinear and irregularly time-varying characteristics of arcs produce current harmonics and voltage ripples in the feeding network, which could be represented approximately.

6.9 ARC INTERRUPTION

The situation with dc is basically different from that under ac.

6.9.1 DC Case

In the dc case the arc current is forcibly brought down to zero. To do that, as will be described in Chapter 11, the arc is driven rapidly by its thermal buoyancy and magnetic field to extend its length and to subdivide into a number of partial arcs. Thus the voltage required to maintain it would increase rapidly. The time for arc interruption could be estimated given the electric circuit and breaker characteristics. The voltage and current oscillograms would look like those represented in Figure 6.10. It should be noted that the circuit inductance acts to delay the arc

Arc Discharge 159

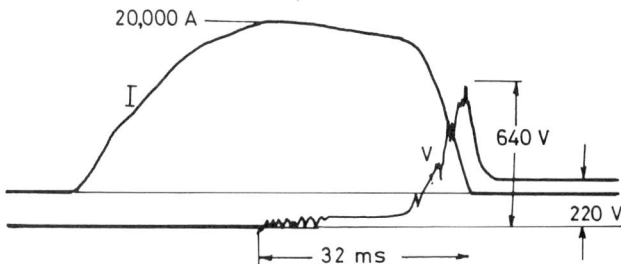

Fig. 6.10 Voltage and current variations during arc interruption. Note the overvoltage caused by the circuit inductance and the capacitance across the arc.

interruption. On the other hand, a resistance or capacitance across the arc would accelerate its interruption, as it would draw an additional current through the circuit resistance.

6.9.2 AC Case

Due to its thermal capacity, the conductance of the postarc gap cannot vanish instantly at current zero. Some time would be needed for the electrode spots and the gases in the postarc gap to cool down and for the electrons and ions of the arc plasma to recombine and/or diffuse. The conductance of the gap drops during the first few microseconds after arc extinction, and the gap's dielectric strength builds up in the subsequent tens of microseconds (Buttler and Whittaker, 1972).

This behavior of the postarc gap was originally represented by Cassie as a resistance increasing exponentially with time. The representation was further improved by Mayr when he considered the arc column to lose heat from its surface rather than from its whole volume. His relation took the form

$$R \frac{d}{dt} \frac{1}{R} = \frac{1}{\theta}\left[\left(\frac{VI}{W}\right) - 1\right] \tag{6.9}$$

where W is a constant. Thus, for a steady arc, its equivalent $d(1/R)/dt = 0$ (i.e., its VI = constant). This tends to corroborate the shape of the V-I characteristics of dc arcs (Fig. 6.4).

This model does represent the arc around the current zero and indicates how the postarc column resistance continues to increase. Some further refinements for this model were introduced by Rizk (1985) and others (Butler and Whittaker, 1972). The constant θ

was shown to depend on the magnitude of the arc current. It could not actually be defined as a time constant in the rigid mathematical sense, realizing how complex the arc phenomena are.

Current Chopping

The circuit to be interrupted by the circuit breaker inevitably has distributed capacitance and inductance. As the power-frequency current through the breaker arc falls toward zero, the arc colum diameter shrinks and exhibits a negative V-I characteristic. This effectively negative resistance of the arc would accentuate negatively damped oscillations of the local LC circuit at a frequency on the order of 10 kHz.

Such oscillations could be initiated by any sudden drop in the arc voltage due to any change in its conditions. With the high-frequency component superposed on that at the power frequency, the total arc current drops rapidly and prematurely to zero (Fig. 6.11). This phenomenon is current chopping. It induces excessive overvoltages in the circuit as discussed in Chapters 11 and 14.

Arc Reignition

The voltage that appears across the postarc gap after its extinction may or may not reignite it. If it does, the current will

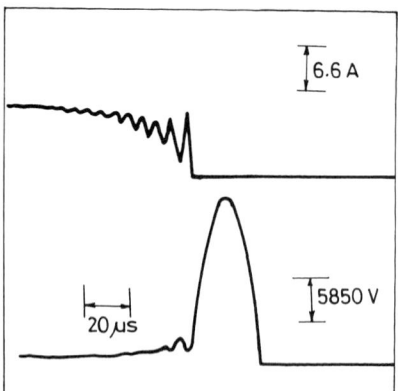

Fig. 6.11 Current and voltage variations during ac circuit interruption by an air-blast circuit breaker. Note the superposed oscillations in the current and its chopping before its natural zero. [From Rizk (1963).]

Arc Discharge

continue to flow for another half-cycle. If it does not, the circuit breaker or switch is considered to have opened the circuit successfully.

Arc reignition is decided by a race with time between the voltage appearing across the postarc gap and the deionization of its gases. Ionization would continue to be partly thermal because of the residual heat in the new cathode spot and in the gases in the gap. The applied electric field would also accelerate electrons, giving them energies that would probably be high enough for ionization by collision, as explained in Chapter 3.

During the first few microseconds after the virtual current zero, reignition is governed by the energy balance. If the applied voltage per unit length of the postarc gap is E', and this causes a current i to flow as a result of the gap's residual conductance, the input power per unit length of the gap is $E'i$. If this exceeds the heat loss, the arc will reignite.

During a subsequent period arc reignition would be dielectric rather than thermal. It will be governed by the applied voltage exceeding the dielectric strength of the gap. As the electrodes have reversed polarity, a dense cloud of positive ions remains near the new cathode. This cloud would severely distort the field near the cathode, and the dielectric strength of the postarc gap would be relatively low. However, after some microseconds the cloud of positive ions would be so diffuse that its influence on the electric field in the postarc gap would become much less pronounced. Thus the breakdown voltage of the gap rises fast initially, but afterward slowly approaches its ultimate value.

6.10 ARC EROSION

Erosion of the contacts by arcing is caused partly by evaporation of metal from electrode spots during the arc. Some droplets of molten metal may also be removed from the electrode spots by high differential gas pressures in the gap and by excessive electric fields, as explained above.

Tests and analyses of the erosion phenomena have revealed that the energy of the arc, rather than its charge, is the deciding factor. The current wave shape also has an effect on the rate of arc erosion (Fig. 6.12). For otherwise similar conditions, the rate of contact erosion is lower for contact metals with high melting points, greater latent heats of evaporation, and higher thermal conductivities.

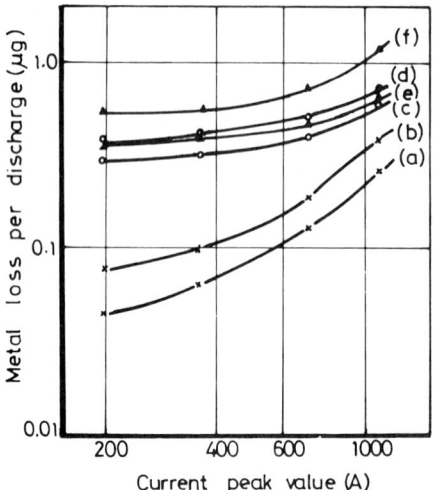

Fig. 6.12 Rates of arc erosion from electrodes of different materials. DC arc charge, 73 mC; arc length, 1 mm: (a) anode and (b) cathode of molybdenum, (c) anode and (d) cathode of steel, (e) anode and (f) cathode of W-Cu sintered mixture.

6.11 APPLICATIONS

The applications of electric arcs include arc-discharge lamps, arc furnaces, arc welders and cutters, and plasma torches. They are also used in the production of high-quality metal strips and for ion implantation on silicon wafers in the production of very large scale integrated circuits.

Arc furnaces are widely used in metal-producing plants. For smelting aluminum, for instance, dc arc furnaces are preferred, particularly when electric power is available economically. Also, furnaces fed from three-phase ac are used for melting and purifying metals.

Glow discharge has been used in a plasma spray for applying a metallic protective coating to important machine parts such as turbine blades, under vacuum (Shankar, 1981). It has also been used more recently for excessive heating of strips of metal powder in a rarefied atmosphere to produce high-quality stainless steel ribbons. This method was proved superior to the conventional method involving successive rolling and heat treatments (Millar, 1987).

Modern techniques for the production of very large scale integrated circuits involve isolating islands of p-type silicon with submicrometer layers of SiO_2. An effective method employs a plasma beam of high-energy positive oxygen ions (O^+) (Dettmer, 1987).

REFERENCES

Barrault, M., Mackburn, T., Edels, H., and Satyanarayana, P. (1972). *IEE conference publication 90*, pp. 221–223.
Buttler, T. and Whittaker, D. (1972). *Proc. IEE*, 119:1295–1300.
Dettmer, R. (1987). *IEE Electronics & Power*, 33(4):273–277.
Flurscheim, C. H. (ed.) (1975). *Power Circuit Breakers—Theory and Design*, Peter Peregrinus, Ltd., Stevenage, Herts, England.
Frind, G. (1960). *Zeitschrift für Angewandte Physik*, 42:515.
Llewellyn-Jones, F. (1957). *The Physics of Electrical Contacts*, Oxford University Press, Oxford.
Millar, R. (1987). *IEE Power Engineering Journal*, 1(5):257–265.
Reece, M. (1975). *Power Circuit Breakers—Theory and Design* (C. H. Flurscheim, ed.), Peter Peregrinus, Ltd., Stevenage, Herts, England, Chapter 2.
Rieder, W. (1967). *Plasma und Lichtbogen*, Friedr Vieweg & Sohn, Verlagsgesellschaft mbH, Braunschweig, West Germany.
Rizk, F. (1963). Ph.D. thesis, Chalmers Technical University, Göteborg, Sweden.
Rizk, F. (1985). *IEEE Trans.*, PAS-104:948–955.
Shankar, S., Koenig, D., and Dardi, L. (1981). *Journal of Metals*, 33(10):13–20.

7
Insulating Liquids

M. KHALIFA *Cairo University, Giza, Egypt*

7.1 INTRODUCTION

Since the turn of the century, oils have been in use for insulating cables, transformers, and circuit breakers. Insulating liquids are broadly classified as organic, mineral, or synthetic. Organic oils started being used late in the nineteenth century, whereas mineral oils were introduced in about 1910 with the development of petroleum refineries. Synthetic liquids with a wide spectrum of properties started being developed by the petrochemical industry in about 1960. They include synthetic hydrocarbons, halogenated hydrocarbons, silicones, and synthetic esters.

7.2 TYPES OF OILS

7.2.1 Organic Oils

This group of liquids includes vegetable oils, rosin oils, and esters. Natural esters are produced by the chemical reaction between a vegetable acid and an alcohol. Molecules of esters contain atoms of carbon, hydrogen, and a considerable number of oxygen atoms, about 20% (Breuer and Hegemann, 1987). The reaction is helped by a catalyst such as sulfuric acid. An inert water entrainer is also included to help in removing the water formed in the reaction.

Phosphate esters are manufactured from such raw material as coal tar "phenol" in a chemical reaction with phosphorus oxychloride, thus:

    phenol + phosphorus oxychloride ⟶

        triphenol phosphate + hydrochloric acid

Phosphate esters have very good fire resistance compared to mineral oils. However, their poor oxidation stability limits their use to special applications. With the inevitable depletion of mineral reserves in the world becoming more of a concern, several attempts have been made recently to use organic liquids in transformers and other applications where mineral oil has been unparallelled and where synthetic oils were being introduced during the last 20 years (Sankaralingam and Krishnaswamy, 1987; Marinho et al., 1987; Chan, 1987).

7.2.2 Mineral Oils

Petroleum is known to comprise a variety of molecular species. It is therefore broadly classified as of a paraffinic, naphthenic, aromatic, or intermediate group of molecules. The paraffin base is characterized by the chemical formula C_nH_{2n+2}, the naphthenes by the formula C_nH_{2n}, and the aromatics by the formula C_nH_n. The number n of monomers per molecule varies over the range of tens to a few hundreds depending on the type of crude, the distillation and purification processes, and its subsequent treatment and life in service.

Oil of the naphthenic group has traditionally been favored for impregnating paper to be used in insulating HV and EHV cables, because of its good gas-absorbing properties. As naphthenic crudes represent only 5 to 10% of the world's total production, which is inevitably declining, alternative liquids have been explored by research laboratories. Synthetic liquids have proved to be the answer. Their properties could in a sense be tailored to match each specific application.

7.2.3 Synthetic Oils

Thanks to intensive research and developments in petrochemical industries, we now have several synthetic liquids covering a wide range of properties, including nonflammability. Available are askarels, olefins, silicon oils, phosphate esters, and other liquids.

Askarels

The synthetic oils commercially known as askarels are manufactured by chlorinating polychlorobiphenyls (PCBs), aromatic hydrocarbons available as byproducts in some petrochemical industries. For example, hexachlorobiphenyl is produced from benzene and has an atomic arrangement of molecules, such as the one shown in Figure 7.1.

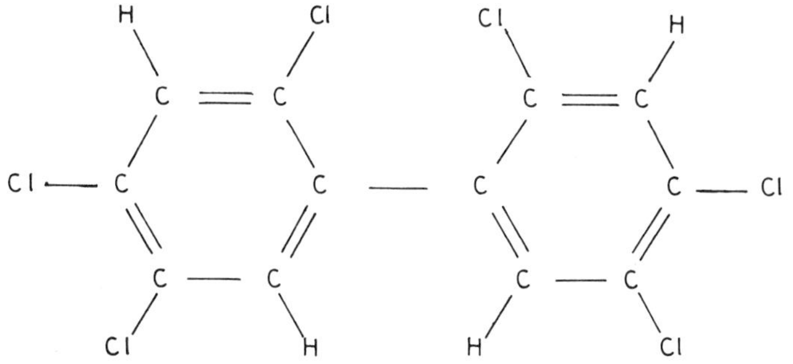

Fig. 7.1 Example of the atomic arrangement in a molecule of hexa-chlorobiphenyl (askarel).

Silicon Oils

The molecules of silicon oils contain silicon and oxygen atoms in addition to their many carbon and hydrogen atoms. As an example, a molecule of poly(dimethyl siloxane) is represented in Figure 7.2, with the number n varying from 10 to over 1000. Typically, n is about 50 for silicon oils used in transformers.

Other Synthetic Liquids

In addition to the phosphate esters mentioned above, in recent years we have witnessed the development of organofluoric liquids of compositions such as $(C_4F_9)_3N$ and $(C_4F_9)_2O$. These have very high chemical stability.

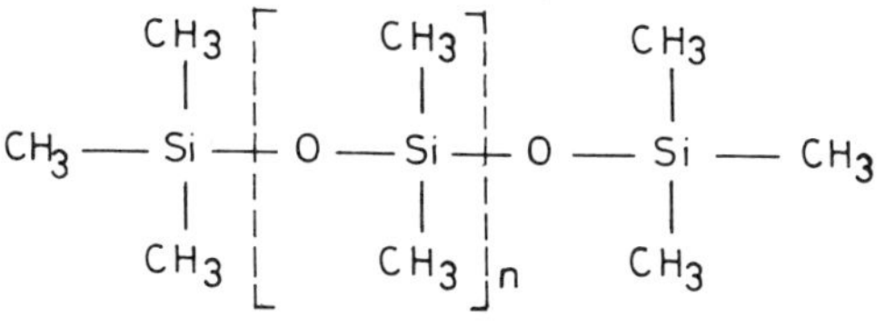

Fig. 7.2 Example of the atomic arrangement in a molecule of poly-(dimethyl siloxane) (silicon oil). n could be 10 to 1000.

7.3 PHYSICAL PROPERTIES

Important physical properties of insulating liquids include specific heat, thermal conductivity, viscosity, and their variation with temperature. These are important for efficient cooling of oil-immersed equipment. Other important qualities are the coefficient of thermal expansion, flammability, and pour point, below which the liquid starts to thicken and solidify and thus would be useless for convection cooling of equipment immersed in it.

The variations in the viscosities of some typical mineral oils—a silicon oil, an ester, an olefin, and chlorinated hydrocarbon—are depicted in Figure 7.3. In the figure are shown three grades (I,

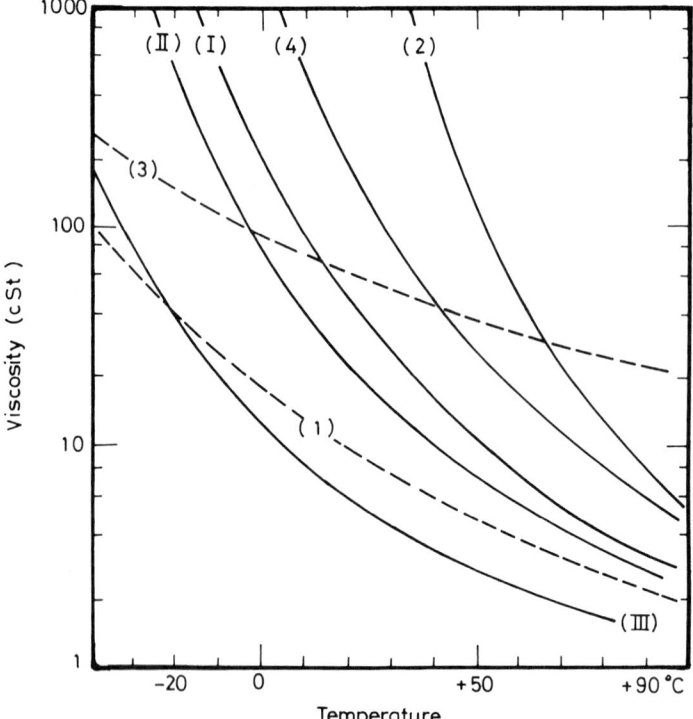

Fig. 7.3 Variation of viscosity with temperature for (1) low-viscosity polybutene (olefin), (2) hexachlorobiphenyl (askarel), (3) silicon oil for transformers, (4) trixylenyl phosphate ester. I, II, III: mineral oils graded by the International Electrotechnical Commission.

Insulating Liquids 169

II, and III) of mineral oil prescribed by the International Electrotechnical Commission for use in transformers and similar equipment (IEC, 1982a). In Figure 7.3, it is noted that the viscosity of hexachlorobiphenyl is considerably higher than those of other synthetic and mineral oils, except at excessive temperatures. The viscosities of silicon oil and olefin are distinctly less sensitive to temperature variation.

In applications where the fire hazard is a concern, mineral oils are usually replaced by askarels or phosphate esters because of their remarkable fire resistance. Esters are preferred because of their low toxicity and the fact that they are biodegradable. Also, organofluoric liquids are stable up to temperatures as high as 500°C and are ideal for cooling electronic equipment of the sealed-off types that have large power ratings. Because of their low boiling points, they are used in evaporative cooling of power equipment.

7.3.1 Water Solubility in Oils

Water is molecularly soluble in small quantities in mineral oils, about 50 ppm. Its solubility depends on the molecular composition and on temperature. Water is more soluble in silicon oils and much more in esters, particularly phosphate esters, where it exceeds 600 ppm. Water can be absorbed by the oil from the ambient, or can be produced by oxidation processes of the oil itself, as explained in Section 7.4.

Above the saturation level extra water gets emulsified as finely dispersed droplets of diameter 1 to 10 μm suspended in the oil. If the water content of the mineral oil exceeds about 400 ppm, the droplets tend to flocculate, becoming larger than 10 μm and settling to the bottom of the container. In liquids such as askarels, which have specific gravities exceeding unity, water would collect at the surface.

The water-absorbing property of phosphate esters presents a serious problem for their users. Unless extreme care is taken in their handling, their water content can reach levels as high as 0.1%, with deleterious effects on their properties, particularly the dielectric strength.

7.4 CHEMICAL PROPERTIES

These include the chemical stability of liquid under thermal and electrical stresses. Usually under such stresses, oils get oxidized in the presence of oxygen. The gas could be either in contact with the oil or actually dissolved in it. The oxidation gets

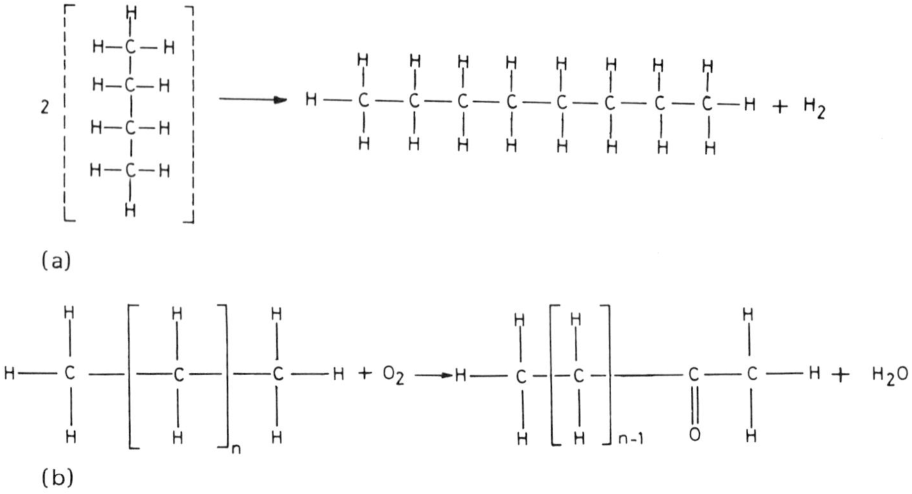

Fig. 7.4 Examples of chemical reactions involved in the oxidation of mineral oils: (a) polymerization, (b) formation of water.

accelerated in the presence of metals such as copper, which act as a catalyst.

The chemical reactions involved may produce lighter molecules (scission) or heavier molecules (polymerization) depending on the type of oil and the reaction conditions. Among the products of oxidation are hydrogen, water, acids, and wax. The wax deposits in the form of sludge. Examples of such reactions are given in Figure 7.4. Some of the oxidation products do inhibit the reaction, which thus stabilizes with time. Some synthetic chemicals have recently been developed (Wilson, 1980; Cookson, 1987), which when added in small quantities to the oil inhibit the oxidation reaction and thus tend to extend the oil's life in service. More will be said about this point in Section 7.8.

Compared to mineral oils, natural esters, askarels, and some olefins oxidize more rapidly. With oxidation, olefins produce acids and polymerize, forming resins. On the other hand, silicon oils at temperatures below 150°C are very stable. Above 200°C, however, they oxidize, forming water, carbon oxides, and acids.

Traces of acid in the oil could be produced by oxidation reactions. Acids could result from the refining process or be present originally in the crude and not completely eliminated during dis-

Insulating Liquids

tillation. In esters, acids form due to their hydrolysis and thermal decomposition. These acids may chemically attack the solid insulation and/or enamel or paint of the immersed equipment.

7.4.1 Chemical Reactions Enhanced by Electrical Discharge

The electrical discharge in oil-immersed equipment can be either a complete or a partial breakdown (i.e., a silent discharge). In the case of silent discharges there is usually a gas bubble, with enhancement of applied electric fields at its boundaries. Under high electric fields some oil molecules at the surface of the bubble get dissociated, releasing hydrogen.

The ionized hydrogen gas produced by the discharge would be chemically active. Some chemical reactions enhanced by gas discharges may include gas absorption, with hydrogen, oxygen, or nitrogen accepted in some unsaturated molecules of aromatic oils. Such degassing action would no doubt help to maintain the stability of the oil and extend its life under silent electrical discharges. Following is an example:

$$(CH_2\text{---}CH_2) + H_2 \xrightarrow{\text{electrical discharge}} (CH_3\text{---}CH_3)$$

The activity of the chemical reaction and its products depend heavily on the type of oil, the gases present, the temperature and pressure, and the presence of catalysts.

In the case of complete breakdown of an oil gap and of immersed solid insulation, an arc is involved. In such cases gases such as acetylene, ethylene, and ethane get evolved, resulting from the chemical decomposition of some oil molecules. Also, carbon monoxide results from decomposition of the solid insulation.

Under severe electrical or thermal stresses, askarels evolve acidic gases, mainly hydrogen chloride. The hydrochloric acid thus formed causes severe electrolytic corrosion of insulation and metals immersed in the askarel. Also, as is known, askarels are nonbiodegradable and are thus ecologically unacceptable. Esters do not suffer from these disadvantages.

7.5 ELECTRICAL PROPERTIES

Electrical properties include the electrical conductivity of the oil, its permittivity, dissipation factor, and dielectric strength.

7.5.1 Electrical Conductivity

It has been noted that when liquids known to be insulants are subjected to direct voltage, they actually conduct a current, however small. Even a simple insulating liquid such as hexane conducts a current that varies with the electric stress applied. It has been suggested that under low electric fields the conduction is due mainly to positive and negation ions belonging to dissociated molecules. Dissociation and recombination are in dynamic equilibrium, thus:

$$A-B \underset{\text{recombination}}{\overset{\text{dissociation}}{\rightleftarrows}} A^+ + B^-$$

However, when an electric field exceeding 1 kV/cm is applied, the rate of dissociation exceeds that of recombination and increases at higher fields (Nelson, Lee, 1987). This state of affairs is basically different from the case of conduction through gases, where below the corona onset level the number of charge carriers is limited and independent of the applied field strength (Chapter 4). This explains why the saturation current is constant in gas gaps, whereas it increases in oil gaps with applied voltage (Cross and Jaksts, 1987).

At still higher fields the current grows at a rapidly increasing rate. This growth is accounted for both by field emission of electrons from the cathode and field-aided dissociation of liquid molecules. As the ions drift through the liquid, they gain energy from the field and lose some of it in collisions with the liquid molecules. The energy thus gained by the liquid molecules accounts for extra vibrations. The vibrations of $C-H$ and $C-C$ bonds among the carbon and hydrogen atoms result in some of them being broken, thus producing extra ions.

Under impulse voltages conduction currents reach much higher values because the choking effect of space charges accumulating at the electrodes would be insignificant compared to the dc case, as in gas discharges.

In liquids of commercial purity, additional conduction currents would be caused by the impurities. These would include a wide variety of particles: droplets of water and acids, resins, and cellulose fibers. Their different conductivities, permittivities, and affinities for ionization would have widely varying effects on conduction through the insulating liquid.

Conductivity rises sharply with temperature due to both the increased dissociation of the liquid molecules and its decreasing viscosity. At a temperature T the number n of dissociated molecules per unit volume could be related to the total number per unit volume N in terms of W, the dissociation energy, as

$$n = N \exp(-W/kT) \tag{7.1}$$

k being Boltzmann's constant. Thus the conductivity of the liquid due to dissociated molecules would be

$$\sigma = ne(\mu_+ + \mu_-) \tag{7.2}$$

Here e is the electronic charge and μ is the ion mobility. It is evident that the conductivity depends exponentially on temperature, as observed experimentally (Fig. 7.5) (Wilson, 1980).

Under ac the liquid conductivity represents a considerable part of its dielectric losses. The other part of the loss under ac is due to hysteresis in the polarization of the liquid molecules. The losses are expressed as a dissipation factor = tan δ, where δ is the loss angle. The dissipation factor increases only slightly with applied electric stress at moderate levels. Under very high stresses, however, the dissipation factor increases considerably with stress, at an ever-increasing rate until the liquid breaks down (Denat et al., 1983).

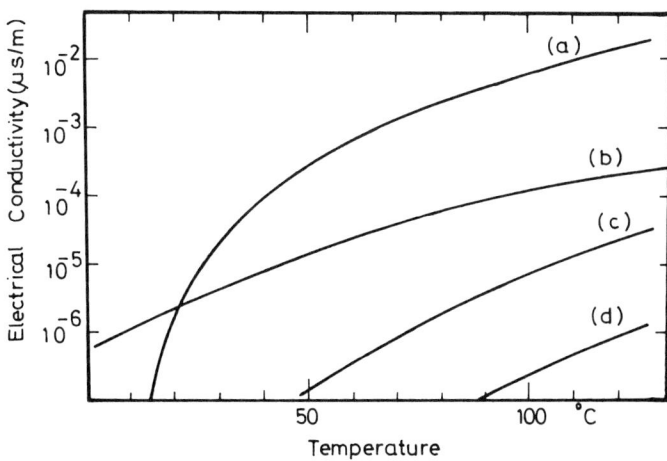

Fig. 7.5 Electrical conductivities of some insulating liquids as functions of temperature: (a) PCB (askarl), (b) commercial transformer oil, (c) purified transformer oil, (d) highly purified transformer oil.

7.5.2 Dielectric Constant

The dielectric constants (i.e., relative permittivities) of insulating liquids range from about 2.2 for mineral oils to about 5.3 for askarels. Askarels have therefore been favored for applications in capacitors. The dielectric constants of mineral and most synthetic oils vary only slightly with temperature.

7.6 THEORIES OF DIELECTRIC BREAKDOWN

Unlike the case of gases, there is no single theory that has been unanimously accepted for the breakdown of insulating liquids. There are two main reasons for the underdeveloped states of theories of liquid breakdown. One is that the liquid phase is much less amenable to theoretical analysis than are gases. Second, insulating liquids are almost never absolutely pure, if not because of the ingress of contaminants, then due to traces left in the liquid by the processes of manufacture and purification. Each of these ingredients plays a role in the breakdown of the bulk of the liquid. Thus it is an accepted fact that the breakdown strength of a sample of insulating oil depends on its impurity content rather than its molecular composition. The three principal theories that have been proposed for the breakdown of insulating liquids are (a) the electronic breakdown theory, (b) the bubble theory, and (c) the suspended particle theory.

7.6.1 Electronic Breakdown Theory

According to this theory, electrons are emitted from the cathode (Kuffel and Zaengl, 1984; Abdel-Bary, 1980). On their way to the anode, the electrons collide with atoms of liquid molecules. If enough energy is transferred during such collisions, some electrons would be knocked off their atoms and drift with the original electrons toward the anode. Thus electron avalanches such as those occurring in gas discharges develop in the liquid and ultimately lead to breakdown. Evidence supporting this theory, even under fast impulse voltages, is rather scarce.

7.6.2 Bubble Theory

Leakage currents are not distributed evenly over electrode surfaces. As directed by the applied electric field, they tend to concentrate at points of microroughness where the field lines converge (Fig. 7.6a). Near such points, the highly concentrated leakage currents with corresponding Joule heating of a microscopic volume of liquid would cause a rapid rise in temperature to high above the boiling

Insulating Liquids

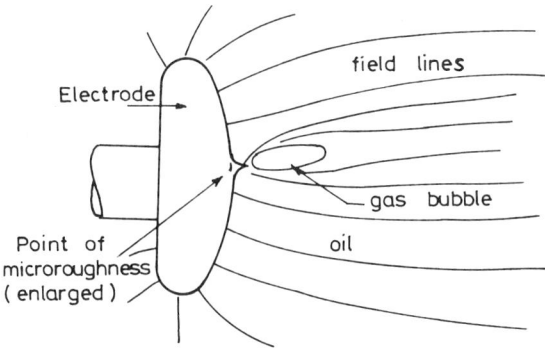

(a)

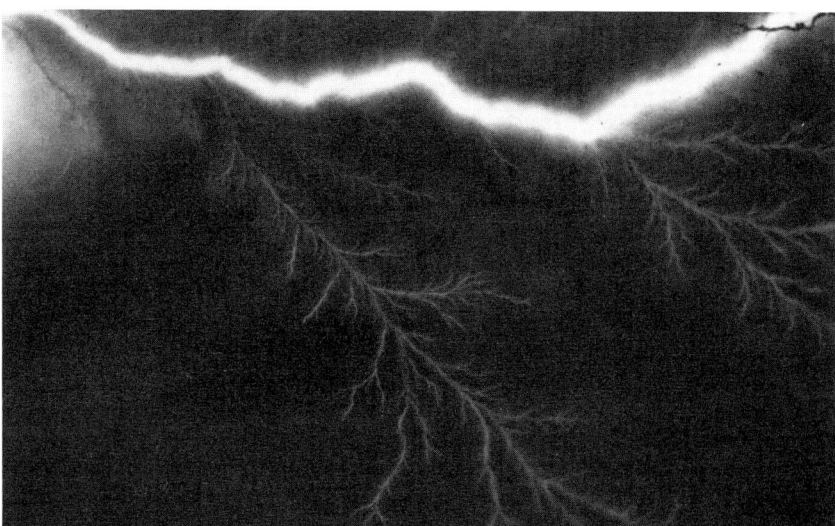

(b)

Fig. 7.6 (a) Formation of a gas bubble at a point of microroughness on the electrode surface in an oil gap, (b) gas discharge in the bubble, developing it into branched streamers with ultimate breakdown.

point. With very limited heat convection and conduction, such a temperature rise could occur over short periods on the order of 1 µs (Abdel-Bary, 1980). Thus a bubble of vapor forms at a point of microroughness on the electrode surface (Fig. 7.6a).

Three other alternatives were proposed to account for formation of the gas bubble (Korobejnikov et al., 1983): release of occluded gases from micropores in electrode surface layers, cavitation caused by mechanical strain of the liquid under the highly concentrated electric field with corresponding electrostrictive pressure differential, and electrochemical dissociation of some liquid molecules with the release of gases. Furthermore, in the case of oil-impregnated paper, as in transformer windings, cables, and capacitors, repeated expansion and contraction under cyclic loads would cause cavitation and form gas bubbles. Under each alternative, the concentrated electric field at points of microroughness on electrode surfaces or somewhere along the gap would play a basic role in the production of gas bubbles.

Tests on oil gaps under high-voltage pulses of nanosecond duration could provide the data on development of gas bubbles at the electrodes and their initiating breakdown (Korobejnikov et al., 1983; Lesaint and Tobazéon, 1987). Because of the difference between the permittivities of the gas and liquid, the electric field inside the bubble would be considerably higher than that in the liquid. When the bubble reaches a critical volume, and under high enough fields, the gas in the bubble would become ionized and electron avalanches would develop into streamers. The space charges formed at the ends of the bubble would act to deform it under the electric field by Coulomb forces. Also, the electrons and ions accelerated by the field inside the bubble would impinge against areas of its walls, helping to decompose some of the liquid molecules and to evaporate others. The gas bubble grows longitudinally (Fig. 7.6a), which would lead eventually to complete breakdown of the entire liquid gap (Fig. 7.6b).

The energy supplied to the gas bubble is consumed in vaporizing some additional liquid (on the order of 10^6 J/m^3), ionizing and expanding the gas bubble (10^5 to 10^8 J/m^3) and decomposing some oil molecules into relatively simpler compounds (10^7 J/m^3). The speed of expansion of gas bubbles was measured in the range 100 m/s to 10 km/s (Felici, 1987).

Negative streamers start from sharp points on the cathode. Their growth is aided by electrons bombarding the gas–oil interface at the streamer's tip. The positive streamer propagates from the anode and electrons fall into its tip from the oil molecules located ahead of it, which get dissociated by the local intense electric field (Chadband and Sadeghzadeh-Araghi, 1987). With the streamer growth, its space-charge field increases exponentially. It may

Insulating Liquids   177

reach the order of 10 MV/cm, exceeding greatly that of the applied Laplacian field at sites away from the electrodes, which explains lateral branching of these streamers (Fig. 7.6b).

The bubble theory is supported by the observed dependence of breakdown strength on the applied static pressure. If the liquid is degassed, its dielectric strength becomes less dependent on the static pressure. Also, gas-absorbing additives cause a rise in the breakdown strength.

7.6.3 Suspended Particle Theory

The particle could well be a fiber, probably soaked with moisture or be even a droplet of water. Under an applied electric field, and because their dielectric constant ε_{rf} is much higher than that of oil ε_{ro}, fiber particles get polarized and move along converging fields (Fig. 7.7).

Assuming hemispherical tips for the fiber particles with radius r, the charge ±q at either of its ends would be

$$\pm q = \pm \pi r^2 (\varepsilon_{rf} - \varepsilon_{ro}) \varepsilon_0 E \qquad (7.3)$$

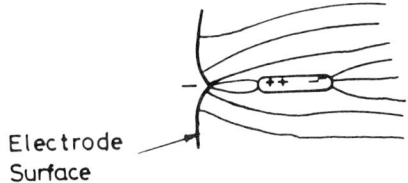

Electrode
Surface

(a)

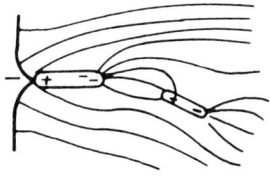

(b)

Fig. 7.7 Steps of collection of fiber or other particles in suspension to eventually bridge the oil gap. They get polarized and directed by the field.

where ε_0 is the permittivity of free space and E is the field strength at either end of the fiber assumed being equal. If the field in the oil gap is nonuniform, as is usually the case, the resultant force will move the fiber (Fig. 7.7). For a short fiber of length equal to radius r, and with $\varepsilon_{rf} \gg \varepsilon_{r0}$, the driving force would be

$$F_E = r^3 \varepsilon_0 E \text{ grad } E \tag{7.4}$$

As this particle moves with velocity v in the oil of viscosity η, there will be an impeding force with a magnitude of $6\pi r \eta v$. When a particle reaches either electrode, its outward tip would act as an extension to the electrode and attract more fibers (Fig. 7.7).

It is noted from equation (7.4) that the driving force on the fiber increases in relation to its size and with the degree of nonuniformity of the field. Thus in a sphere gap they will migrate toward the higher field at its axis and gradually line up and bridge the gap (Fig. 7.8) (Kind, 1978). Such a relatively conducting path would in effect short circuit the gap, resulting in breakdown. Figure 7.9

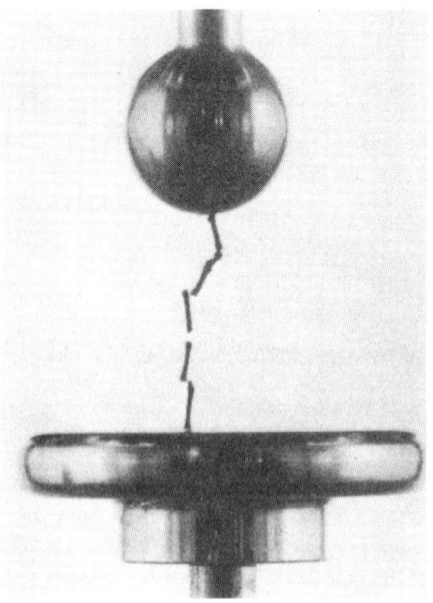

Fig. 7.8 Fiber particles bridging a gap under high voltage. (Courtesy of D. Kind and F. Vieweg & Sohn.)

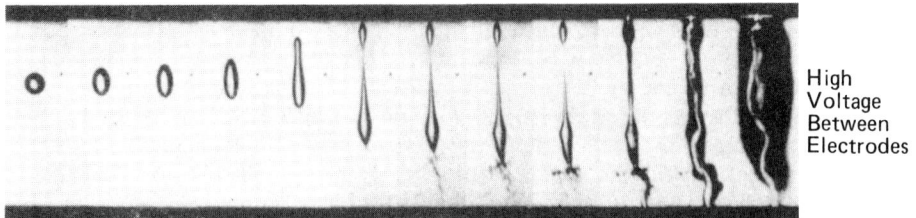

Fig. 7.9 Water drop being gradually elongated by an electric field and ultimately causing breakdown. The water is evaporated by the discharge.

shows how a water drop gets distorted under the electric field in an oil gap and ultimately leads to its breakdown.

The evidence in support of this theory includes the increased time required to reach breakdown of the liquid with increased viscosity. Evidently, this phenomenon could occur under dc and power-frequency ac, but not under high-frequency or fast impulse voltages.

7.7 FACTORS INFLUENCING THE DIELECTRIC STRENGTH OF INSULATING LIQUIDS

The dielectric strength of insulating liquids has been observed to depend on the liquid temperature, applied static pressure, size of the gap, electrode material and surface conditions, and impurities contained in the oil.

7.7.1 Temperature and Pressure

The effect of temperature on the dielectric strength of insulating liquids depends on their type and degree of purity (Calderwood and Corcoran, 1987). For example, the dielectric strength of dry transformer oil is insensitive to temperature except slightly below the boiling point. There, the dielectric strength decreases drastically, probably because of the formation of vapor bubbles and their growth, which are aided by the decrease at such temperatures of the oil's viscosity and surface tension. The dielectric strengths of oils that have a trace of moisture are sensitive to temperature variations over the full range from about $-20°C$ up to the boiling point of about $250°C$.

The dielectric strength of insulating liquids under dc and power-frequency ac increase significantly with applied static pressure. Raising the pressure from atmospheric to 10 times higher increases the dielectric strength by about 50%, depending on the type of liquid (Abdel-Bary, 1980). Another effect of pressure is the suppression of pre-breakdown discharges. These observations support the bubble theory of liquid breakdown. Under very fast impulse voltages of duration less than 0.05 μs, breakdown voltage is insensitive to both pressure and temperature, as observed by Korobejnikov and his colleagues (1983).

7.7.2 Electrode and Gap Conditions

The breakdown voltage of an oil gap depends on its width as well as the electrode shape and material. For gaps with highly nonuniform fields, for example, such as that of a point-to-sphere gap, there is a polarity effect. The negative dc breakdown voltage is lower than the positive voltage up to a critical gap length above which the relation reverses (Yehia, 1984). This critical gap length depends on the liquid and the electrode material. There seems to be no simple explanation for these phenomena. However, the material of the cathode surface layer determines the electric stress necessary for electron emission. These electrons play a decisive role in the conduction and breakdown processes.

The size and shape of electrodes determine the volume of liquid subjected to high electric stress and the degree of field nonuniformity. The bigger this volume, the higher the probability of its containing impurity particles. The more of these particles that are present, the lower would be the breakdown voltage of the liquid gap. The effect of moisture content is shown in Figure 7.10.

The sensitivity of liquid breakdown to these factors is logically higher under dc and power-frequency ac than under fast impulse voltages (Yehia, 1984; Krueger, 1987). Thus, the impulse ratios of highly nonuniform gaps of contaminated or technically pure liquids can reach about 7, much higher than the gas gaps of similar geometries.

It has also been shown that stressing the oil gap under high voltage for a long time, and repeated sparks of limited energy, tend to raise the breakdown voltage of the gap (Dawoud, 1986). This is called conditioning the oil gap. Particles in suspension collect at zones of field concentration. Points of microroughness on the electrodes get eroded by concentrated discharge currents. A film of discharge byproducts gradually covers the discharge areas of both electrodes. In the case of silicon oil, repeated breakdowns tend to cover the electrodes with a film of gel and solid decomposition products (Krueger, 1987). If a high-energy arc is allowed

Insulating Liquids

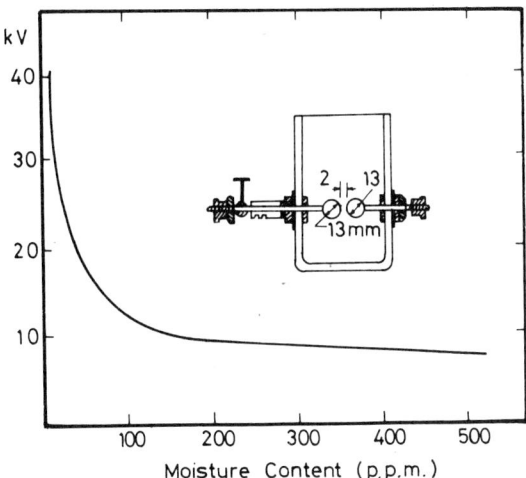

Fig. 7.10 Dielectric strength of mineral oil as a function of its moisture content.

to take place in the liquid gap, the arc products would cause the liquid properties to deteriorate.

7.7.3 Impurities

Impurities include solid particles of carbon and wax, by-products of aging and discharges, cellulose fibers, residues of filtration processes, water, acids, and gases. Impurities usually cause a reduction in the dielectric strength of an insulating liquid, the largest effect being that of the simultaneous presence of moisture and fibers. Cellulose fibers are known to be hygroscopic. Thus, floating moist fibers would tend to bridge the oil gap, as explained earlier.

Under both ac and dc the effect of a trace of moisture is drastic on meticulously dried liquids, much greater than that of commercial liquids (Fig. 7.10). The effect of moisture is less pronounced in the case of oil gaps with strongly nonuniform fields and with liquids containing no fibers (Abdel-Bary, 1980). Because water solubility is considerably higher in silicon oil and phosphate esters than in mineral oil, they need to be much more carefully dried and kept.

7.8 AGING

When an insulating liquid has been kept at elevated temperatures and/or under electric stress and exposed to oxygen, its properties get degraded. The amount of moisture content and the acidity level rise, while resistivity and dielectric strength drop. Oxidation of the insulating liquid is accelerated at higher temperatures, with more oxygen available and in the presence of a catalyst such as copper. As low-viscosity highly refined oils have higher tendencies to dissolve air and oxygen, they become oxidized faster than do oils of lower grades and higher viscosities. Among petroleum oils, those of the naphthenic groups are considerably more stable than the paraffinics and aromatics, and therefore have dominated applications with transformers, switchgear, and cables.

While the insulating liquid is in service or being stored in contact with oxygen, some impurity particles which inevitably are present gradually flocculate. Thus particle complexes grow in size. Silent discharges of the corona type, and concentrated leakage currents, help the formation of water, resins, acids, and the evolution of hydrogen. Disruptive discharges such as arcs, and intensive localized heating of the liquid and solid insulation produce particles of carbon, and gases such as carbon monoxide, carbon dioxide, and acetylene. Wax often forms by polymerization of oil at the walls of gas bubbles when the bubbles get ionized. The aggressive acidic products of oxidation and discharges attack the solid insulation, iron, and copper immersed in the liquid. Also, incompletely cured varnishes on oil-immersed windings dissolve in the oil and polymerize. Solid particles of carbon, wax, corroded iron, and polymerized liquid settle as sludges. Thus the physical, chemical, thermal, and electrical properties of the insulating liquid deteriorate.

To maintain the qualities of an insulating liquid in service, its important physical, chemical, and electrical properties have to be regularly checked. Before the qualities of the insulating liquid change beyond permissible levels, special measures should be taken to reclaim the liquid.

7.8.1 Additives

To prolong the life of insulating liquids in service, action can be taken along two fronts. First, the actual process of aging—which is mainly oxidation—should be inhibited. Second, the oxidation products should be treated so as to minimize their deleterious effects on the liquid properties.

To start with, to minimize the rate of oxidation, the amount of oxygen dissolved in, or in contact with, the liquid should be

minimized. An example is the case of sealed equipment with a
nitrogen cushion above the oil. Otherwise, oxidation inhibitors are
dissolved in the liquid. These are also scavengers that react with
oxidation products and thus break the oxidation chain reaction.
There are passivators that react with metal salts which otherwise
would act as catalysts for the oxidation reaction. The amount of
salt added to the liquid is on the order of 0.1%, although the exact
amount for optimal results depends on the salt and base liquid
composition (Kamath and Murthy, 1987).

Oxidation inhibitors react with the free radicals and peroxides
produced by the oxidation process and thus break their chain reaction mechanism, which would otherwise give momentum to the oxidation process (Dawoud, 1986). By reacting with peroxides of radicals, passivators would prevent the formation of naphthanates of
copper and iron which are the usual catalysts for the oxidation
reaction. For example, for petroleum oils, established additives
include di-*tert*-butyl-*para*-cresol (DBPC), dimethylaniline (DMA),
quinones, anthracenes, and phenyls, whereas for askarels, anthraquinone acts as a scavenger for the hydrogen chloride, which is its
most chemically aggressive decomposition product.

7.9 TESTS ON INSULATING LIQUIDS

To ensure the qualities of insulating liquids and their compatibility
with the equipment in which they are to be used, numerous tests
for these liquids have been prescribed. The International Electrotechnical Commission has issued publications describing testing methods and the criteria to be used in accepting insulating liquids (e.g.,
IEC, 1978, 1982a,b). The electrical tests include measurement,
under controlled conditions, of the dielectric strength, dielectric
dissipation factor, resistivity, and permittivity. Tests for the
liquid's physical characteristics include measurement of its viscosity,
pour point, flash point, and moisture content. Chemical tests include measuring its degree of acidity, oxidation stability, and
gassing characteristics (IEC 1963, 1974a,b).

The moisture content of an insulating liquid is measured by
heating a sample to a set temperature in a rarefied atmosphere.
The water vapor developed in the test vessel is a function of the
water content in the liquid. Also, the degree of acidity of oil
could be measured by noting the amount of potassium hydroxide
that just neutralizes the acids in the sample. It has been observed
that the acidity of oil affects its dissipation factor much more
sensitively than it does its dielectric strength (Krueger, 1987).

There are other tests for measuring and identifying the gases dissolved in the liquid. Equipment has been developed, based on the principle of fuel cells, for on-line monitoring of the amount of hydrogen dissolved in transformer oil while in service (Webb, 1987). Identifying the gases dissolved in the insulating liquid of equipment and measuring their relative quantities could assist greatly in monitoring the performance of equipment and in diagnosing any incipient fault at an early stage. Thus a major equipment fault could be prevented and the corresponding capital loss could be avoided.

7.10 RECONDITIONING OF INSULATING LIQUIDS

Insulating liquids normally remain in service as long as their qualities have not deteriorated beyond levels set by standard specifications and general experience. For instance, their breakdown voltage, across the specified test gap of 2 mm, should not decrease below 30 to 55 kV; the magnitude depends on whether the liquid is used in low-voltage or high-voltage equipment. For the latter, insulation is usually designed to undergo higher stresses. The corresponding limits for the resistivity of the liquid are 3 and 10 $G\Omega \cdot m$, and for the dissipation factor are 0.2 and 0.05 at 90°C. In more highly stressed equipment such as cables, the limit for the dissipation factor is set as low as 0.001.

The level of acidity neutralization factor of insulating liquids could be measured as the amount of potassium hydroxide sufficient to neutralize the acids in 1 g of the liquid. The acceptable level is 0.5 mg KOH/g. The acceptable level of water content is 15 to 30 mg/kg; the lower figure is for power equipment. The level is set even lower (0.1 mg/kg) for EHV cables, as their working electrical stresses are very high (IEC, 1982a).

As the level set for any of the foregoing characteristics is approached, measures for reconditioning the liquid should be taken. There is portable equipment now available for reconditioning insulating oils while in service (Fig. 7.11). A method well known and in use is that of filtering and vacuum drying. While under a vacuum of about 1 kPa, the liquid is heated to about 30 to 60°C, high above the boiling point of water at such a reduced pressure. With large oil surfaces exposed to the vacuum, it becomes freed from its moisture content and dissolved gases. In the process it also gets freed from acids and particulate matter. The process could include replenishing inhibitors and other additives in the liquid. Reconditioned oils usually have about half the life of new ones.

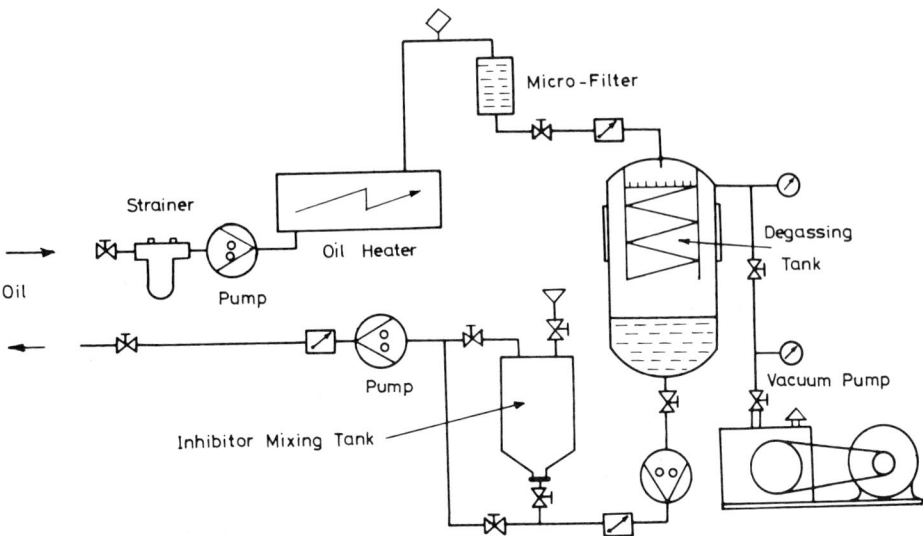

Fig. 7.11 Setup into which insulating oil is drawn, filtered, dried under vacuum, degassed, and its inhibitors replenished before it is returned to the equipment.

REFERENCES

Abdel-Bary, M. (1980). M.S. thesis, Ain-Shams University, Cairo.
Breuer, W. and Hegemann, G. (1987). *Proc. CIGRE Symposium on New and Improved Materials for Electrotechnology*, Vienna, May, Report 500.09.
Calderwood, J. and Corcoran, P. (1987). *Proc. 9th Int. Conference on Conduction and Breakdown in Dielectric Liquids*, Salford, England, pp. 124–128.
Chadband, W. and Sadeghzadeh-Araghi, M. (1987). *Proc. 9th Int. Conference on Conduction and Breakdown in Dielectric Liquids*, Salford, England, pp. 325–330.
Chan, J. (1987). *IEEE Trans.*, EI-3(3):10–11.
Cookson, A. H. (1987). *Proc. CIGRE Symposium on New and Improved Materials for Electrotechnology*, Vienna, May, Report 500.00.
Cross, J. and Jaksts, A. (1987). *Proc. 9th Int. Conference on Conduction and Breakdown in Dielectric Liquids*, Salford, England, pp. 271–279.

Dawoud, R. (1986). M.S. thesis, University of Alexandria, Alexandria, Egypt.
Denat, A., Cosse, B., and Cosse, J. (1983). *Proc. 4th Int. Symposium on High Voltage Engineering*, Athens, Paper 24.02.
Felici, N. (1987). *Proc. 9th Int. Conference on Conduction and Breakdown in Dielectric Liquids*, Salford, England, pp. 30–35.
IEC (1963). *Method for the Determination of the Electrical Strength of Insulating Oil*, Publication 156, International Electrotechnical Commission, Geneva.
IEC (1974a). *Methods for Assessing the Oxidation Stability of Insulating Liquids*, Publication 474, International Electrotechnical Commission, Geneva.
IEC (1974b). *Methods for Sampling Liquid Dielectrics*, Publication 475, International Electrotechnical Commission, Geneva.
IEC (1978). *Measurement of Relative Permittivity, Dissipation Factor and D.C. Resistivity of Insulating Liquids*, Publication 247, International Electrotechnical Commission, Geneva.
IEC (1982a). *Specification for Unused Mineral Insulating Oils for Transformers and Switchgear*, Publication 296, International Electrotechnical Commission, Geneva.
IEC (1982b). *Determination of Water in Insulating Oil, Oil-Impregnated Paper and Pressboard*, Publication 733, International Electrotechnical Commission, Geneva.
Kamath, K. and Murthy, T. (1987). *Proc. CIGRE Symposium on New and Improved Materials for Electrotechnology*, Vienna, May, Report 500.07.
Kind, D. (1978). *An Introduction to High Voltage Experimental Technique*, Friedr Vieweg & Sohn, Verlagsgesellschaft mbH, Braunschweig, West Germany.
Korobejnikov, S., Yanshin, K., and Yanshin, E. (1983). *Proc. 4th Int. Symposium on High Voltage Engineering*, Athens, Paper 24.09.
Krasucki, Z. (1962). *Proc. IEE, 109B*(Supplement 22):435–439.
Krueger, M. (1987). *Proc. 9th Int. Conference on Conduction and Breakdown in Dielectric Liquids*, Salford, England, pp. 487–501.
Kuffel, E. and Zaengl, W. (1984). *High Voltage Engineering Fundamentals*, Pergamon Press Ltd., Oxford.
Lesaint, O. and Tobazéon, R. (1987). *Proc. 9th Int. Conference on Conduction and Breakdown in Dielectric Liquids*, Salford, England, pp. 343–347.
Marinho, A., Sampaio, E., and Monteiro, M. (1987). *Proc. CIGRE Symposium on New and Improved Materials for Electrotechnology*, Vienna, May, Report 500.06.
Nelson, J. and Lee, M. (1987). *Proc. 9th Int. Conference on Conduction and Breakdown in Dielectric Liquids*, Salford, England, pp. 298–302.

Sankaralingam, S. and Krishnaswamy, K. (1987). *Proc. CIGRE Symposium on New and Improved Materials for Electrotechnology*, Vienna, May, Report 500.01.
Theoleyre, S. and Tobazéon, R. (1983). *Proc. 4th Int. Symposium on High Voltage Engineering*, Athens, Paper 24.05.
Webb, N. (1987). *IEE Power Engineering Journal*, 1(5):295-298.
Wilson, A. (1980). *Insulating Liquids, Their Uses, Manufacture and Properties*, Peter Peregrinus Ltd., Stevenage, Herts, England.
Yehia, S. (1984), Ph.D. thesis, University of Alexandria, Alexandria, Egypt.

8
Solid Insulating Materials

M. KHALIFA   *Cairo University, Giza, Egypt*

8.1   INTRODUCTION

Solid insulating materials are encountered in every electric and electronic device and piece of equipment, both large and small. They are vital for isolating conductors and for their proper performance. Solid insulants cover a very wide range, including organic and inorganic, natural and synthetic, and simple, bonded and impregnated materials. Their common feature is that they are compounds—not pure chemical elements. They have extremely high electrical resistivities and high dielectric strengths below certain temperatures. They differ in their electrical, physical, and chemical properties, which are used as a guide in selecting the material appropriate for each application. After discussing the important electrical, physical and chemical properties common for solid insulating materials, and the theories of their electrical breakdown, some of the most widely used, will be briefly described.

8.2   ELECTRICAL PROPERTIES

These electrical properties are essentially the dielectric constant, electrical resistivity, and dielectric strength of the material. The dielectric constant (i.e., the relative permittivity) is determined by the phenomenon of polarization that takes place inside the material when under an electric field. The polarizability of a dielectric is an intrinsic property that depends on its molecular composition.

The electrical resistivity of the material, together with any hysteresis that might take place under ac fields, determines its dielectric losses, expressed by power engineers as the loss factor or dissipation factor, and by electronics engineers in terms of the quality factor of the dielectric. The dielectric strength of an insulator depends on its material, its shape and size, its ambient, the type and duration of the applied voltage, the presence of field concentration, and other factors as discussed in Section 8.4.

Solid dielectrics are compounds as a rule; their properties may acquire magnitudes, depending on the direction in which they are measured. In the special cases where this happens the material is nonisotropic. This phenomenon of nonisotropicity appears in some crystalline materials, including quartz.

8.2.1 Relative Permittivity

Most dielectrics are chemical compounds of positive and negative ions. Thus, under electric fields, each ion tends to migrate toward the electrode of opposite polarity, as is well known (e.g., Seely and Poularikas, 1979). This polarization induces bound charges on the electrodes (Fig. 8.1). The electric flux density D is related

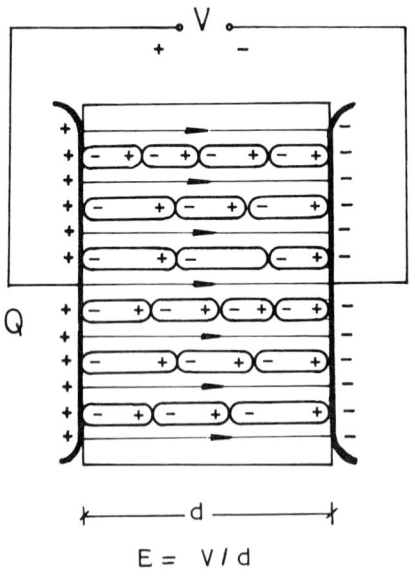

(a)

Fig. 8.1 (a) Polarization of a dielectric block under dc electric field: (b) current components fed to a capacitor as they vary with time from the instant it is connected to a dc source, (c) field caused by surface charge on a spherical cavity within the dielectric block of (a).

Solid Insulating Materials

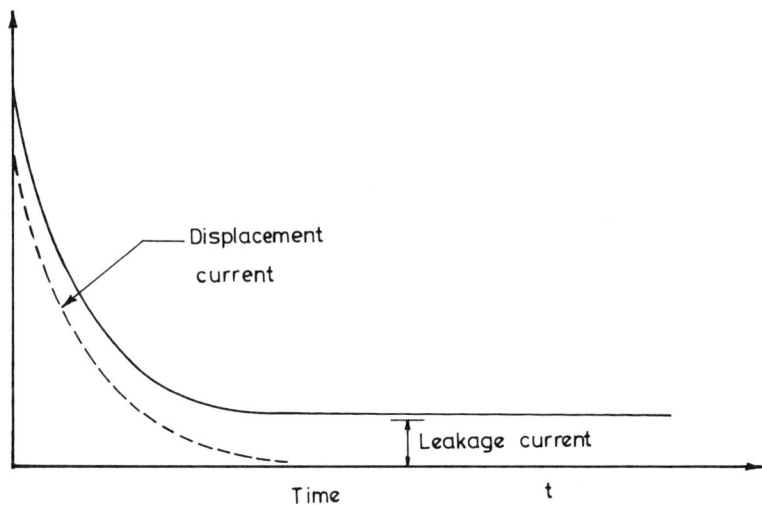

(b)

(c)

Fig. 8.1 (Continued)

to the field strength E by the well-known relation $D = \varepsilon E = \varepsilon_0 \varepsilon_r E$, where ε_0 and ε are, respectively, the permittivities of free space and the dielectric. The relative permittivity (i.e., the dielectric constant) of the material $\varepsilon_r = \varepsilon/\varepsilon_0$. Thus

$$D = \varepsilon_0 E + P \tag{8.1}$$

where P is the partial charge density on the dielectric surface resulting from its polarization. P is also defined as the dipole moment per unit volume of the dielectric. From equation (8.1) it is evident that

$$\varepsilon_r = 1 + \frac{P}{\varepsilon_0 E} = 1 + \chi \qquad (8.2)$$

χ being the susceptibility of the dielectric.

It can be easily shown that when a dielectric is charged, as in a capacitor, its energy stored per unit volume is $(1/2)\varepsilon_0 \varepsilon_r E^2$. Therefore, dielectrics with higher permittivities are favored by capacitor manufacturers. Typical values of the dielectric constants of insulating materials in common use are listed in Table 8.1.

The Field Inside a Cavity

The insulator body may contain particles of impurities or gas voids (in, e.g., plastics, glass, or paper). In such a case the field strength inside the void is bound to differ from that outside, depending on the relative permittivities of the dielectric and the void. Take, for example, the enlarged spherical gas void inside the dielectric shown in Figure 8.1. The field inside it would exceed that outside it (E) by the component E_c produced by the charges on its surface. The density of this surface charge would vary with θ according to a cosine curve, with a minimum = P. Thus

$$E_c = \oint d(E_c \cos \theta) = \int_0^\pi \frac{P(\cos^2 \theta) 2\pi r^2 \sin \theta}{4\pi \varepsilon_0 r^2} d\theta \qquad (8.3)$$

$$= \frac{P}{3\varepsilon_0} \qquad (8.3)$$

Substituting for P from equation (8.2), the total field in the void would be E_t:

$$E_t = \frac{2 + \varepsilon_r}{3} E \qquad (8.4)$$

In case the void has a different shape, this expression would differ. For example, if the void approaches a capillary tube along the field extending to the electrodes, $E_t = E$. On the other hand, if the void approaches a thin disk normal to the field, $E_t = \varepsilon_r E$, as can easily be shown.

8.2.2 Dielectric Polarization

The majority of dielectrics are nonpolar; that is, the centers of charges of the positive and negative ions in the molecule coincide. In a polar dielectric such as poly(vinyl chloride) (PVC) C_2H_3Cl, the size and charge of the chlorine atom are quite different from those of hydrogen atoms (Fig. 8.2). Thus a net dipole forms different from the case of polyethylene (C_2H_4) where the four hydrogen atoms of the monomer are balanced. Under an electric field, the dipoles of a polar dielectric tend to get oriented along

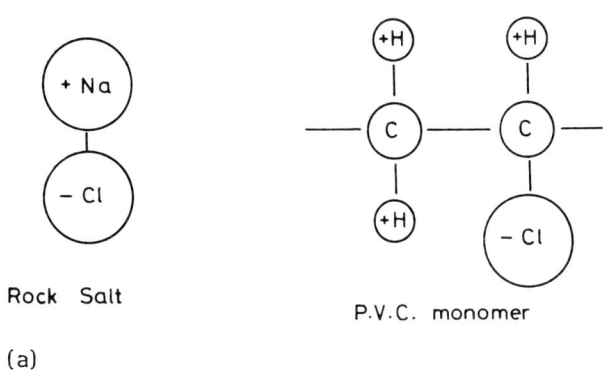

(a)

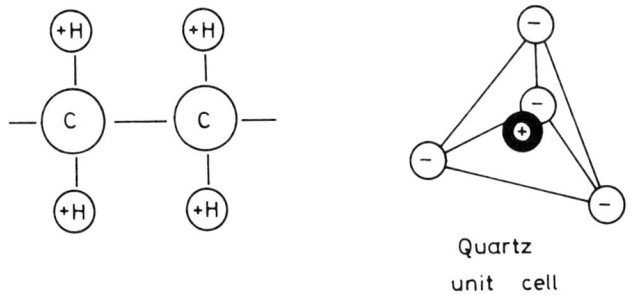

(b)

Fig. 8.2 Molecular representation of some polar (a) and nonpolar (b) dielectrics.

Table 8.1 Electrical Properties of Some Insulating Materials

Material	ε_r	tan δ at 25°C, 1 MHz (average)	ρ_V at 25°C ($\Omega \cdot$cm)	Breakdown stress (kV/mm)
Quartz glass	3.8	5×10^{-4}	10^{17}	
Lead glass	9			
Soda-lime glass	7	0.01		10–40
Glass ceramic	5–7	0.001–0.1	10^{12}–10^{14}	20–80
Porcelain	5	0.04	10^{12}–10^{15}	20–30
Zirconium ceramic	7–12	0.003		20–30
Alumina ceramic	10	0.0005		
Barium titanate ceramic	2000–8000	0.1		

Solid Insulating Materials

Micanite	5–11	0.003	10^{14}–10^{16}	50 (for 1-mm specimen) 100–200 (for 0.1-mm specimen)
PVC	6	0.1		
PE	2.3	0.0001	10^{15}–10^{18}	40
Polystyrene	2.6	0.0001	10^{16}–10^{18}	30–40
PTFE	2	0.0002	10^{16}–10^{18}	25
Bakelite	4.5	0.1	10^{13}	
Vulcanized natural rubber	3.5	0.05		
Methyl methacrylate	3.6	0.01		

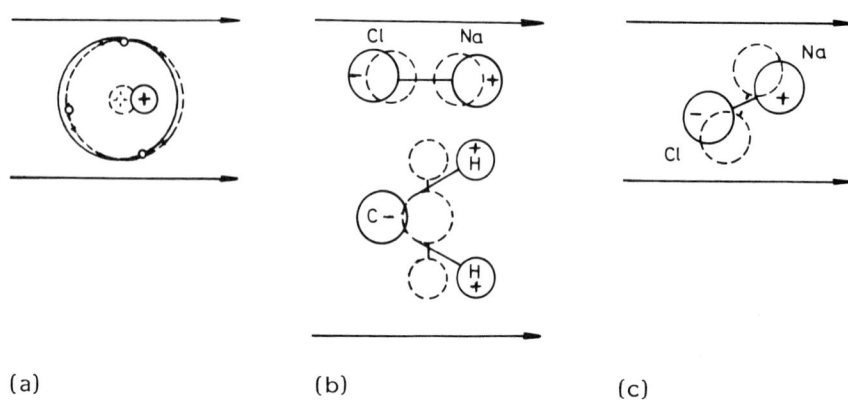

Fig. 8.3 Displacement polarization of an atom (a) and a molecule (b), and orientational polarization of a molecule (c). Arrows indicate field direction.

the field. In addition to this orientational polarization, the applied electric field exerts a force on each positive charge (be it a positive ion or even a nucleus of an atom) and on each negative charge (negative ions and electrons) and thus slightly displaces them along the field. Displacement polarization results in all dielectrics, both polar and nonpolar (Fig. 8.3). Also, impurity particles with permittivities and conductivities that differ from those of the parent material would get charged and tend to migrate along the applied field (migrational polarization).

Thus, in a unit volume of the dielectric, there are N_e atoms in N_m molecules, each with a dipole moment p_e and p_m, respectively, they add up to

$$P = N_e p_e + N_m p_m = (N_e \alpha_e + N_m \alpha_m) E_t \tag{8.5}$$

where α is the polarizability of each under the local field E_t. From equations (8.4) and (8.5) it could easily be shown that

$$\frac{3 \varepsilon_0 (\varepsilon_r - 1)}{\varepsilon_r + 2} = N_e \alpha_e + N_m \alpha_m \tag{8.6}$$

which is Clausius-Mosotti equation (Hummel, 1985).

It is obvious that each component of polarization needs time to materialize fully (Figs. 8.1 and 8.3). The fastest is electronic

Solid Insulating Materials

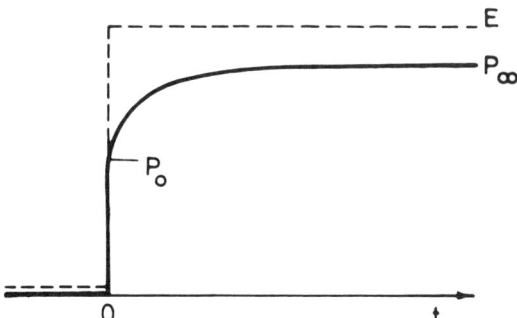

Fig. 8.4 Buildup of polarization P in a dielectric under the effect of a step-function-imposed electric field E. P_0 initial polarization; P_∞, polarization value after infinite time.

polarization, with a relaxation time of the order of 10^{-16} s, whereas the slowest is migrational polarization for which the relaxation time extends to seconds, minutes, hours, or in some cases, even longer. The relaxation time for orientational polarization reaches the order of 10^3 s for glass.

The relaxation time for polarization manifests itself in the fact that if a step-function voltage is applied across a capacitor with a solid dielectric, its polarization will take some time, however small, to build up, as shown in Figure 8.4. In studies concerned with periods much longer than the electronic and ionic relaxation times, and if the other polarizations are assumed to have one equivalent relaxation time τ, the polarization of the dielectric under study would vary with time according to a relation of the form

$$P(t) = P_\infty - (P_\infty - P_0) \exp\left(-\frac{t}{\tau}\right) \tag{8.7}$$

If the initial and ultimate magnitudes of polarization P_0 and P_∞ are significantly different, this would mean that the relative permittivity of the material has two correspondingly different values, according to equation (8.2), ε_{r0} and $\varepsilon_{r\infty}$.

Now, if a capacitor with such a dielectric is subjected to a sinusoidally alternating voltage with angular frequency ω, the polarization will vary with the voltage but lagging slightly behind it depending on the difference $(P_\infty - P_0)$ and on ω as compared to $1/\tau$. The polarization and permittivity would thus be complex quantities, that is,

$$P = P' - jP'' \tag{8.8}$$

and

$$\varepsilon_r = \varepsilon_r' - j\varepsilon_r'' \tag{8.9}$$

It could easily be shown that

$$\varepsilon_r' = \varepsilon_{r\infty} - \frac{\varepsilon_{r\infty} - \varepsilon_{r0}}{1 + \omega^2 \tau^2} \tag{8.10}$$

and

$$\varepsilon_r'' = \frac{\omega \tau}{1 + \omega^2 \tau^2} (\varepsilon_{r\infty} - \varepsilon_{r0}) \tag{8.11}$$

These two components depend on the dielectric material, its relaxation time, and the frequency of the applied electric field. Both are shown schematically in Figure 8.5.

Under dc and also at very low frequencies, $\varepsilon_r' \longrightarrow \varepsilon_{r0}'$ and ε_r'' almost vanishes, as noted from Figure 8.5 and equations (8.10) and (8.11). With increase in frequency, the inertia of the dipoles will cause the lag of the orientational polarization behind the field alteration to become more and more significant, until this type of

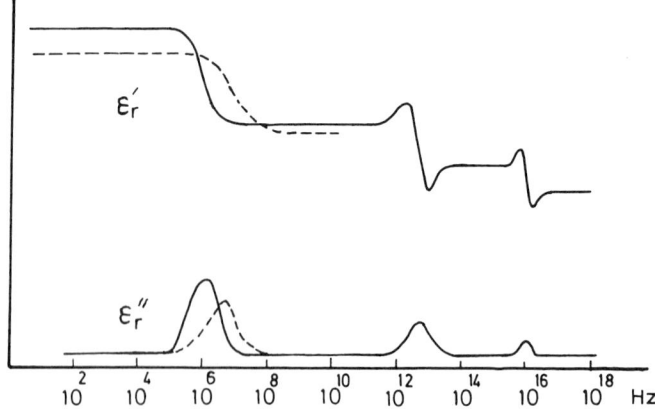

Fig. 8.5 Schematic showing the components ε_r' and ε_r'' of the complex permittivity of a dielectric as they vary with the frequency of the applied electric field: ─────, at a temperature T_1; ─ ─ ─ ─ ─ ─, at a higher temperature T_2. [From Tareev (1980).]

polarization ceases to contribute to the whole. Thus ε'_r drops. It can easily be shown that at $\omega\tau = 1$, $d\varepsilon'_r/d\omega$ is maximum, and ε''_r reaches a peak. This explains the variations of the two curves in Figure 8.5 at the first and subsequent resonant frequencies. The variation of ε'_r with frequency is usually termed the "dispersion" curve, while the variation of ε''_r is known as the "resonance absorption characteristic," as ε''_r represents dielectric loss (i.e., energy absorption).

The dispersion of dielectric materials could be measured by tests described by national and international standard specifications (IEC, 1973). A capacitor containing the dielectric is charged from a dc source with a voltage V_0 and then momentarily short circuited and its residual open-circuit voltage V_1 is measured; the dispersion is expressed as $V_1/(V_0 - V_1)$.

The effect of temperature on the permittivity and polarization of nonpolar dielectrics is negligible. In polar materials, however, the random thermal motion of the dipoles is heavily influenced by temperature. The higher the temperature, the lower will be the relaxation time τ. Hence the magnitudes of ε'_r will be lower and the resonant frequencies of ε''_r will be higher (Fig. 8.5).

8.2.3 Electrical Conduction

Practical insulating materials have admittedly high resistivities but are, nonetheless, finite. For wood, marble, and asbestos the volume resistivity ρ_v is in the range 10^6 to 10^8 $\Omega \cdot m$, whereas for polystyrene and polyethylene it is in the range 10^{14} to 10^{16} $\Omega \cdot m$. Quartz crystals are nonisotropic; ρ_v is about 10^{12} $\Omega \cdot m$ along its optical axis and about 10^{14} $\Omega \cdot m$ in directions normal to it. The volume and surface resistances of an insulator are two paths in parallel for the leakage currents I_v and I_s it draws when connected to a power source (Fig. 8.6). Their relative values can vary drastically depending on the insulator material and its surface conditions.

Electrical conduction through dielectrics is undertaken by ions, basically different from conductors where electrons are responsible for the conductivity. The reason is that in dielectrics the energy necessary for dislodging ions from their positions in the atomic lattice is on the order of that of their thermal vibrations at normal room temperature. On the other hand, the energy needed to liberate electrons in a dielectric is at least one order of magnitude higher. For example, in a simple dielectric such as rock salt (NaCl), the energy required to remove an ion from its lattice position is 0.8 eV, whereas electrons need at least 6 eV for their liberation.

It is evident that with increased temperature, more ions could be dislodged from their positions in the atomic lattice and could

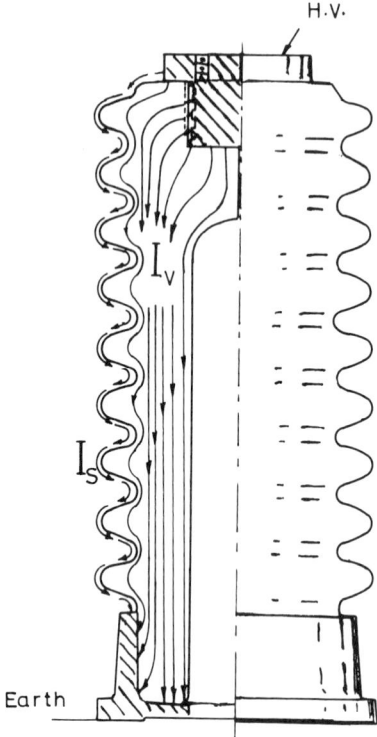

Fig. 8.6 Insulating bus support in half section showing the two components of its leakage current: I_V, through its volume, and I_S, creeping on its surface.

contribute to the electrical conduction. Also, their mobilities increase with temperature exponentially. Therefore, for insulating materials the volume resistivities decrease sensitively with increased temperature.

8.2.4 Dielectric Losses

Practical insulating materials draw leakage currents, however small. There is additional energy loss in the process of polarizing the material. The latter loss is significant under ac with frequencies approaching the resonant frequencies shown in Figure 8.5.
Thus, under ac the current drawn by a practical capacitor leads the applied voltage by an angle slightly less than 90°; the difference δ is termed the loss angle. Power engineers refer to the

Solid Insulating Materials

factor (tan δ) as the loss factor, whereas electronics engineers prefer to use (cot δ) the quality factor of the dielectric.

Effect of Temperature

There are two factors to be considered here. First, in a homogeneous dielectric the conductivity contributed by the jumping of ions from one site to another under the applied electric field becomes easier with higher vibrational energies (i.e., at higher temperatures). The volume resistivities of insulating materials therefore drop considerably with temperature. Distinguished examples are plastics and glass. With PVC, for example, the resistivity drops and the loss factor rises about 10-fold for a temperature rise from 80°C to 120°C. Second, in a heterogeneous material such as paper, a trace of moisture accentuates the effect of temperature. For oil-impregnated paper, for instance, 3% by weight of moisture causes the loss factor to rise from 0.5% at 20°C to about 30% at 99°C.

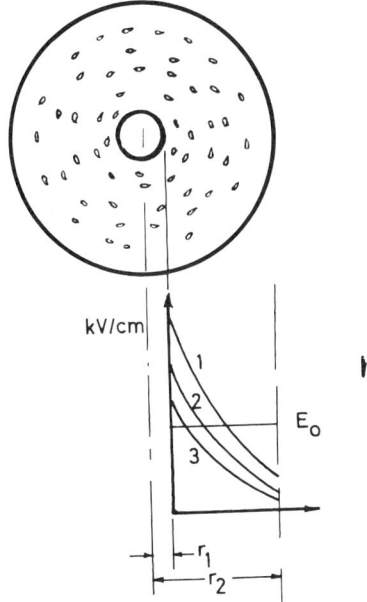

Fig. 8.7 Cross section of a coaxial paper-insulated cable with gas voids scattered in its insulation. E_0 is the gas discharge onset field strength. The curves represent the radial field distributions at different applied voltages $V_1 > V_2 > V_3$.

Effect of Electric Stress

The loss factor for an insulating material is not much affected by the magnitude of the applied electric stress except at levels high enough for secondary phenomena to set in. A well-known example is the ionization that occurs in gas voids within the material. The corresponding power loss is analogous to the corona loss (Chapter 5). It causes a considerable rise in the loss factor above the discharge-onset voltage (Fig. 8.7). At the higher voltages the gases in additional voids get ionized. A well-designed and well-made cable should have the minimum number of gas voids (Chapter 12). Measuring the loss angle δ is one of the important tests to be carried out on test samples of high-voltage cables (Chapter 18).

8.3 THERMAL, MECHANICAL, AND CHEMICAL PROPERTIES

Heat is conducted through insulating materials by phonons. The thermal vibrations in a crystal or a solid continuum could be represented as elastic waves propagating through the solid with the velocity of sound. Thermal conduction of insulating materials is the one important factor determining heat transfer in cables, capacitors, and some electronic devices. In transformers and machines, on the other hand, heat is removed primarily by convection. Evaporative cooling is sometimes used to cope with hot spots.

Thermal conductivities of plastics range between 0.15 and 0.3 W/m°K, whereas for porcelain and glass it is about 1.2 to 1.7 W/m°K. For alumina and magnesia it is as high as 35 W/m°K, which recommends both materials for making special ceramics.

8.3.1 Thermal Classes

Insulating materials commonly used in machines, transformers, capacitors, and cables have been classified according to the maximum safe temperature each material or combination of materials can endure for a very long time under normal service conditions (IEC, 1984). The code letter and temperature limit assigned to each class are given in Table 8.2. Examples of materials belonging to each class are given below.

Class Y: paper, cotton, and natural silk, when not impregnated with oil, varnish, or other insulating liquid. This class also

Table 8.2 Thermal Classes of Solid Insulating Materials

Class code[a]	Y	A	B	E	F	H			
Maximum temperature for continuous operation (°C)	90	105	120	130	155	180	200	220	250

[a]Before 1984, the IEC had grouped the three top classes in one group termed class C, withstanding temperatures >180°C.
Source: IEC (1984).

includes vulcanized natural rubber and thermoplastics limited by their softening temperature to be within this class, such as polyethylene (PE), cross-linked polyethylene (XLPE), and poly(vinyl chloride) (PVC).

Class A: paper, cotton, and natural silk when impregnated with oil, varnish, or resin. This class also includes some organic enamels, varnishes, and thermoplastics with suitable softening temperature.

Class E: laminated and molded plastics with thermosetting binders, epoxy enamels, and varnishes of suitable thermal endurance (e.g., phenol-formaldehyde, melamine-formaldehyde, and polyethylene terephthalate fibers).

Class B: inorganic fibrous and flexible materials (e.g., mica, glass, and asbestos) when bonded or impregnated with organic binder such as shellac, alkyd, and rosin with mineral fillers.

Class F: materials of class B but bonded or impregnated with resins of suitable thermal endurance (e.g., epoxy and silicon alkyd).

Class H: materials of class B but bonded or impregnated with suitable inorganic resin, such as silicon resin. This class includes silicon rubber.

Class C: inorganic materials (e.g., mica, ceramics, fused quartz) when used without binders or with binders suitable for the required thermal endurance. This class also includes polytetrafluoroethylene (PTFE). In 1984 this class was split (IEC, 1984) into three subclasses, coded 200, 220, and 250, corresponding to their safe operating temperatures.

It is realized that insulating materials in general can withstand temperatures significantly higher than the limit assigned to their class but only for a very short duration—a fraction of a second for a short circuit cleared rapidly by a circuit breaker. At too high a temperature, insulating materials inadmissibly degrade in electrical and mechanical properties. Their electrical conductivities and losses increase. Their mechanical strength and elasticity limit decrease considerably. Some insulating materials, such as plastics and cellulosic materials, are apt to melt, decompose, shrink, craze, char, or even burn at elevated temperatures.

8.3.2 Effects of Moisture

Paper and other organic fibrous materials are bound to contain at least traces of moisture. The cellulose they contain, just like hydroxyl carboxyl and amine groups, has an affinity for water. Also, being polymers, most plastics have a measurable permeability for water. Moisture has been found to permeate through micropores

Solid Insulating Materials

in plastic sheaths of underground cables. Once inside the cable and under electric stress, these water droplets would aim at sites of higher field concentration by electrophoresis. At their new sites they would highly distort the field, become ionized, and start water treeing, which may eventually cause breakdown of the cable, as discussed in Section 8.4. The deleterious effects of moisture are worsened by salts, acids, and such easily ionizable ingredients if dissolved in water. A trace of 3% moisture absorbed in dry paper reduces its resistivity by about six orders of magnitude. Therefore, meticulous care is taken during the manufacture of paper and similar cellulosic materials when designed for use as electrical insulation.

8.3.3 Tests

Tests have been prescribed for ensuring various qualities of insulating materials. Thermal endurance tests are by no means straightforward, because the degradation of an insulating material due to exposure to a high temperature cannot be expressed by a single variable. However, a thermal endurance test that has been widely adopted is to measure the loss in weight of a specimen after exposure to a certain high temperature for several hundred hours (Steffens, 1986; IEC, 1981, 1983, 1987).

8.4 DIELECTRIC BREAKDOWN

The process of breakdown of solid dielectrics is much more complex than that of gases, for example. Although solid insulating materials have been the subject of numerous investigations, no single theory fully explains the process of breakdown and predicts the breakdown stress of a given insulator. One of the principal reasons for this state of affairs is the dependence noted for the breakdown on numerous factors, including the material's temperature, voltage duration, and ambient conditions. However, three primary processes of dielectric breakdown have been discerned: intrinsic, thermal, and electrochemical. Each occurs under a certain set of conditions conducive to it. They are all discussed briefly in this section.

8.4.1 Intrinsic Breakdown

Intrinsic breakdown occurs purely due to the electronic behavior of the dielectric, with no effect of ambient or temperature rise. It is therefore sometimes called electronic breakdown. It is an ideal, extremely difficult to identify in practice by eliminating all secondary causes of breakdown. Care should be taken not to allow breakdown due to thermal instability of the dielectric with temperature rise caused by leakage currents, nor should the dielectric fail under electromechanical stresses, nor should there be impurities or gas discharges to initiate failure due to electrochemical decomposition.

It is thus evident that intrinsic breakdown would be possible for idealized crystalline materials. In 1947, Fröhlich proposed a criterion for electronic breakdown as "the electric stress at which the dielectric acquires a field-enhanced conductivity." He developed a theory for crystalline dielectrics, such as potassium chloride, relating the breakdown stress to six constants of the crystal lattice. A detailed discussion of his theory is beyond the scope of this section. As his crystal constants are not measurable with any accuracy for simple crystalline dielectrics, to say nothing of practical amorphous materials, a brief qualitative review will suffice.

Cosmic and other radiations would set a few electrons free in the crystalline dielectric. These would be accelerated by the applied electric field and would collide with some lattice sites on their way. The energies of these electrons differ according to Boltzmann's distribution (Chapter 3). Depending on the energy of each individual electron, its energy gain from the electric field and loss during the interaction with the lattice site it collides with vary as shown schematically in Figure 8.8. Curves 1, 2, and 3 represent the

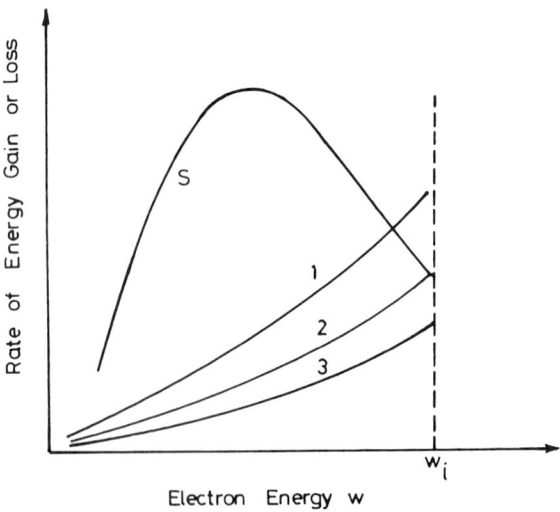

Fig. 8.8 Rates of energy gain by electrons from the applied electric field E and the rate of their energy loss S to the crystal lattice as functions of the electron energy. W_i is the ionization energy for the atoms in the lattice: $E_1 > E_2 > E_3$.

Solid Insulating Materials 207

rates of energy gain by an electron from fields with different strengths $E_1 > E_2 > E_3$, while curve S represents the energy loss. The limit W_i is the energy sufficient for the electron to ionize the atom it interacts with at the lattice site.

For intrinsic breakdown to occur, the energy gain should just exceed the loss. This would result in the field-enhanced conductivity suggested by Fröhlich. Thus the critical field strength is E_2, as shown in Figure 8.8. Attempts have been made to measure intrinsic breakdown voltages of dielectric specimens in the laboratory under carefully controlled conditions. Care was taken to eliminate or greatly suppress secondary causes of breakdown. The highest breakdown stress reached for mica, for example, was 10 MV/cm and for polyethene was 8 MV/cm, obtained with extreme care to prevent secondary effects from causing breakdown at much lower stresses.

8.4.2 Thermal Breakdown

Thermal breakdown would result from disruption of thermal equilibrium in the dielectric: The rate of heat generation within the material exceeds that of cooling. In such a case the temperature starts to rise monotonously instead of leveling off at a steady value (Fig. 8.9a). With temperature rise, the volume resistivity of the insulating material decreases and the dielectric losses increase. Thus with a monotonous temperature rise, a runaway process sets in, leading eventually to breakdown.

Attempts at mathematical analysis of this situation were made by Whitehead in 1951 (Kuffel and Zaengl, 1984). For simplicity, consider the linear heat flow to the ambient together with the dielectric losses in a capacitor of a large area A under dc (Fig. 8.10). The heat balance of the elemental volume ($\Delta A\ \Delta x$) at a distance x from the center plane is

$$\frac{E^2(x)}{\rho_v(x)} \Delta A\ \Delta x = -k\ \Delta A\ \Delta x\ \frac{d^2 T}{dx^2} \qquad (8.12)$$

where $E(x)$ is the electric field strength at x, k the thermal conductivity of the dielectric; and ρ_v its volume resistivity, known to decrease almost exponentially with temperature. Realizing that the density of leakage current through the dielectric $i = E(x)/\rho_v(x)$ is independent of x, equation (8.12) could be rearranged to

$$-k\ \frac{d^2 T}{dx^2} = iE(x) \qquad (8.13)$$

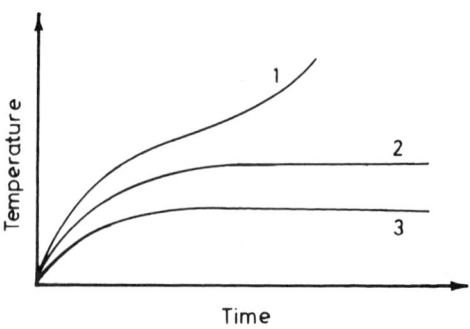

(a)

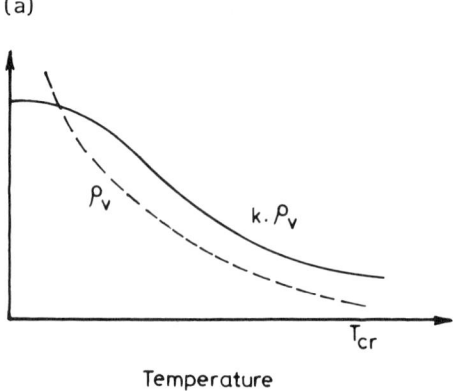

(b)

Fig. 8.9 (a) Temporal increase of temperature of dielectric in a capacitor when subjected to different step voltages: $V_1 > V_2 > V_3$. (b) Schematic showing the variations of the volume resistivity ρ_V of a dielectric and of the product of its volume resistivity and its thermal conductivity k. T_{cr} is the critical temperature at which the dielectric fails.

Solid Insulating Materials

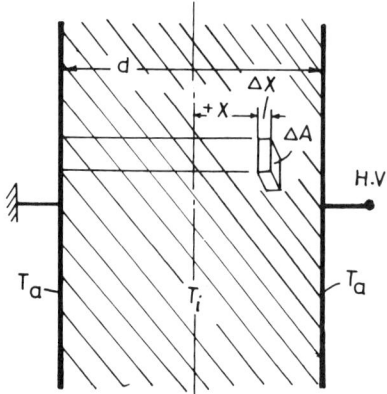

Fig. 8.10 Heat flow to the ambient from a dielectric block. The heat is caused by the dielectric loss under voltage. The maximum temperature T_i occurs at the center plane, whereas T_a is the ambient temperature.

Integrating both sides over the distance $0 - x$, we get

$$-k \frac{dT}{dx} \Big|_0^x = i\,[V(x) - V(0)] \qquad (8.14)$$

as $dT/dx = V(0) = 0$ at $x = 0$, the center plane of the dielectric block. Thus, rearranging and integrating equation (8.14), we get

$$\frac{V^2}{8} = -\int_{T_i}^{T_a} k\rho_v\, dT \qquad (8.15)$$

where V is the total voltage across the dielectric block.

The critical conditions are reached when the temperature at the center of the dielectric $T_i = T_{cr}$, the critical temperature at which the thermal runaway process sets in. Thus, according to Whitehead's theory, the thermal breakdown voltage of a dielectric block is independent of its thickness and could be obtained by integrating the product ($k\rho_v$) over the range from the ambient to the material's critical temperature (Fig. 8.9b). Computations of this area were made for numerous insulating materials and their voltages for thermal breakdown ranged between 1 and 5 MV (Alston, 1968). Experimental measurements, however, yielded voltages up to only 400 kV. The discrepancy is explained at least partially by the nonhomogenicity of most dielectric materials and inaccurate values of their physical characteristics at higher temperatures.

One other important point to note here is that there is a limit for the thermal breakdown voltage, no matter how thick the insulation is made. This is to be kept in mind when designing high-voltage machines and EHV cables. The relation (8.15) eventually does not hold for very small thicknesses. A theoretical account can be developed if the surface temperature of the specimen is allowed to exceed that of the ambient, which is more realistic.

In this case the heat balance equation per unit area with the entire thickness of the specimen (Fig. 8.10) takes the form

$$\frac{V^2}{\rho_v d} = HF(T_i)$$

or

$$\frac{V^2}{d} = H\rho_v F(T_i) \qquad (8.16)$$

The function $F(T_i)$ correlates the surface temperature to the highest temprature inside the specimen, and the product $HF(T_i)$ gives the rate of heat dissipation to the electrodes and ambient. As the dielectric resistivity ρ_v is another function of temperature, the right-hand side of equation (8.16) could be considered as a constant for a given critical temperature at which the dielectric fails thermally. Thus this simple approximate approach yields the relation between the thermal breakdown voltage and the small thickness d of a dielectric specimen as

$$V = cd^{1/2} \qquad (8.17)$$

c being a constant.

Although the discussion above was concerned with a dc voltage applied across the dielectric specimen, the same result could be reached had we considered instead an ac voltage. In that case the dielectric loss would be expressed as ($\omega CV^2 \tan \delta$).

8.4.3 Electrochemical Breakdown

With electric field concentration at some points (e.g., near the edge of the HV electrode), corona discharges would occur. Also, discharges would take place within gas voids inside such insulation as multilayer paper or plastic insulation. The gas ions driven by the electric field would impinge on the wall of the insulation and react chemically with some of its surface layer molecules. Thus Chemical and thermal degradation of the insulating material occurs at these microscopic sites. The field strength at the discharge tip would be much higher than the macroscopic field strength on the specimen.

The vulnerability of solid insulation under surface discharges increases with the proportion of interatomic bonds that produce free carbon atoms on pyrolysis. The alternative to carbonization is erosion of the insulator surface. Erosion causes roughness of plastic surfaces, and tubules formed slowly penetrate the insulator body. Each tubule may have a diameter of a few free paths of the gas molecule (Ichinose, 1988). Extension of tubular cavities inside plastic insulation may be ascribed to chemical decomposition, volatilization of the polymer molecules under the discharge, or fracture at the microscopic site under electromechanical forces.

Treeing in Plastics

Treeing began to receive wide attention in the mid-1970s as polymers started replacing oil-impregnated paper to insulate high-voltage cables. Electric treeing initiated by gas discharges at microscopic sites, if not inhibited, was seen to lead to slow, yet complete breakdown.

Although specimens of polyethylene and cross-linked polyethylene tested under ac of short duration break down at no less than 700 to 800 kV/mm, and the macroscopic working stresses in cables with such an insulation range from about only 3 kV/mm (in low-voltage cables) to 20 kV/mm (for 250-kV cables), some cables have actually failed in service as a result of treeing (Deschamps et al., 1984; Hosokawa et al., 1982). Treeing begins at microscopic sites of imperfections in the molecular structure of polymers. There may be metallic particles, cavities, or thermo-oxidated or deteriorated particles of material, all of which are extremely difficult to eliminate. Also, water can diffuse through microscopic pores and cracks in the insulation and sheaths of polymer-insulated cables. Such cracks might be 10 nm in width in the amorphous regions between crystalline lamellae of semicrystalline polymers (Sletbak and Ildstad, 1984; Marton, 1987; Filippini et al., 1987). Water, with its soluble ions, would migrate by electrophoresis under the electric field to sites of higher intensity. When water is involved the phenomenon is termed water treeing.

A concentrated electric field at the tip of a tubule, pulsating under ac, combined with hydrostatic pressure caused by water

Fig. 8.11 Tree developed at the tip of a needle electrode inserted into an epoxy dielectric. (Courtesy of A. Rasmy and O. Gouda.)

inside it being heated by concentrated leakage currents and chemical decomposition, would result in pulsating mechanical forces at the tip of the tubule. This would cause extension and branching of the tubule in a treelike shape in steps of microfractures of the material (Fig. 8.11).

Several attempts have been made to explain water treeing (Meyer and Filippini, 1979; Sletbak, 1979). It was recently suggested by Fischer and his co-workers (1987) that water treeing involves decomposition of hydrogen peroxides with probable catalytic action of some metallic ions. They thus could retard water treeing in cross-linked polyethylene (XLPE) by including some antioxidants, (e.g., barbituric acid). However, further intensive research needs to be carried out before conclusive results can be reached. Techniques for testing the vulnerability of polymers to water treeing were described by Saure and Kalkner (1987).

The time to complete electrochemical breakdown depends on the number and intensity of discharges within cavities or on the surface. These discharges are generally independent of the duration of the half-cycle of the applied ac voltage, as long as the dielectric's properties remain fairly constant. For most polymers their loss factor and other properties do not change significantly over the frequency range 50 to 2000 Hz. This has encouraged the development of accelerated life tests for such dielectrics at the higher frequencies, the life being counted in cycles of the test ac voltage.

8.5 INSULATING MATERIALS

Traditional materials include cellulosics such as paper and organic fabrics. They also include mica, glass, and ceramics. Cellulosic materials suffer from hygroscopicity and are bound to contain gas voids. Therefore, to be usable as good insulating materials, they have to be specially treated and impregnated with oil or varnish. The insulating characteristics could be improved drastically by combining paper with polypropylene (PPP) (Samm, 1987; Schaible, 1987; Mark, 1984).

Mica has traditionally been used where high temperatures and/or surface discharges are experienced (e.g., in vacuum tubes, heaters, and dc machine commutators). Mica powder mixed with finely ground glass in a hot-pressing process yields a dense machinable material with excellent electrical qualities, particularly suitable for insulators in high-voltage high-frequency equipment. Recently, synthetic mica could be developed where the OH groups are replaced by fluorine atoms, enabling the mica to withstand much higher temperatures.

Porcelain and glass have traditionally been used in insulating overhead power lines and busbars (Chapter 9). However, since the 1960s much lighter insulators with excellent electrical and mechanical characteristics could be made of epoxy resins reinforced with glass fibers. Depending on the composition of ceramics, they could be manufactured for use as insulators with almost any applicaation, as magnets, or as semiconductors with resistivities sensitively varying with temperature, humidity, or electric stress.

Several newly developed insulating materials fall within the classification as polymers. Their molecular arrangements are either linear or cross-linked, the former including elastomers and thermoplastics and are generally very flexible and yield with time under a given mechanical load. The higher the temperature, the greater the yield. They include polyethylene (PE), poly(vinyl chloride) (PVC), and polyamides (nylons). Cross-linking has considerably improved the electrical qualities of PE. Thus XLPE has been favored as an insulant in both low-voltage and medium-voltage cables. As an exception to the rule, polytetrafluoroethylene (PTFE), which is a linear polymer, is not flexible nor does it yield and melt at high temperatures. Because of the high binding energies of fluorine atoms, the polymer is about 95% crystalline. At temperatures as high as 400°C, it decomposes. It enjoys thermal and mechanical

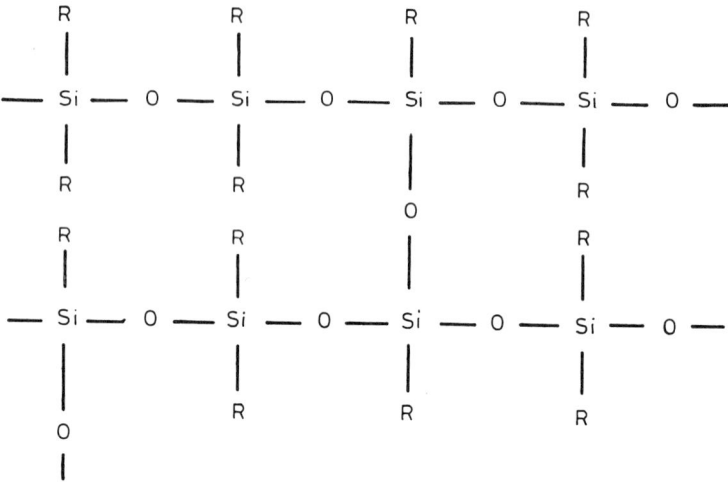

Silicone Resin

Fig. 8.12 Representation of molecular arrangements of a silicon resin.

endurance and chemical stability, much higher than that of PE and PVC (Margolis, 1985).

Depending on their production process, polysiloxanes (silicon resins) can be either elastomers or thermosetting. The latter has a molecular arrangement such as that represented in Figure 8.12. The organic radical R could be a methyl (CH_3), ethyl (C_2H_5), or phenyl (C_6H_5). The silicon's link with oxygen Si—O is stronger than that of carbon C—C, which accounts for the considerably higher thermal endurance for silicon polymers.

8.6 ACTIVE DIELECTRICS

In addition to the passive dielectrics employed solely as electrical insulators, there are dielectrics that have other applications. In such "active" dielectrics there are mutual effects between imposed electric fields and mechanical stresses, heat influx or penetrating light. They are used in piezoelectric and pyroelectric transducers, nonlinear capacitors, ferroelectric memory devices, electro-optic voltage sensors, electric light shutters, vidicons, and electrets.

Their far-reaching applications are extending rapidly in many areas of engineering (CIGRE, 1987).

REFERENCES

Alston, L. (1968). *High Voltage Technology*, Oxford University Press.
Böttger, O., Gölz, W., and Saure, M. (1987). *Proc. CIGRE Symposium on New and Improved Materials for Electrotechnology*, Vienna, May, Report 620.07.
CIGRE (1987). *Proc. CIGRE Symposium on New and Improved Materials for Electrotechnology*, Vienna, May.
Dechamps, J., Michel, R., and Lepers, J. (1984). *CIGRE report 21-08*.
Filippini, J., Poggi, Y., Rahamarimalala, V., de Bellet, J., and Matey, G. (1987). *Proc. CIGRE Symposium on New and Improved Materials for Electrotechnology*, Vienna, May, Report 620.04.
Fischer, P., Peschke, E., Schroth, R., and Farkas, A. (1987). *Proc. CIGRE Symposium on New and Improved Materials for Electrotechnology*, Vienna, May, Report 620.06.
Hosokawa, K., Kojma, K., Toshida, N., Kasahara, T., and Kaneko, R. (1982). *CIGRE report 21-09*.
Hummel, R. (1985). *Electronic Properties of Materials*, Springer-Verlag, Berlin.
Ichinose, N. (1988). *IEEE Electrical Insulation Magazine*, 4(3):24–30.
IEC (1973). *Specifications of Insulating Materials Based on Mica—Definitions and General Requirements*, Publication 371-1, International Electrotechnical Commission, Geneva.
IEC (1981). *Methods of Tests*, Publication 371-2, International Electrotechnical Commission, Geneva.
IEC (1983). *Specifications for Individual Materials*, Publication 171-3, International Electrotechnical Commission, Geneva.
IEC (1984). *Recommendations for the Classification of Materials for the Insulation of Electrical Machines and Apparatus in Relation to Their Thermal Stability in Service*, Publication 85, International Electrotechnical Commission, Geneva.
IEC (1987). *Guide for the Determination of Thermal Endurance Properties of Electrical Insulating Materials—General Guidelines for Ageing Procedures and Evaluation of Test Results*, Publication 261-1, International Electrotechnical Commission, Geneva.
Kuffel, E. and Zaengl, W. (1984). *High Voltage Engineering Fundamentals*, Pergamon Press Ltd., Oxford.
Margolis, J. (1985). *Engineering Thermoplastics: Properties and Applications*, Marcel Dekker, Inc., New York.

Mark, R., ed. (1984). *Handbook of Physical and Mechanical Testing Paper and Paperboard*, Marcel Dekker, Inc., New York.
Marton, K. (1987). *Proc. CIGRE Symposium on New and Improved Materials for Electrotechnology*, Vienna, May, Report 620.08.
Meyer, C. and Filippini, J. (1979). *Polymers*, 20:1186–1187.
Samm, R. (1987). *IEEE Electrical Insulation Magazine*, 3(4):41–42.
Saure, M. and Kalkner, W. (1987). *Proc. CIGRE Symposium on New and Improved Materials for Electrotechnology*, Vienna, May, Report 620.10.
Schaible, M. (1987). *IEEE Electrical Insulation Magazine*, 3(1):8–12.
Seely, S. and Poularikas, A. (1979). *Electromagnetics—Classical and Modern Theory and Applications*, Marcel Dekker, Inc., New York.
Sletbak, J. (1979). *IEEE Trans.*, PAS-98:1358–1365.
Sletbak, J. and Ildstad, E. (1984). *Rec. IEEE Int. Symposium on Electrical Insulation*, Montreal, pp. 29–32.
Steffens, H. (1986). *IEEE Electrical Insulation Magazine*, 2(6):39–40.
Tareev, B., ed. (1980). *Electrical and Radio Engineering Materials*, Izdatelstvo Mir, Moscow.
Valuzat, P. and Goddet, T. (1987). *IEEE Electrical Insulation Magazine*, 3(6):24–26.

Part II

9
High-Voltage Busbars

A. El-MORSHEDY *Cairo University, Giza, Egypt*

9.1 INTRODUCTION

The substation or switching station functions as a connection and switching point for transmission lines, subtransmission feeders, generating units, and transformers. The substation design aims to achieve a high degree of continuity, maximum reliability, and flexibility, and to meet these objects with the highest possible economy.

The substations used in distribution systems operate at voltage levels up to 69 kV. Transmission substations serving bulk power sources operate at voltages up to 765 kV. Voltage classes used for major substations include 69, 115, 138, 161, 230, and 287 kV considered as high voltage, and 345, 500, and 765 kV, considered as extrahigh voltages (EHV). Higher voltage classes, including 1100 and 1500 kV, are in various stages of planning or construction. These are referred to as ultrahigh voltages (UHV).

In this chapter we deal with the conventional type of substations where atmospheric air is the insulating medium. Gas-insulated switchgear (GIS) is discussed in Chapter 10.

Substation buses are the most important part of the station structure since they carry high amounts of energy in a confined space and their failure would have very drastic repercussions on the power supply continuity. Therefore, the bus system must be built to be electrically flexible and reliable enough to give continuous service. It must have adequate capacity to carry all loads and robust construction to withstand foreseeable abnormal electromechanical forces.

9.2 BUSBAR ARRANGEMENTS

Most substations conform to one or the other of the following basic arrangements as a result of different factors: (a) single bus, (b) double bus with double breaker, (c) double bus with single breaker, (d) main and transfer bus, (e) ring bus, and (f) breaker-and-a-half with two main buses. The choice of the arrangement to be used depends on the relative importance assigned to such items as safety, reliability, simplicity of relaying, flexibility of operation, first cost, ease of maintenance, available ground area, location of connecting lines, ease of rearrangement, and provision for expansion.

9.2.1 Single Bus

The single bus is the one in common use and has the simplest design. It is generally used in small outdoor stations having relatively few outgoing or incoming feeders and lines. The single-bus design (Fig. 9.1) is not normally used for major substations. Dependence on one bus can cause a serious outage in the event of bus failure.

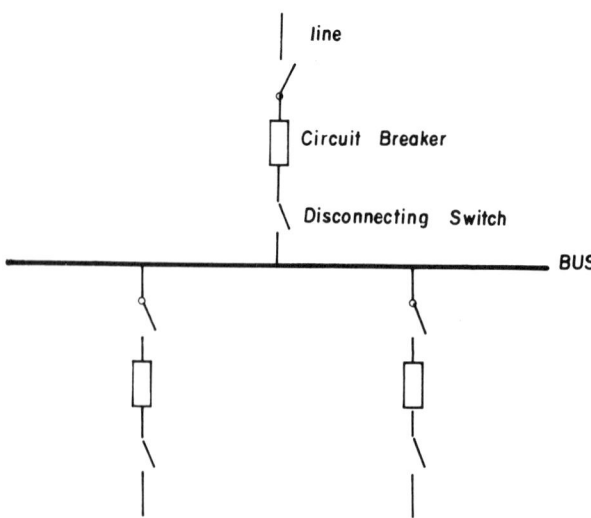

Fig. 9.1 Single busbar arrangement.

High-Voltage Busbars

9.2.2 Double Bus with Double Breaker

The double bus with double breaker design (Fig. 9.2) is used with large switching stations. It requires two circuit breakers for each feeder circuit. Normally, each circuit is connected to both buses, which presents a high order of reliability.

9.2.3 Double Bus with Single Breaker

This scheme (Fig. 9.3) uses two main buses, and each circuit includes two disconnecting switches. A bus-tie circuit breaker is connected to the main buses. When closed it allows transfer from one bus to the other without deenergizing the feeder circuit; only the bus disconnecting switches need be operated then.

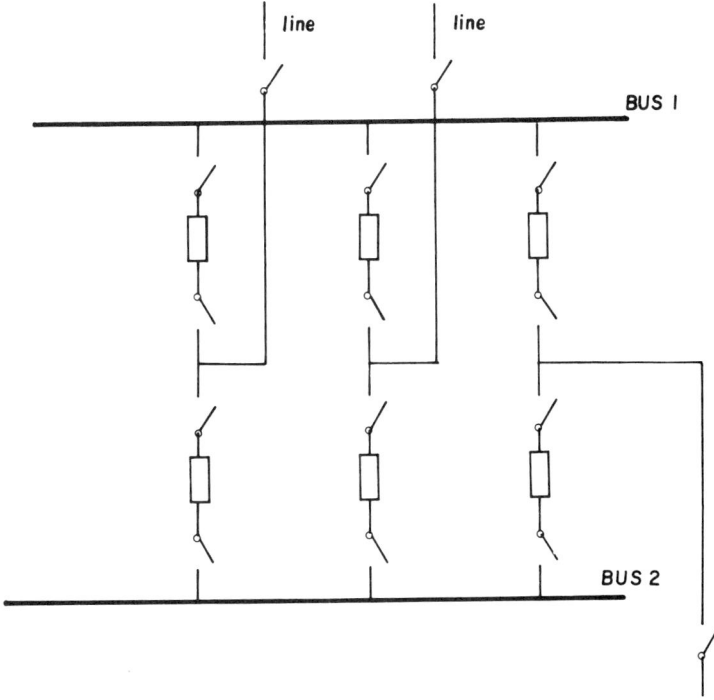

Fig. 9.2 Double busbars with double breakers.

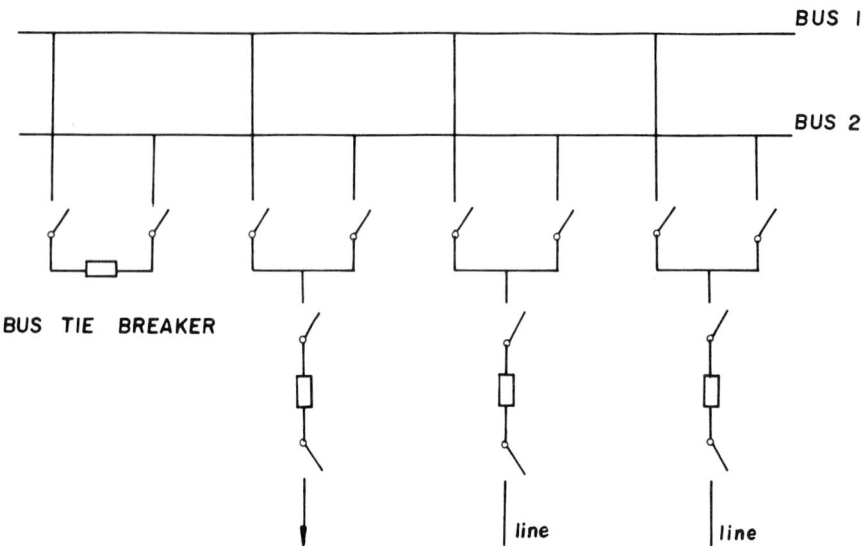

Fig. 9.3 Double busbars with single breaker.

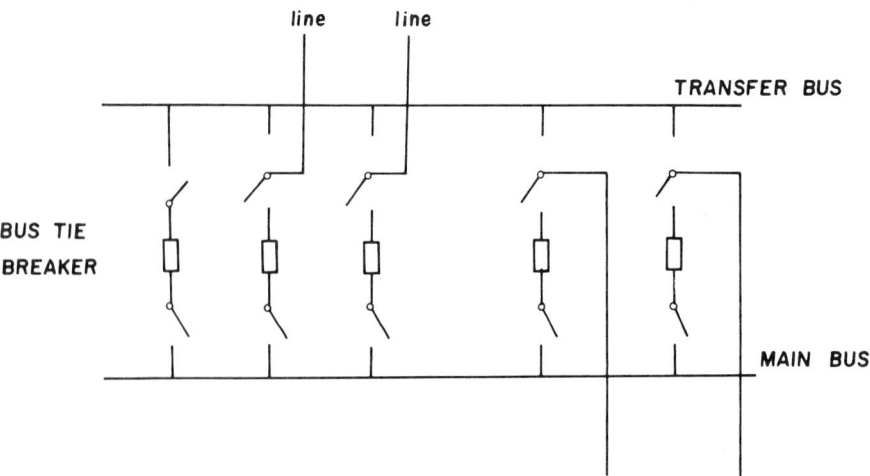

Fig. 9.4 Main and transfer busbars.

High-Voltage Busbars

9.2.4 Main-and-Transfer Bus

The main-and-transfer bus (Fig. 9.4) is used for most distribution work. One circuit breaker serves each circuit. The transfer bus is a standby for emergency use. Also, a bus-tie circuit breaker is provided to tie the main and transfer buses together when the need arises.

9.2.5 Ring Bus

The advantage of the ring bus scheme (Fig. 9.5) is that it requires the use of only one circuit breaker per circuit. In addition, each outgoing circuit has two sources of supply. This system is economical in cost, reliable, and very flexible.

9.2.6 Breaker-and-a-Half with Two Main Buses

This scheme (Fig. 9.6) has three breakers in series between the main buses. Two circuits are connected between the three breakers, hence the term breaker-and-a-half. Under normal operating conditions all breakers are closed and the two main buses are energized. To trip a circuit, the two associated circuit breakers must be opened.

It has been found that the two commonest busbar arrangements at all voltage levels are the single busbar and the double busbar

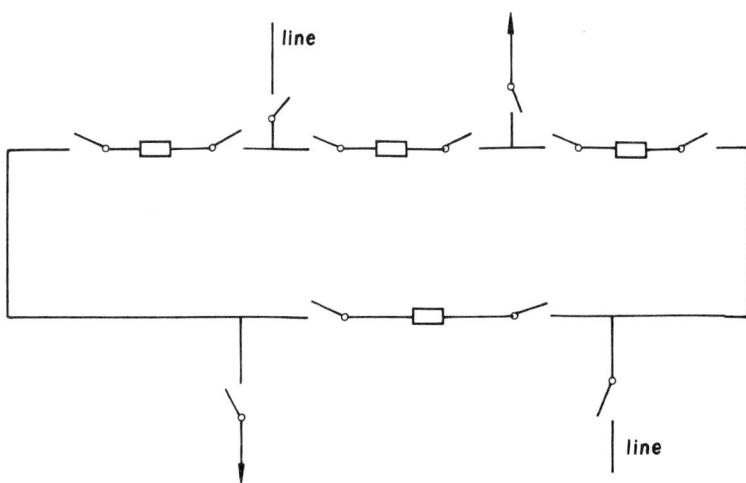

Fig. 9.5 Ring bus.

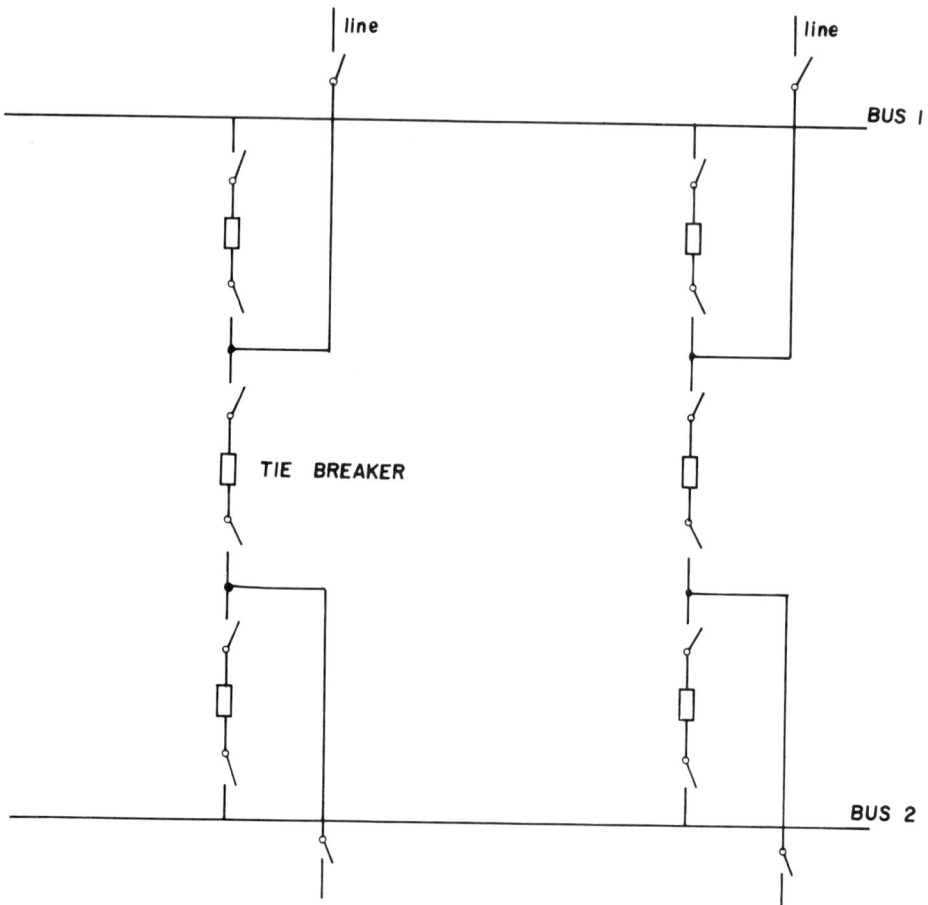

Fig. 9.6 Breaker-and-a-half with two main buses.

with single breaker. The single-busbar arrangement predominates up to 70 kV, while the double busbar with single breaker is more commonly used above 70 kV. The ring bus and the breaker-and-a-half arrangements have been gaining popularity among designers for voltages above 160 kV.

9.3 RIGID BUS VERSUS STRAIN BUS

Usually, low-voltage and medium-voltage busbars are of the rigid bus type, using copper or aluminum bars or tubings as the phase conductors, with pedestal-type insulator supports. Those of higher voltages use strain insulators and stranded aluminum (ACSR) or copper conductors.

Substations of the EHV class are usually of the strain bus design. Medium-voltage substations normally use the rigid-bus approach and enjoy the advantage of low station profile and ease of maintenance and operation.

Combinations of rigid and strain bus constructions are sometimes employed in conventional arrangements, up to 765 kV, to benefit from both types.

The advantages of the rigid bus compared with the strain bus design are as follows:

1. The rigid bus design employs less steel and simple, low-level structures.
2. The rigid conductors are not under constant strain.
3. The pedestal-type bus supports are usually more accessible for cleaning.
4. The rigid bus has low profiles, which provides good visibility of the conductors and apparatus.

The disadvantages of the rigid bus design are:

1. It is comparatively expensive, due to the higher cost of tubing and connections.
2. It usually needs more supports and insulators.
3. It is more sensitive to structural deflections, which may lead to possible damage.
4. It usually requires more ground space than the strain bus.

9.4 BUS CONDUCTOR MATERIALS

Several factors determine the proper choice of busbar conductor material. These factors are determined largely by the type of installation and the size of electrical load to be carried. Factors to be considered when making this choice are voltage drop, power loss, current-carrying capacity, ice and wind loads, short-circuit loads, and conductor corrosion.

The busbar materials in general use are aluminum and copper. Heat-treatable aluminum alloys, especially in tubular shapes, are most widely used in HV and EHV outdoor substations. They combine high strength and good conductivity.

Aluminum is one-third the weight of copper for a specified length. Aluminum and its alloys require little maintenance. For a given current rating and for equal temperature rise, the aluminum bus would have a 33% larger cross section than the equivalent copper bus. The resulting deflection for the copper bus is about 31% greater than for the aluminum bus. As aluminum has several advantages over copper, most rigid bus installations use tubings of aluminum or its alloys.

9.5 BUSBAR CLEARANCES

For safety and reliability it is essential that adequate clearances to live parts be provided. The minimum phase-to-ground and phase-to-phase air clearances in substations are prescribed to guarantee the withstand levels of switching and lightning impulses considered necessary by the system designer (see Chapter 14).

In fact, in substations up to 200 kV, air insulation is predominantly dictated by lightning overvoltages. There the minimum phase-to-phase clearances are usually taken about 15% longer than those to grounded metal frames. In certain cases larger phase-to-phase clearances have been preferred as dictated merely by the requirements of equipment maintenance.

The dielectric stresses in the air gaps between phases and to ground due to switching overvoltages have, on the other hand, very different characteristics (see Chapters 4 and 14). In particular, the stress between the phases is greater than the stress to ground. Moreover, the behavior of air gaps under switching impulse is affected by electrode shapes, which in the case of clearances between phases are usually different from the case of clearances to ground.

9.5.1 Phase-to-Ground Clearances in Substations

The insulation levels for systems and equipment with voltages rated at 245 kV and above are combinations of two components: the rated switching impulse withstand voltage and the rated lightning impulse withstand voltage (Section 18.3).

The air clearances between phase conductors and between them and grounded metal frames could be estimated knowing the dielectric strengths of such air gaps under ac and impulse voltages. Numerous tests have been carried out in high-voltage testing stations in many countries in dry air and under simulated rain conditions for air gaps between rods and planes, between parallel conductor bundles, between conductors and planes, and between conductors and portal frames. The results have been expressed by empirical relations and could also be predicted by a generalized formula (see Section 4.8). For rod-to-plane and similar gaps, positive impulse flashover voltages are lower than those with negative polarity. Rain has no significant influence on positive impulse flashover voltages (Rizk, 1976).

These relations could be used for estimating the 50% breakdown voltages of such air gaps under switching and lightning impulses. The standard deviations σ of these breakdown voltages were observed, by experiment, to be about 6% for switching surges and 3% for lightning surges. Based on the evaluated 50% impulse breakdown voltages of air gaps, their impulse withstand voltage V_w could be estimated by the formula

$$V_w = V_{50\%}(1 - 1.3\ \sigma) \tag{9.1}$$

This withstand voltage represents the level at which the probability of flashover is less than 10%. In substations three main categories of phase-to-ground clearances are present:

1. Distances between conductors and portals
2. Distances between live parts of apparatus and portals
3. Distances between conductors and ground

Table 9.1 gives the air clearances according to the recommended rated impulse withstand voltages.

9.5.2 Clearances between Conductors

The distances between different live conductors are determined not only by their voltages but also by the necessity in some cases to carry out maintenance work while neighboring conductors are alive. Lightning surge stresses between phases will normally be within the magnitude of stresses to ground. On the other hand, the phase-to-phase overvoltage peaks are subjected to considerable variations, depending on the wave shapes of the two phase-to-ground components possibly having opposite polarities and on the ratio and time delay between these two components (El-Morshedy, 1982). The 50%-flashover voltage of phase-to-phase air clearances is dependent on the wave shapes used in laboratories for the two phase-to-ground components, their times to crest and the time delay between their crest values representing expected field

Table 9.1 Phase-to-Ground Air Clearance Dependent on the Rated Switching and Lightning Impulse Withstand Voltages

Rated switching impulse withstand voltage (kV)	Air clearance (m)	Rated lightning impulse withstand voltage (kV)	Air clearance (m)
650	1.15	750	1.35
750	1.45	850	1.55
850	1.79	950	1.73
950	2.16	1050	1.92
1050	2.55	1175	2.14
1175	3.07	1300	2.37
1300	3.64	1425	2.60
1425	4.24	1550	2.83
1550	4.87	1800	3.28
		1950	3.56
		2100	3.83
		2400	4.38

conditions. It has been recommended that phase-to-phase insulation tests be carried out with two equal switching impulses of opposite polarities, equal crest values, and synchronous peaks. The crest value of each impulse should be half the rated switching impulse withstand voltage between phases. Besides being representative for the real stresses, only this method allows testing of phase-to-phase insulation under actual geometrical substation conditions without causing undue flashover on phase-to-ground insulation.

The air clearances in substations may be generally identified as:

1. Clearances between conductors
2. Clearances between conductors and apparatus
3. Clearances between poles of the same phase

Most electrode configurations are characterized by highly inhomogeneous and sometimes symmetric field distributions. Rod gaps, rod-to-plane gaps, and gaps between ring-shaped electrodes have

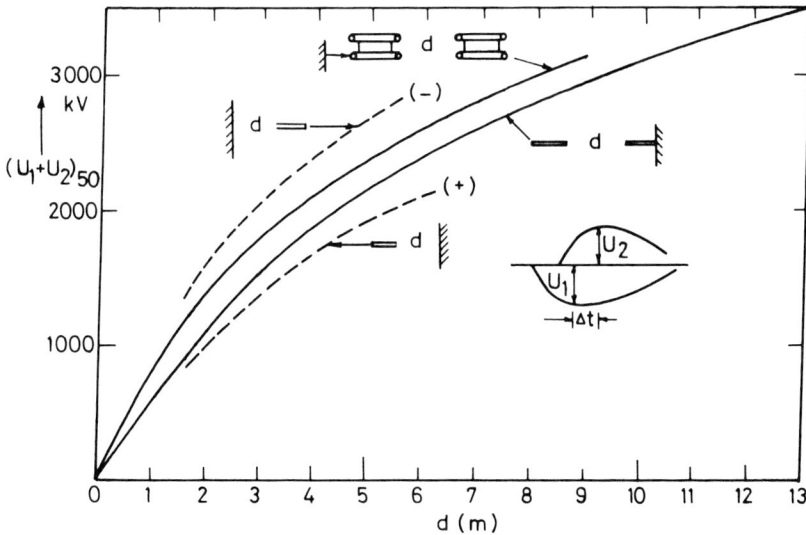

Fig. 9.7 Fifty percent flashover voltages of a gap d between conductors of the shapes shown, under two simultaneous equal and opposite switching surges, U_1 and U_2 ($\Delta t = 0$).

generally been accepted as representing practical geometrical conditions, typical for the high-voltage and extra-high-voltage (EHV) ranges.

Figure 9.7 shows the test results for the 50% flashover voltages using two synchronous ($\Delta t = 0$) switching impulses with equal crest values U_1 and U_2. The time to crest for the positive component has been equal to that value which gives the lowest flashover voltage (Sections 4.8 and 15.2). Since the electrical field distribution is symmetric for ring and rod gaps, no polarity effect was observed. Only for nonsymmetric electrode configurations is the influence of the polarity remarkable. For a rod-plane gap, Figure 9.7 shows that the lower values are obtained with the positive polarity at the rod, where the field concentration is highest.

Table 9.2 gives the recommended minimum clearances between phases as represented by the rod—rod and ring—ring configurations. These clearances were calculated from the 50%-interphase flashover voltages.

As mentioned above, the degree of inhomogeneity of the electric field varies with the voltage level under consideration. Therefore, the determination of the necessary interphase clearance is

Table 9.2 Correlations between the Switching Impulse Withstand Voltage-to-Ground and between Phases and Minimum Phase-to-Phase Air Clearances

Rated switching impulse withstand voltage to ground (kV)	Ratio between phase-to-phase and phase-to-ground switching impulse withstand voltage	Switching impulse withstand voltage between phases (kV)	50% switching impulse flashover voltage between phases (kV)	Minimum phase-to-phase air clearance (m)	
				Ring–ring	Rod–rod
650	1.62	1050	1139	1.50	2.05
750	1.57	1175	1274	1.75	2.35
850	1.53	1300	1410	2.05	2.65
950	1.63	1550	1681	2.65	3.35
1050	1.71	1800	1952	3.40	4.20
1175	1.66	1950	2115	3.90	4.75
1300	1.73	2250	2440	5.20	6.20
1425	1.79	2550	2766	6.80	7.95
1550	1.74	2700	2928	7.65	8.95

more related to the results for the rod—rod configuration in the lower voltage range. In the EHV range, however, the ring—ring configuration becomes more and more representative because of the sizes of the hardware used at such voltage levels.

Some trends in the design practices of North American 500-kV and 750-kV ac substations are indicated by the clearances in Table 9.3 (IEEE Working Group, 1983). It is noted that the minimum phase-to-ground and phase-to-phase clearances are greater than those recommended in Tables 9.1 and 9.2, and that in EHV substations rigid, strain, or a combination of both bus constructions are in use.

9.6 THERMAL RATING

In the design of buses the temperature rise of conductors above ambient while carrying current is very important. Buses are generally rated on the basis of the temperature rise that can be permitted without danger of overheating equipment terminals, bus connections, and joints.

The permissible average temperature rise for plain copper and aluminum buses while carrying their normal loads is usually limited to 30°C above an ambient temperature of 40°C. This value is the accepted standard for IEEE, NEMA, and ANSI. This is the overall temperature rise and a maximum or hot-spot temperature rise of 35°C is permissible. Above this temperature oxidation increases rapidly and would give rise to cumulative and excessive heating at joints and contacts. Many factors influence the heating of the bus, such as the type of material used and their size and shape (Conway, 1979). The maximum continuous current-carrying capacity is important in selecting the proper conductor material, size, and shape of cross section.

9.6.1 Conductor-Rated Current

The ampacity (i.e., current-carrying capacity) of a busbar is determined by its ratio of surface area (for heat dissipation) to its cross-sectional area. This means that for a single shape there are limits to the current-carrying capacity with respect to its dimensions. For large currents it is better to provide several parallel on-edge flat bars than to use a single bar of equivalent total area and the same height. With the arrangement of multiple bars, the proximity and skin effects should be considered in estimating their ac losses (Weiss and Csendes, 1982). Figure 9.8 shows comparative ampacity ratings at power frequency for various conductor arrangements, each with a cross-sectional area of 25.8 cm^2 (Simpson and Greenfield, 1971). Evidently, the thick single bar is impracticable

Table 9.3 Examples of Electrical Station Design

	American Electric
	500 kV
Minimum clearance (metal to metal)	
Phase-to-phase (m)	6.1
Phase-to-ground (m)	3.7
Above roadway inside substation (m)	7.9
Main bus parameters	
Current rating	3360 A
Type (rigid/strain)	Strain
Phase spacing (center to center, m)	7
Span (maximum distance between supports, m)	30.5 and 110
Basic insulation level (BIL) (kV)	1550
Conductors	
Number per phase	2
Diameter (cm)	4.19
Type	ACSR
Subconductor spacing (cm)	45.7
Height above ground at support (m)	8.9
Bay bus	
Current rating	3360 A
Type (rigid/strain)	Strain
Bay span (m)	30.5
Bay conductors	
Number per phase	2
Diameter (cm)	4.19
Type	ACSR
Subconductor spacing (cm)	45.7
Height above ground at support (m)	8.9

Power (AEP) 765 kV	Ontario Hydro 500 kV		Hydro Quebec 735 kV
10.7	6.5		12.2
6.1 (horizontal)	4		5.6
23.5 to roadway, 11 min. clearance to grade	10		16.7
3730 A	2770 A		2000 A
Rigid	Rigid		Strain/rigid
13.7	7.6		15.3
13.7	10.2		10
2050	1800		2200
1	1		2
14.7	11.43		5.87
Al tube	Al tube		Al
			38.1
23.5	14.3		25.9 and 10.7
	High level	Low level	
5350A	2860 A	1930 A	2000 A
Strain	Strain	Rigid	Strain/rigid
32.9	30.5		79.3
2	2	1	2
6.1	4.45	7.62	5.87
Al conductor	Al conductor	Al tube	Al conductor
45.7	33		38.1
38.1 and 12.2	28.4	7	25.9 and 10.7

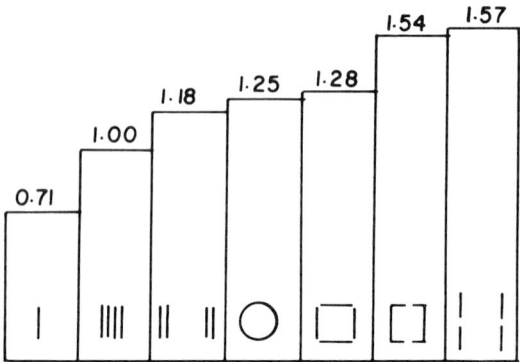

Fig. 9.8 Comparative ampacity ratings for various arrangements and shapes of busbar conductors.

for high currents, especially for ac use. The ampacity of a given busbar can be estimated using the following simple relation:

$$I^2 R = (W_c + W_r) A_s \quad \text{Watts} \quad (9.2)$$

where I is the conductor current (A), R the conductor ac resistance (Ω/m), W_c heat dissipated by convection (W/cm^2), W_r heat dissipated by radiation (W/cm^2), and A_s surface area (cm^2 per meter of busbar length).

For circular conductors, the skin effect causes its ac resistance to exceed its dc resistance, their ratio being

$$\frac{R_{ac}}{R_{dc}} = k \sqrt{\frac{\mu_r f}{\rho}} \quad (9.3)$$

where k is a constant listed in tables (Weiss and Csendes, 1982), f is the frequency (Hz), μ_r the relative permeability, and ρ resistivity (Ω/m^2). A more elaborate method is followed for computing the proximity effect using the same basic physical phenomena (Salama and Hackam, 1984).

Evidently, tubes have lower ac resistance than solid and flat conductors of the same net cross section. Tubes with thin walls are affected the least by skin effect. Because of the higher resistivity of aluminum, aluminum conductors are affected less by skin effect

High-Voltage Busbars

than are copper conductors of similar cross sections, as seen from equation (9.3).

The heat dissipated by convection W_c and radiation W_r can be estimated from the following relationships

$$W_c = \frac{5.73 \times 10^{-3} \sqrt{p\nu}}{T_a^{0.123} \sqrt{d}} \Delta\Theta \qquad \text{W/cm}^2 \qquad (9.4)$$

$$W_r = 5.7 \, E \left[\left(\frac{T}{1000}\right)^4 - \left(\frac{T_o}{1000}\right)^4 \right] \qquad \text{W/cm}^2 \qquad (9.5)$$

where

p = pressure in atmospheres ($p = 1.0$ for atmospheric pressure)

ν = surrounding air velocity (m/s)

T_a = average absolute temperature of conductor and ambient air (K)

d = outside diameter of cylindrical conductor (cm)

T = absolute temperature of the conductor (K)

T_o = absolute temperature of the ambient (K)

$\Delta\Theta$ = temperature rise of conductor above ambient ($T - T_o$) (K)

E = relative emissivity of conductor surface

= 1.0 for black body and about 0.5 for average oxidized copper and aluminum

For other conductor shapes, d is the outside diameter of an equivalent circular conductor of the same surface area. By calculating ($W_c + W_r$), A_s, and R it is then possible to determine I from equation (9.2). This method is generally applicable to both copper and aluminum conductors. Tests have shown that aluminum conductors dissipate heat at about the same rate as copper conductors of the same outside diameter when the temperature rise is the same.

9.6.2 Considerations for Short-Circuit Currents

An adequate conductor size is necessary for carrying the fault current caused by an external short circuit without its temperature exceeding the safe limit during the period from the instant of initiation of the fault until its interruption. For both copper and aluminum bus connectors carrying dc, the following equation was suggested:

$$I_f = \frac{\kappa A}{6.45 \sqrt{t}} \quad A \qquad (9.6)$$

where I_f is the dc fault current, t the time (s) from initiation to clearing of the fault, and A the sectional area of the busbar (mm^2). κ is a coefficient ranging between 600 and 750, depending on the conductor material, its maximum allowable temperature, and the ambient temperature (Simpson and Greenfield, 1971). For the case of copper busbars, the value of coefficient κ is about 70% higher. For ac the value of I_f obtained should be multipled by the ratio $(R_{ac}/R_{dc})^{-1/2}$.

9.7 MECHANICAL STRESSES ON BUSBAR CONDUCTORS

A bus conductor carrying a short-circuit current is subjected to electromagnetic forces that tend to pull adjacent bars to each other if they carry currents in the same direction and push them apart if the currents are in opposite directions (Fig. 9.9). This force increases quadratically with the current magnitude and depends on the shape and arrangement of conductors and the natural frequency of the complete assembly, including mounting structure and insulators (Craig and Ford, 1980; Awad and Huestis, 1980).

The electromagnetic force on either of the two conductors carrying currents of instantaneous values I_1 and I_2 and separated by a distance d could easily be evaluated as

$$F = \frac{2 I_1 I_2}{10^5 d} \quad N/m \qquad (9.7)$$

whether both forces are those of attraction or repulsion depends on the relative directions of the currents I_1 and I_2 (Fig. 9.9).

As a result of the electromagnetic forces between the bus conductors there will be a mechanical reaction on the bus supports. These forces are far greater than their ordinary loading caused by the conductors' weights and wind. The insulators should have enough mechanical strength and resilience.

The short-circuit current may reach excessive peak values, depending on the instant the short circuit occurs with respect to the ac voltage wave and the X/R ratio of the faulted circuit. This causes an offset in the current wave (Fig. 9.10). For example,

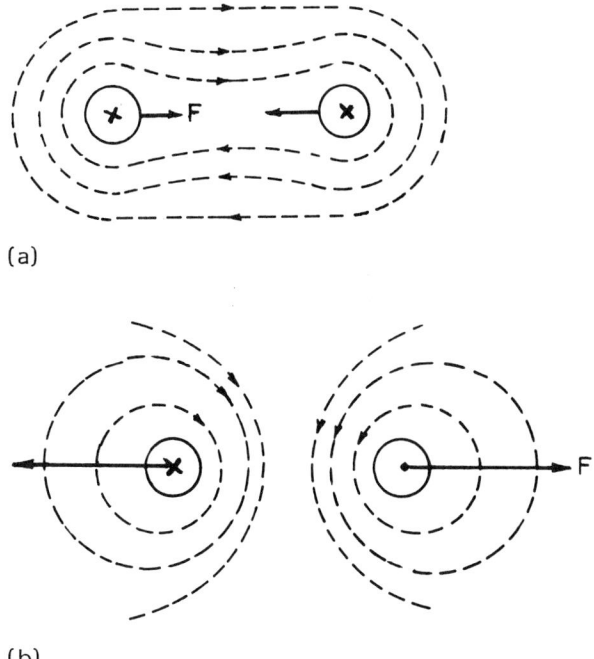

Fig. 9.9 Magnetic fields and electrodynamic forces acting on two parallel conductors carrying currents in the same direction (a) and when the currents are in opposite directions (b).

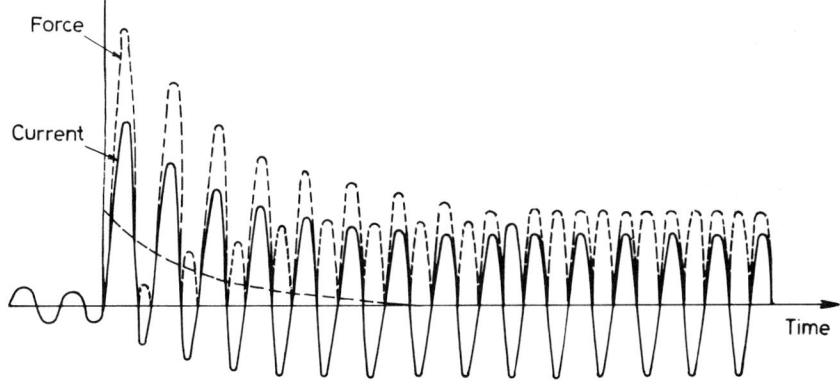

Fig. 9.10 Asymmetrical short-circuit current in a circuit with a high X/R ratio, and the corresponding electrodynamic force acting on the busbars.

Table 9.4 Maximum Instantaneous Short-Circuit Lateral Force Acting upon Bus Conductors

Bus bar arrangement	Type of circuit and kind of fault	Force, N/m
	DC or single-phase AC. Symmetrical	$F = 4\,I_{rms}^2/(10^5 d)$
	Single phase ac. Fully offset asymmetrical wave	$F = 16\,I_{rms}^2/(10^5 d)$
	3-phase AC conditions of asymmetry to give maximum force on any phase	$F = 13.9\,I_{rms}^2/(10^5 d)$
	3-phase AC	
	1. Fully offset asymmetrical wave in phase A	$F = 12\,I_{rms}^2/(10^5 d)$
	2. Conditions of asymmetry to give maximum force on A or C	$F = 13\,I_{rms}^2/(10^5 d)$
	3. Conditions of asymmetry to give maximum force on B	$F = 13.9\,I_{rms}^2/(10^5 d)$

I in amp., d in cm.

High-Voltage Busbars

if the X/R ratio is 20, the first current peak will be about 2.183 times the sustained rms fault current.

The electromagnetic forces set up between parallel conductors in which current varies as in Figure 9.10 produce force pulsations; the component at power frequency dominates for the first few cycles, but afterward the dominant component is at double the power frequency. A bus installation should be designed so that its natural frequency of oscillation or that of any of its parts or supporting structures is not close to the power frequency or its multiples, to evade resonant pulsations that might otherwise endanger the mechanical structure.

9.7.1 Electrodynamic Force Magnitudes

There are considerable differences among short-circuit forces depending on the type of fault and the bus conductor arrangement. The probability of a short circuit occurring on all three phases is very remote, yet it should be considered in the design of the busbars. The phase-to-ground and double-phase-to-ground faults are more likely. Also, during the operation of a circuit breaker clearing a short circuit, one phase current gets interrupted first and thus a three-phase fault turns into a double-line fault. The maximum instantaneous short-circuit lateral forces acting on bus conductors in various arrangements are listed in Table 9.4.

9.8 INSULATORS

Air-insulated busbars are supported if rigid, or if a strain type suspended by insulators. These insulators have traditionally been made of special porcelain or toughened glass. With the developments in polymeric materials, traditional insulators have for several decades been experiencing serious competition from insulators made of epoxy reinforced with glass fibers. Polymeric insulators usually weigh less than half or even a fourth of their corresponding porcelain insulators for the same mechanical and electrical strengths (Karady, Limmer, 1979).

Busbar insulators should have adequate electrical and mechanical strength under the expected ambient and operating conditions of temperature variations, rain, pollution, and normal and abnormal loadings. The abnormal loadings include electrodynamic forces occurring during short circuits on the busbars as described above. Acceptable electrical and other qualities of insulators are usually set by national and international standard specifications (see, e.g., IEC, 1979).

Insulators for outdoor use should pass high-voltage tests under simulated rain conditions. Also, insulators to be employed in sites of significant atmospheric pollution should be designed and tested accordingly, as discussed below.

For the higher operating voltages, more insulator units are stacked or connected in a chain, whereas for greater mechanical strength more insulator chains are bolstered together in "parallel." "Long rod" insulators can replace chains of short "cap-and-pin" units.

9.8.1 Insulator Pollution

When pollution accumulates on insulator surfaces and its soluble ingredients dissolve in the moisture accumulated from fog or dew, a conducting path, however resistive, gets established on the insulator surface all the way from the high-voltage conductor to the grounded metal frame. Thus leakage currents flow. With non-uniformities in pollution and leakage currents naturally occurring over insulator surfaces, hot spots form. A localized spot dried up by concentrated leakage currents pushes the current paths sideways and thus extends laterally to form a dry band. Such a dry band, with its drastically increased resistance, takes a major fraction of the total insulator voltage. A partial arc could strike across the band. With partial arcs burning on one unit of an insulator chain, additional voltage stresses appear over other units. The voltage distribution over the chain comes to be far from uniform (Aoshima et al., 1981). This may eventually develop into a complete flashover of the whole insulator chain under normal operation voltage. Alternatively, the current through the partial arcs may be too small to maintain them. In such a case, the wet contaminated areas of the insulator surfaces dry up gradually and the partial arcs get extinguished. Intensive research has been going on in many countries for several decades (Khalifa et al., 1988; Houlgate et al., 1982; Pei-Zhong and Cheng-Dong, 1982).

Rain does not always provide the cleaning needed to combat pollution, particularly under conditions of heavy marine or industrial pollution, and when rain is not regular enough; desert pollution is a case in point (El-Koshairy et al., 1982). Measures usually taken to improve the insulation strength in polluted atmospheres include using "anti-fog" insulators with extended leakage paths and insulators with semiconducting glaze. Also, greasing their surfaces prevents the formation of continuous conducting films. Combating pollution of high-voltage insulators has also been achieved successfully by regular washing of the insulators while under high voltage, thus maintaining power supply continuity. To help improve system performance, pollution monitors have been developed so that the insulator cleaning operation would be carried out only when necessary (Khalifa et al., 1988).

REFERENCES

Aoshima, Y., Tatsuya, H., and Keiji, K. (1981). *IEEE Trans.*, *PAS-100*: 948–955.
Awad, M. B. and Huestis, H. W. (1980). *IEEE Trans.*, *PAS-99*: 480–487.
Conway, B. J. (1979). *IEEE Trans.*, *PAS-98*: 1384–1402.
Craig, B. and Ford, G. L. (1980). *IEEE Trans.*, *PAS-99*: 434–442.
El-Koshairy, M. A. B., El-Sharkawi, E., Awad, M., Zarzoura, H., Khalifa, M., and Nosseir, A. (1982). *CIGRE paper 33-09*.
El-Morshedy, A. M. K. (1982). *Proc. IEE*, *129*: 199–205.
Houlgate, R. G., Lambeth, P. J., and Roberts, W. J. (1982). *CIGRE paper 33-01*.
IEC (1979). *Tests on Indoor and Outdoor Post Insulators for Systems with Nominal Voltages Greater Than 1000 V*, Publication 168, International Electrotechnical Commission, Geneva.
IEEE Working Group (1983). *IEEE Trans.*, *PAS-102*: 513–520.
Karady, G., Limmer, H. D. (1979). *Proc. American Power Conference*, pp. 1120–1125.
Khalifa, M., El-Morshedy, A., Gouda, O. E., and Habib, S. E.-D. (1988). *Proc. IEE*, *135C*: 24–30.
Pei-Zhong, H. and Cheng-Dong, X. (1982). *CIGRE paper 33-07*.
Rizk, F. (1976). *IEEE Trans.*, *PAS-95*: 1892–1900.
Salama, M. and Hackam, R. (1984). *IEEE Trans.*, *PAS-103*: 1493–1501.
Simpson, T. W. and Greenfield, E. W. (1971). *Electrical Conductor Handbook*, The Aluminum Association, New York.
Weiss, J. and Csendes, Z. (1982). *IEEE Trans.*, *PAS-101*: 3796–3803.

10
Gas-Insulated Switchgear

H. ANIS *Cairo University, Giza, Egypt*

10.1 INTRODUCTION

Gas-insulated switchgear (GIS) that uses compressed SF_6 gas overcomes many of the limitations of the conventional open-type HV switchgear, as it offers the following advantages:

1. The space occupied by the switchgear is greatly reduced.
2. It is totally unaffected by atmospheric conditions such as polluted or saline air in industrial and coastal areas, or desert climates.
3. It possesses a high degree of operational reliability and safety to personnel.
4. It is easier to install in difficult site conditions (e.g., on unstable ground or in seismically active areas).
5. In addition to having a dielectric strength much greater than that of air, SF_6 has the advantages of being nontoxic and nonflammable.

10.2 CHEMICAL AND PHYSICAL PROPERTIES OF SF_6

At atmospheric pressure, SF_6 has a dielectric strength two to three times that of air, and this ratio increases with increasing pressure. At 3 atm, the dielectric value is about the same as that of transformer oil. In addition to this superior dielectric property, SF_6 has the following basic physical and chemical properties:

1. It is nontoxic, having almost the chemical properties of a noble gas. It can, however, cause suffocation if it displaces oxygen

Table 10.1 Physical Properties of SF_6

Molecular weight	146.05
Melting point	−50.8°C
Sublimation temperature	−63.9°C
Density (liquid)	
At 50°C	1.98 g/mL
At 25°C	1.329 g/mL
Density (gas at 1 bar and 20°C)	6.164 g/L
Critical temperature	45.6°C
Critical pressure	36.557 atm
Critical density	0.755 g/mL
Surface tension (at −50°C)	11.63 dyn/cm
Thermal conductivity ($\times\ 10^4$)	3.36 cal/s/cm^2/°C/cm
Viscosity (gas at 25°C $\times\ 10^4$)	1.61 poise
Boiling point	−63.0°C
Specific heat (at 30°C)	0.143 cal/g
Relative density (air = 1)	5.10
Vapor pressure (at 20°C)	10.62 bar

Source: Maller and Naidu (1981).

    from an area because of its much higher density than that of air.
2. It is nonflammable.
3. It is noncorrosive. It does not react with other materials, as it is inert. Furthermore, it inhibits surface erosion and oxidation. Only when SF_6 gas is heated above 500°C will it decompose, and then its decomposition products react with hydrogen and other materials. With special filters, however, it is possible to alleviate this effect (Maller and Naidu, 1981).
4. It has a high partial vapor pressure at both normal and low temperatures.

Gas-Insulated Switchgear

5. It has excellent heat transfer characteristics. Its high molecular weight and low viscosity enable it to transfer heat by convection more effectively than do common gases.

Furthermore, SF_6 gas exhibits excellent properties for arc quenching and is therefore also used as an interrupting medium in circuit breakers instead of oil or air. The use of SF_6 in circuit breakers is discussed in Chapter 11. Table 10.1 summarizes some of the basic physical properties of SF_6. Mixing a small amount of SF_6 gas with a relatively inexpensive gas such as nitrogen, hydrogen, carbon dioxide, or air increases the dielectric strength of the latter substantially.

10.3 LAYOUT OF GAS-INSULATED SWITCHGEAR

In GIS all live parts are enclosed in a compressed-gas system which is divided into a number of compartments. This division enables the isolation of one compartment for maintenance or repair purposes while the other compartments remain pressurized. In Figure 10.1a the single-line diagram of a double-busbar arrangement is shown. In Figure 10.1b the diagram is redrawn in the form of a typical gas GIS circuit breaker bay. Basic components that make up any one GIS bay are the circuit breaker, disconnectors, earthing switches, busbars, and current and voltage transformers. The implementation of the foregoing arrangement into a real GIS depends on the voltage level. Figures 10.2 and 10.3 show sectional views of the general arrangements of two different gas-insulated switchgear bays belonging to the ranges 145 kV and 220 to 800 kV, respectively.

10.3.1 GIS Enclosure Configuration

The GIS enclosure forms an electrically integrated earthed casing for the entire switchgear. The GIS enclosure can be either of the three-phase type, as in Figure 10.2, or the single-phase type, as in Figure 10.3. The advantages of the three-phase common enclosure design are:

1. A smaller number of enclosures is required per feeder (one-third).
2. In the case of a three-phase common enclosure, an arc between phase and ground will within a few milliseconds evolve into a phase-to-phase fault between conductors, owing to ionization

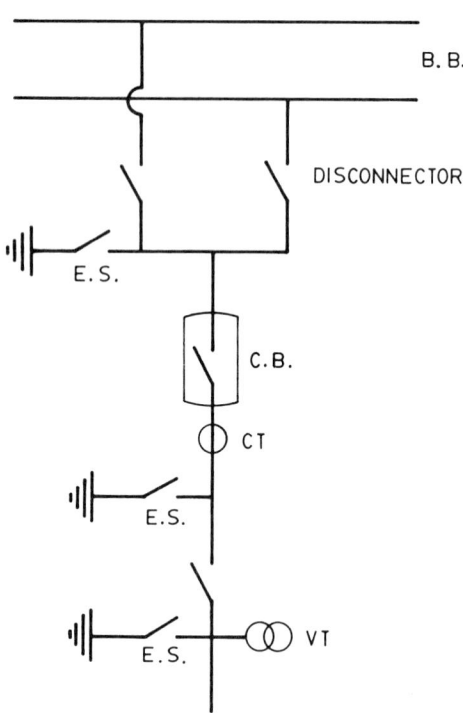

(a)

Fig. 10.1 (a) Single-line diagram of a feeder bay of a double-busbar system; (b) diagram of a typical GIS feeder bay showing gas-filled compartments. 1, Busbar i; 2, busbar ii; 3, disconnector; 4, current transformer; 5, circuit breaker; 6, voltage transformer; 7, maintenance earthing switch; 8, high-speed earthing switch; 9, gas connector; 10, density monitor.

of the gap, and at the same time the phase-to-ground arc will extinguish. Consequently, an enclosure burn-through is not possible.

3. For the same parameters (voltage level, conductor size, clearances between phases and phase-to-ground) the resultant field stress in a common three-phase enclosure is approximately 30% less than that in a single-phase enclosure and hence less probable of failure.

Gas-Insulated Switchgear

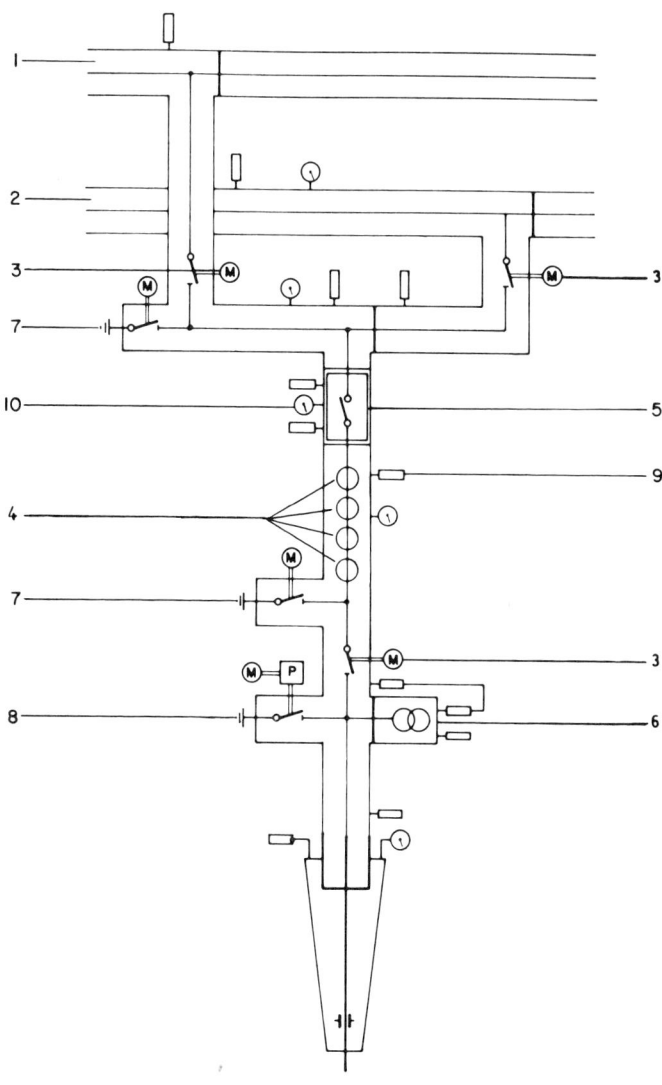

(b)

Fig. 10.1 (Continued)

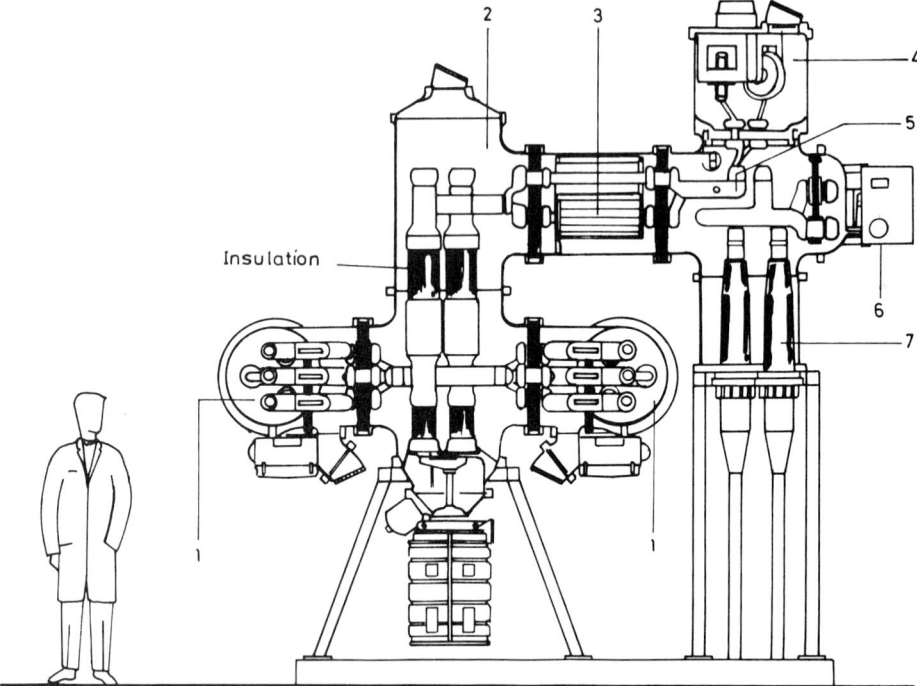

Fig. 10.2 Sectional view of an SF_6 gas-insulated, 145-kV circuit breaker bay using one enclosure for the three phases. 1, Busbar with combined disconnector/earthing switch; 2, circuit breaker; 3, current transformer; 4, voltage transformer; 5, combined disconnector/earthing switch; 6, high-speed earthing switch; 7, cable-end unit.

4. The absence of complicated tie rods and linkages between poles for the circuit breaker, isolator, and grounding switch drives simplifies the drive system.

However, three-phase enclosures are used only for voltages below about 200 kV. Above that level, insulation requirements necessitate the use of single-phase enclosure types.

10.3.2 Enclosure Material

Both aluminum and steel are used for SF_6 GIS enclosures. Both materials fulfill specific purposes. However, the use of steel for

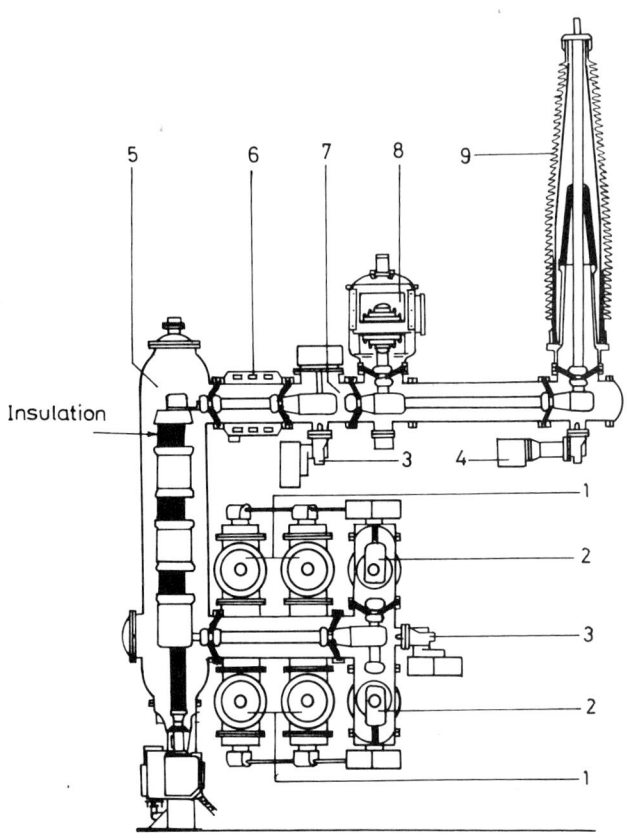

Fig. 10.3 Sectional view of a typical SF_6 gas-insulated EHV (220–800 kV) bay using single-phase enclosures. 1, Busbar; 2, busbar disconnector; 3, maintenance earthing switch; 4, high-speed earthing switch; 5, circuit breaker; 6, current transformer; 7, disconnector; 8, voltage transformer; 9, SF_6/air bushing.

single-phase enclosure GIS is, for example, limited to lower current densities because of the problem of electrical losses and heating in the enclosure resulting from induced circulating and eddy currents. In the case of a three-phase or single-phase enclosure high-pressure GIS, steel has the major disadvantage that it cannot be shaped and welded into the same homogeneous form as can be achieved with aluminum castings to ensure optimum field stress distribution for a given gas pressure and conductor configuration.

10.3.3 SF_6 Gas-Insulating Pressure

With respect to the SF_6 insulating gas pressure, there are two principal GIS designs: high-pressure GIS operating at about 4 bar (405 kPa), and low-pressure GIS, which operates at 1.2 bar absolute. While for a rated voltage of 72.5 kV and below, the low-pressure GIS is the more suitable design, at higher voltage levels the advantages of high-pressure GIS become dominant.

A higher SF_6 gas pressure results in higher dielectric strength. Consequently, for a given voltage level that required conductor spacing is reduced, resulting in a more compact design. On the other hand, the dielectric strength increases at a lower rate than the gas pressure. The reason behind this is the increasing sensitivity of SF_6 insulating gas to the roughness of conductor and enclosure surfaces and to contamination at higher gas pressures, which in turn can reduce the reliability and increase the requirement for servicing and maintenance. As the gas pressure is increased, so are the requirements for a homogeneous electric field distribution if the higher pressure is to be fully utilized. Consequently, for higher SF_6 gas pressures it becomes increasingly necessary to ensure a careful homogeneous shaping of conductors, components, and enclosures.

From the above it can be deduced that there is a limit to the SF_6 insulating gas pressure above which the economics of a design decrease as the disadvantages of the higher gas pressure dominate. By the same reasoning, for voltages 110 kV and above there is also a limit to the SF_6 insulating gas pressure below which the increasing production cost for larger enclosures and greater material expenditure as well as larger buildings make the design uneconomical.

Another important consideration when selecting the SF_6 insulating gas pressure is the behavior of the GIS when a leak occurs. A leakage resulting in a drop in the SF_6 insulating gas pressure automatically means a reduction in the dielectric strength and therefore the integrity of the switchgear. In the case of high-pressure GIS with a rated insulating gas pressure of approximately 4 bar, a severe gas leakage with a pressure drop down to 1 bar will result in a reduction of the dielectric strength, and therfore the BIL, by about 75% of the rated value. On the other hand, in the case of low-pressure GIS with a rated insulating gas pressure of 1.2 bar, the same leak will result in only a 15% deterioration in dielectric strength.

10.3.4 Conductor System

The conductors of a GIS normally consist of aluminum tubes, the diameter and wall thickness of which depend on voltage and rated current (Sections 10.6 and 10.6.3). Spring-loaded copper contact fingers constitute the female contacts and copper plugs the male contacts. The contact surfaces are silver plated and the contacts are welded to the aluminum conductors. The conductor system, together with supporting insulators, must be properly designed to withstand the electrical, thermal, and mechanical stresses that arise during normal service and during short-circuit conditions.

10.3.5 Solid Spacers

Solid insulators are used in gas-insulated apparatus for physical support of high-voltage conductors and for mechanical operation of the switchgear. They take various shapes, such as annular disks, truncated cones, and post supports. The presence of a solid insulating material in the compressed SF_6 medium distorts the field distribution in the GIS duct (Stone et al., 1987). Several problems are introduced by the use of spacers in GIS:

1. Spacers can limit the operating gradient (in kV/m) of the system, due to aging of the solid material.
2. Spacers can affect the short-term behavior of the GIS (i.e., its dielectric strength), due to a number of factors, as detailed in Section 10.7.3.

10.4 COMPONENTS OF GIS

In addition to circuit breakers (discussed in Chapter 11), a GIS bay includes disconnectors, voltage and current transformers, surge arresters, bushings and cable-end boxes, and gas density monitors.

10.4.1 Disconnectors

Disconnectors are made up from insulators, enclosures, and conductors of different geometrical shapes to give an optimum layout, as shown in Figures 10.2 and 10.3 and in detail in Figure 10.4. They are equipped with copper contacts that are spring loaded to give the disconnector high electrical efficiency and high mechanical reliability. Disconnectors must be carefully designed and tested to be able to break small charging currents without generating too-high overvoltages, otherwise a flashover to earth. The operating mechanisms of the disconnectors and earthing switches are of the same design for most GIS. The main features are motorized or manual operation, electrical interlocking against incorrect operation, and mechanically lockable end positions.

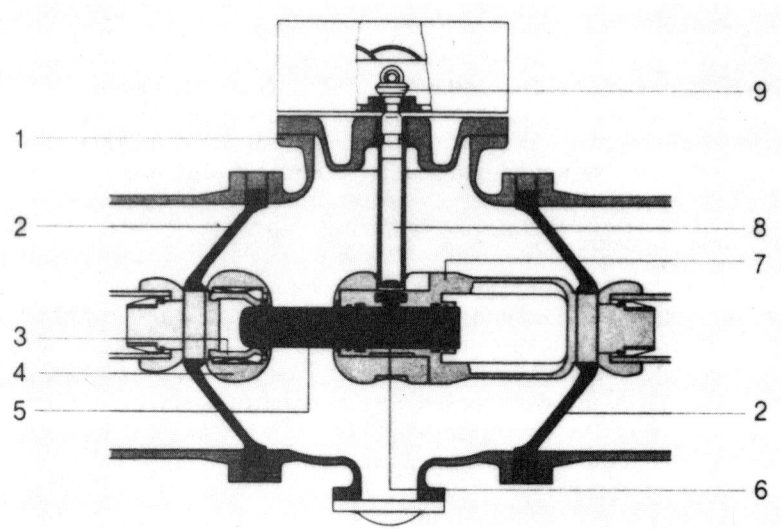

Fig. 10.4 GIS disconnector: (1) enclosure, (2) barrier insulator, (3) fixed contact, (4) shielding for fixed contact, (5) moving contact, (6) rack-and-pinion drive, (7) contact support, (8) insulated drive shaft, (9) driving motor.

10.4.2 Earthing Switches

Two different types of earthing switches are normally used, the slow-operating earthing switch and the fast-closing (high-speed) earthing switch (see Figs. 10.2 and 10.3). Slow-operating earthing switches are used for protection purposes when work is being done in the substation, but it is operated only when it is certain that the high-voltage system is not energized. The fast-closing earthing switch can close against full voltage and short-circuit power. The fast-closing operation is achieved by means of a spring-closing device.

10.4.3 Voltage Transformers

The most commonly used voltage transformer is of the inductive type (Fig. 10.5). In the three-phase enclosed GIS designs, three voltage transformers are placed in one enclosure. It is also possible to design a voltage transformer consisting of a low-capacitive voltage divider connected to an electronic amplifier. The capacitance between the inner conductor and a concentric measuring electrode

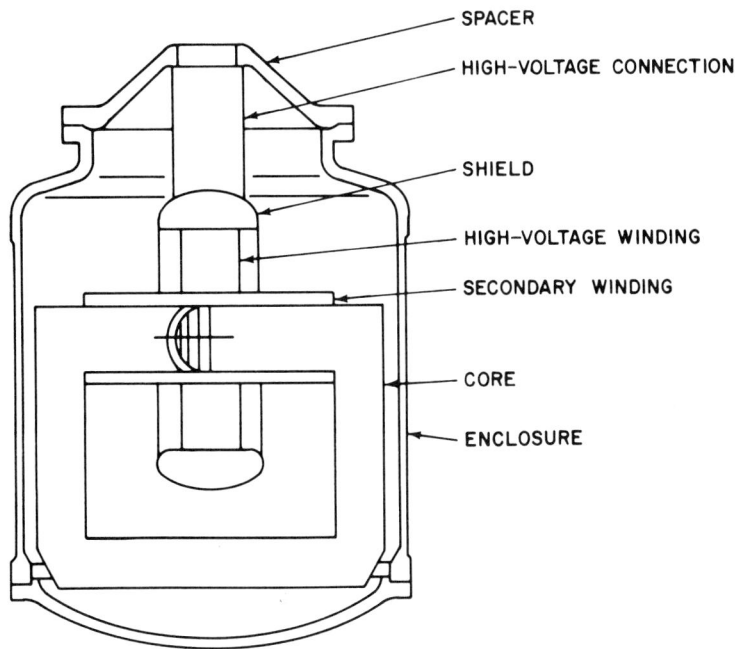

Fig. 10.5 GIS voltage transformer.

near the enclosure is then used as the high-voltage capacitor. This design is suitable only for the highest system voltages (Reisinger et al., 1987).

10.4.4 Current Transformers

In the single-phase enclosed GIS, the core of a current transformer is located outside the enclosure, thus ensuring a completely undisturbed electrical field between the enclosure and the conductor. The return current in the enclosure is broken by an insulating layer. Figure 10.6 shows the single-phase enclosed arrangement. In the three-phase enclosed GIS design the cores of the current transformers are normally located inside the enclosure (Fig. 10.2). They are preferably placed outside on the SF_6 bushings or on cables.

10.4.5 Surge Arresters

The SF_6 gas-insulated arresters are based on the same active parts as conventional arresters (i.e., varistors and spark gaps) but with

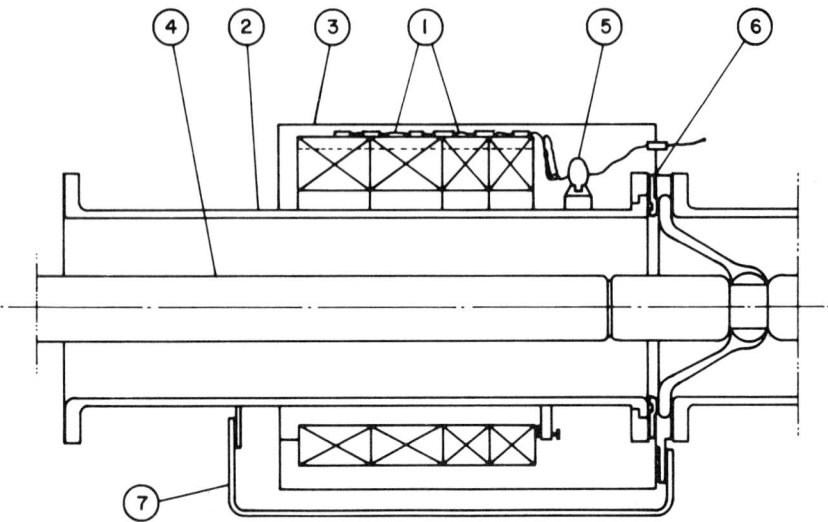

Fig. 10.6 GIS current transformer: (1) core unit, (2) enclosure, (3) cover, (4) HV conductor, (5) secondary connection, (6) insulation layer, (7) short-circuiting bar.

very compact designs. The spark gap elements being sealed off from the atmosphere, and the entire arrester being insulated with dry compressed gas creates a highly consistent performance within tolerances. It is therefore possible to use arresters with lower sparkover voltages, which provides a closer margin for protecting the system insulation. The reduction can reach 10% compared to conventional arresters.

In SF_6 gas-insulated arresters the metallic earthed parts are much closer to live parts, exploiting the high insulation strength of compressed gases. Thus capacitances to earth of live parts are far greater than in conventional arresters. Therefore, extra precaution is needed to compensate the resulting nonlinearities in voltage distribution along the components of gas-insulated arresters. This is achieved in their design (e.g., by including a metallic hood) (Fig. 10.7). The shape and dimensions for such a hood necessary for linearizing the field distribution along the arrester elements can be determined by field computations (Chapter 2). Figure 10.7 shows an arrester with a rated voltage of 120 kV. In arresters with higher rated voltages the hood is supplemented with a number of grading metal rings fitted at certain points along the arrester's active parts. SF_6 gas-insulated arresters can, of course, be integrated into the GIS in any desired position, depending on the protection requirements.

Gas-Insulated Switchgear

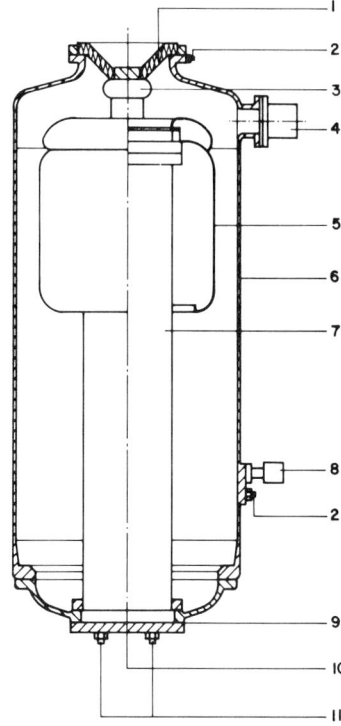

Fig. 10.7 GIS-enclosed 120-kV arrester: (1) support and barrier-insulator, (2) SF$_6$ gas connections, (3) HV connection, (4) bursting plate connection, (5) hood for field control, (6) metal cladding, (7) insulating tube, (8) manometer, (9) insulation, (10) earth connection, (11) N$_2$ gas connection.

10.4.6 Bushings

Overhead lines and all-air insulated components are connected to the SF$_6$ GIS by air/SF$_6$ gas-filled bushings. These bushings (Fig. 10.3) employ capacitive grading and are divided into two independent gas compartments by a barrier insulator.

The space surrounded by the porcelain insulator is filled with SF$_6$ gas to slightly above atmospheric pressure. If the porcelain is damaged, this reduces the risk down to a minimum. The gas space on the switchgear side of the barrier insulator has the same SF$_6$ gas pressure as the switchgear. Oil-filled condenser bushings can also be used for high voltages. Bushings used for the direct connection of a GIS to a transformer may also be of the oil-filled type.

10.4.7 Cable-End Boxes

High-voltage cables of the various types are connected to the SF$_6$ switchgear via cable-end boxes (Fig. 10.2), which consist of the cable-end bushing with connecting flange, the enclosure, and the barrier insulator with female plug contacts. The pressure-tight bushing separates the SF_6 gas compartment from the insulating medium of the cable. A completely dry cable termination for connecting XLPE cables to GIS has the advantages of smaller dimensions and better thermal properties.

10.4.8 Gas Density Monitors

The dielectric strength of the switchgear insulated with SF_6 gas and the breaking capacity of the SF_6 circuit breaker depend on the density of the gas. Since the pressure varies with temperature, it is the gas density that is monitored. For this, a density monitor is employed. The gas compartments separated by barrier insulators (Fig. 10.1b) are each monitored by its own density monitor.

10.5 COMPRESSED GAS-INSULATED CABLES

Although overhead transmission still represents the most economical solution to bulk power transmission, it is becoming more difficult to construct overhead lines in densely populated areas, and underground cables are then required (Chapter 12). SF_6 compressed gas-insulated cables present possibilities for underground transmission of high power with rated voltage above 123 kV. The range of application of compressed gas-insulated cables extends from short links, to transmitting energy from cavern-type power stations over long transmission distances. Conventional cables are adequate for rated voltages up to about 150 kV over short distances and without extremely high rated currents. The application range of these cables can be expanded only by forced cooling.

For higher voltages it will become increasingly more difficult to dissipate the heat losses since the insulation thickness, and hence the internal thermal resistance, will be necessarily increased. Also, the dielectric loss increase with voltage—raised to an exponent of about 1.4—can hardly be contained within acceptable limits. The compressed-gas cable is an attractive alternative in this case.

10.5.1 Comparative Conductor Size

The active losses of a transmission line depend primarily on the cross section of the conductor. Additional active losses occur in cables and compressed gas-insulated cable systems due to loss currents in the metallic enclosure and to dielectric losses (Chapter 12). The cross sections of individual transmission media that can be utilized in practice are determined largely by their active resistance, which also influences their range of application. Figure 10.8 shows a comparison of cross sections that can be produced in practice.

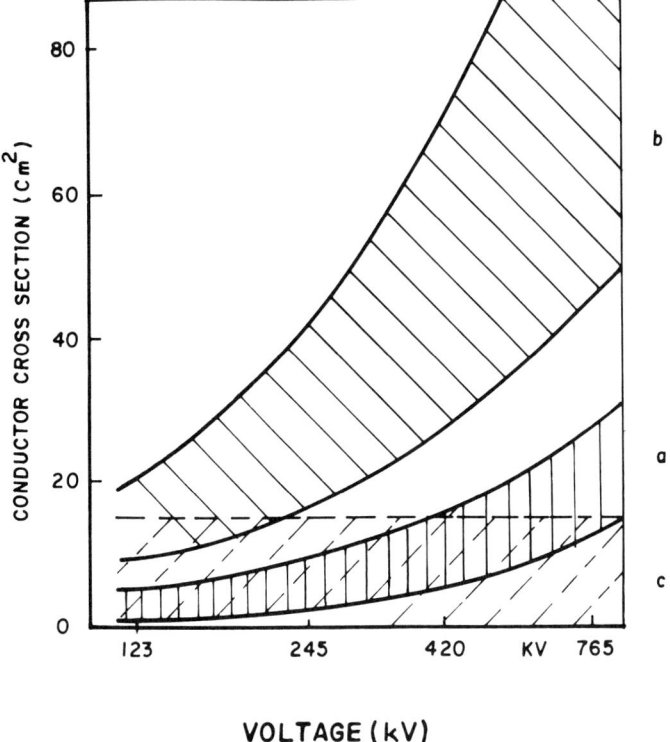

Fig. 10.8 Comparative conductor cross sections as functions of rated voltage: (a) overhead lines, (b) compressed-gas cables, (c) conventional cables.

Overhead lines are generally aluminum-stranded conductors with cross sections of 240 to 500 mm^2. This relatively small range is, however, increased using bundled conductors for higher voltages. Furthermore, the amount of reactive losses is also superior to those of compressed gas-insulated cables, which is particularly important when the transmission distance is long.

Whereas conventional cables show heavy capacitive losses that only slightly changes with loading, a compressed gas-insulated cable changes from capacitive to inductive losses as loading is increased. A point is passed where the ideal operating condition is reached (i.e., at zero reactive losses). The surge impedance load of compressed gas-insulated cables is three to four times smaller than that of overhead lines. Therefore, the transmission properties of a compressed gas-insulated cable correspond to those of three to four overhead lines operating in parallel (Chapter 14).

10.6 DIMENSIONING OF COMPRESSED GAS ENCLOSURE

The dimensions of the enclosure are generally decided by considering the duct to be a coaxial cylinder arrangement with conductor of outer radius r and enclosure of inner radius R. As explained in Chapter 2, the electric field at the surface of the conductor is

$$E_r = \frac{V}{r \ln(R/r)} \tag{10.1}$$

where V is the applied phase voltage.

As shown in Chapters 2 and 12, E_r is minimum (i.e., ionization is least likely to develop) when

$$\frac{R}{r} \sim e \sim 2.72 \tag{10.2}$$

If this relation is combined with knowledge of the breakdown threshold of the gas (i.e., the maximum allowable value for E_r), the enclosure dimensions would be determined. For coaxial cylinders, several formulas for the breakdown threshold have been published. One formula sets the threshold under positive impulse voltages at

$$E_{th} = 81.2p + 10 \quad \text{kV/cm} \quad 0.1 < p < 0.4 \text{ MPa} \tag{10.3}$$

For a 420-kV GIS whose basic impulse insulation level (BIL) is 1425 kV and operating at 0.35 MPa, equations (10.1) to (10.3) give r = 4.8 cm and R = 13.2 cm. Values obtained in this way are somewhat less than what should be used in practice where particle contamination and surface roughness have their adverse effects.

10.6.1 Condensation Threshold

Another important factor in the dimensioning of GIS ducts is that condensation of the gas should be avoided, typically at the lowest ambient temperature. According to Figure 10.9, for a given minimum temperature there exists an upper limit of gas density above which condensation is sure to occur (e.g., a gas density of 60 g/L at −20°C). This maximum allowable gas density, in turn, corresponds to a maximum allowable withstand electric field as shown

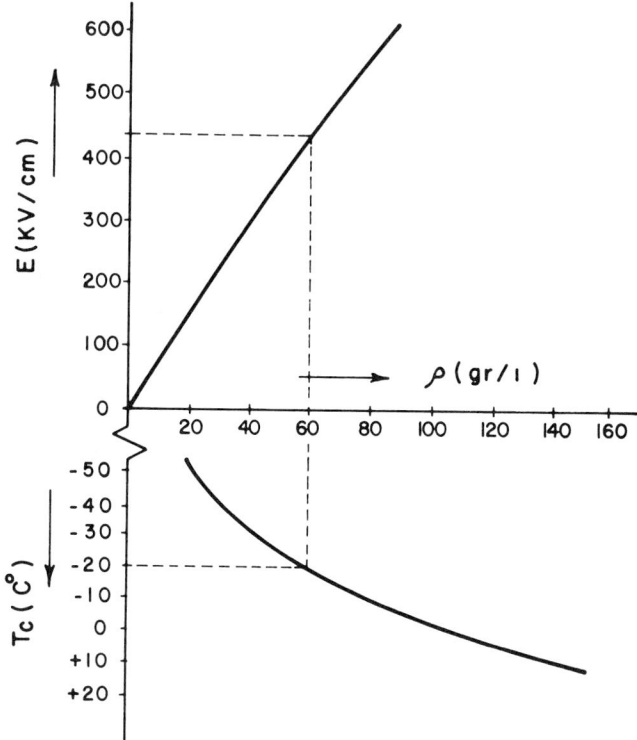

Fig. 10.9 Relations among the impulse breakdown strength E of SF_6, the gas density ρ, and condensation temperature T_c.

in Figure 10.9 (e.g., 430 kV/cm at 60 g/L). Finally, the electric field in the gas can be ensured not to exceed this maximum only if a lower limit is imposed on the duct dimensions according to equation (10.1). For the current example the field is maintained below 430 kV/cm for a typical system BIL of 1425 kV if the conductor and inner duct radii are larger than 3.3 and 9.0 cm, respectively.

10.6.2 Spacer Flashover

Solid support insulators and spacers represent a parallel insulation to the compressed gas insulation, and hence breakdown may, in general, materialize in either insulation. Under lower gas pressures a flashover across a solid insulator is highly unlikely to occur prior to a breakdown in the gas and therefore poses no design problem. At higher gas pressures, in contrast, a flashover across an insulator is likely since the dielectric strength of the gas is much increased while that of a solid insulator is nearly unaffected by increasing gas pressure.

The surface flashover strength of cast resin, the material of which solid supports are usually made, is about 30 kV/cm (rms). The relation between the applied phase voltage and the maximum field on a supporting insulator depends largely on the insulator shape. Efforts have been made to reduce the maximum field down to only 1.2 of its average value along the insulator (Stone et al., 1987). Therefore, the maximum field on the insulator is related to the phase voltage by

$$E_m = \frac{1.2 V_{ph}}{R - r} \qquad (10.4)$$

If E_m is equated to the reported dielectric strength of 30 kV/cm, for a 420-kV GIS,

$R - r = 9.7$ cm

If the optimal relation between the inner and outer radii, equation (10.2), is considered, the minimum dimensions of the system can be determined as

$r = 5.6$ cm

$R = 15.3$ cm $\qquad (10.5)$

10.6.3 Heat Dissipation Considerations

While the GIS duct diameter computed on an insulation basis increases more or less linearly with rated system voltage, the rated current increases much less than linearly with voltage. At lower system voltages, therefore, the duct diameter may be required to be larger than that determined on an insulation basis to limit the temperature rise in the duct.

The heat generated by the "copper" losses in both the conductor and the duct wall will be dissipated to the atmosphere by radiation and convection. By increasing the thicknesses of both the hollow conductor and the duct wall, more current can be passed in the system without excessive temperature rise.

Gas-Insulated Switchgear

10.7 FACTORS AFFECTING INSULATION STRENGTH

The level of dielectric strength of compressed SF_6 outlined in the foregoing sections may not be fully attainable in reality. Several factors are known to have an effect on the dielectric strength of SF_6 when used in an actual GIS. In the following sections the effects of those factors are discussed.

10.7.1 Effect of Electrode Material

Under very high pressures when the breakdown strength exceeds about 200 kV/cm, the material of the electrodes in an SF_6 gap begins to have an influence over breakdown. The uniform field breakdown strength is larger with steel electrodes than with copper. The dependence of breakdown strength on electrode material is especially evident in uniform and quasi-uniform field gaps.

10.7.2 Conductor Conditioning and Surface Roughness

Particularly at high SF_6 gas pressures, repeated breakdowns affecting the electrode surfaces have their influence on the strength of the gap. The dielectric strength increases with the number of previous breakdowns until it levels off at an ultimate value. This phenomenon is explained by the conditioning of the electrodes. The effect of conditioning on breakdown in SF_6 was reported to depend on the gas pressure and the electrode area (Anis and Ward, 1988).

The breakdown strength of an SF_6 gap no longer has a single value at a given gas pressure. The fluctuation in the breakdown strength is thus better expressed by its statistical distribution. The dispersion is larger at the higher pressures. The breakdown strength is also reduced by the roughness of the electrode surface. It has been found that the reduction in the breakdown voltage due to electrode roughness depends on the type of gas insulation as well as the product of the protrusion height R_{max} and gas pressure p. When the product of the gas pressure and the protrusion height exceeds 0.8 kPa·cm, the protrusion causes a decrease in the breakdown voltage. However, the effect is negligible if the product is less than 0.4 kPa·cm. The reduction in the maximum withstand stress E related to the gas pressure p as a function of the product pR_{max} is shown in Figure 10.10.

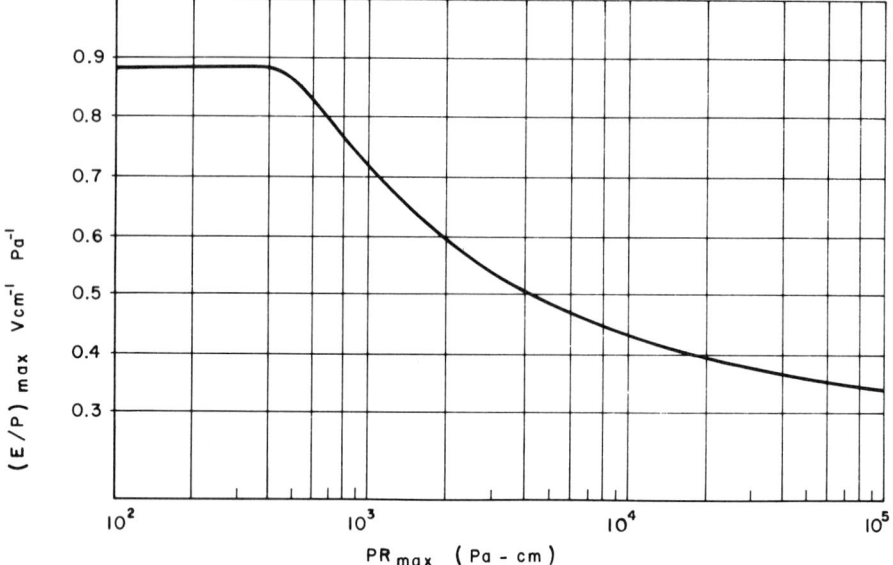

Fig. 10.10 Effect of surface protrusions on the strength of SF_6.

10.7.3 Problems Associated with Solid Spacers

Several factors contribute to the reduction of spacers' dielectric strength:

1. *Imperfect solid conductor adhesion.* Narrow gaps may exist between electrodes and spacers due to imperfect casting and/or mechanical stresses at the solid/gas interface. Studies of the effect of small gas gaps between the spacer and conductors on breakdown performance showed that the reduction in breakdown strength is more pronounced at higher gas pressures. As explained in Chapters 2 and 8, a small gas gap stressed in series with insulation of a dielectric constant ε_r would undergo an electric stress ε_r times higher than the average. Thus a microdischarge is then sure to occur in the gap.
2. *Moisture content.* The presence of some degree of moisture adversely affects the dielectric strength of the spacer/gas interface. It was reported that the breakdown voltage dropped by more than 50% when the humidity was increased from zero to 4000 ppm (Masetti et al., 1982).

3. *Contaminating particles*. Contaminating particles may eventually adhere to the solid spacer by electrostatio forces, causing local field enhancement. In the following section the subject of particle contamination in GIS is discussed in more detail.

As mentioned in Section 10.6.2, the breakdown in a high-pressure GIS is more likely along the gas/spacer interface, unlike low-pressure GIS, where breakdown is purely gaseous. Efforts have been made to optimize the design of post-type spacers based on minimizing the field enhancement in the gas at spacer boundary while accounting for resultant electrodynamic forces on the conductors (Stone et al., 1987; Trinh et al., 1984).

10.7.4 Particle Contamination in GIS

The presence of particle contaminants in gas-insulated systems is by far the most significant factor responsible for the deterioration of insulation integrity. The effect of metallic particles on the SF_6 breakdown voltage is more pronounced at high gas pressures, as shown in Figure 10.11 for a 150/250-mm coaxial system and 0.4-mm-diameter wire contaminants.

In the case of free particles, the particle motion depends largely on the type of applied voltage. Under ac voltage, for a wire particle of given radius, the activity increases with particle length since the particle charge-to-mass ratio at lifting increases with length. In addition, lateral particle movement along the central conductor is possible. Under dc voltages, the particles oscillate between the electrodes, and wire particles may exhibit intense "firefly" activity.

10.7.5 Particle-Initiated Breakdown

Under the influence of the applied field, free conducting particles become charged and oscillate in the interelectrode gap. As a charged particle approaches either electrode, it may lose its charge to the electrode through a microdischarge in the gas. This microdischarge may well trigger the breakdown of the main gap (Cooke, 1978).

Particle-initiated breakdown in compressed gas insulation generally depends, among other parameters, on the position of the particle in the gap. It has been reported that for a given particle-contaminated GIS system at some particle positions in the gap, the breakdown voltage is lower than for others (Cooke et al., 1977). An attempt was made to establish an analytical "breakdown voltage profile" that relates the instantaneous breakdown voltage magnitudes to the instantaneous particle position in the interelectrode gap (Anis

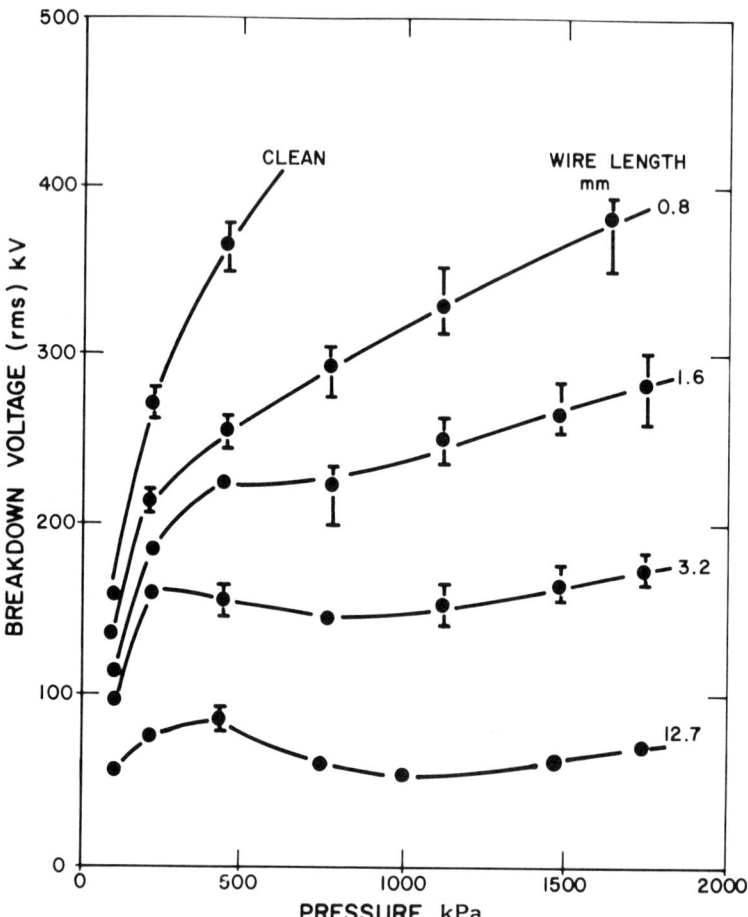

Fig. 10.11 Effects of conducting wire particles of different lengths on the strength of an SF_6 gap at different pressures. [From Cookson and Farish (1973).]

Gas-Insulated Switchgear

and Srivastava, 1981a). For a 1-mm-radius spherical particle in a parallel-plane SF_6 gap at a gas pressure of 5 atm, the breakdown voltage profile is illustrated in Figure 10.12. The profile explains the experimentally discovered existence of a critical particle position at which the insulation displays minimum dielectric withstand. This "critical" breakdown voltage may be taken as the insulation withstand limit under transient as well as steady-state applied voltages. The breakdown voltage profile helped explain the significant difference between the breakdown voltages of fixed and free particles, and the general effects of particle size, gas pressure, and electrode configuration.

10.7.6 Particle Control Techniques

For systems to be reliable and economical, the problem of particle contamination should be overcome. Contamination control in GIS can be achieved either by designing the system with a certain degree of immunity against the presence of particles, or by providing designated low field areas in the system in the form of "particle traps" where the particles can be safely trapped and contained. An electrostatic particle trap could be designed such that a region of low or zero electric field is provided at the outer enclosure by a slightly elevated metal shield. Particles can enter the low-field region created by the trap through slots in the shield. There are several important factors to be considered for particle control with such traps:

1. Particle contaminants have to be moved through the GIS system to the location of the traps. This must be achieved without risking damage to the system from flashovers.
2. The trap must provide rapid capture of the particles without any restriction on particle size and shape.
3. The trap must have positive retention characteristics under ac, dc, and transient voltages as well as mechanical vibrations.

10.7.7 Conductor Coating

Conductors in a gas-insulated system may be coated with a dielectric material to restore some of the dielectric strength of the compressed gas that is lost due to surface roughness and contamination with conducting particles. The improvement in dielectric strength of the system, due to coating, can be attributed to the following effects:

1. Coating reduces the degree of surface roughness on conductors, thus decreasing the high local electric fields (Endo et al., 1983).

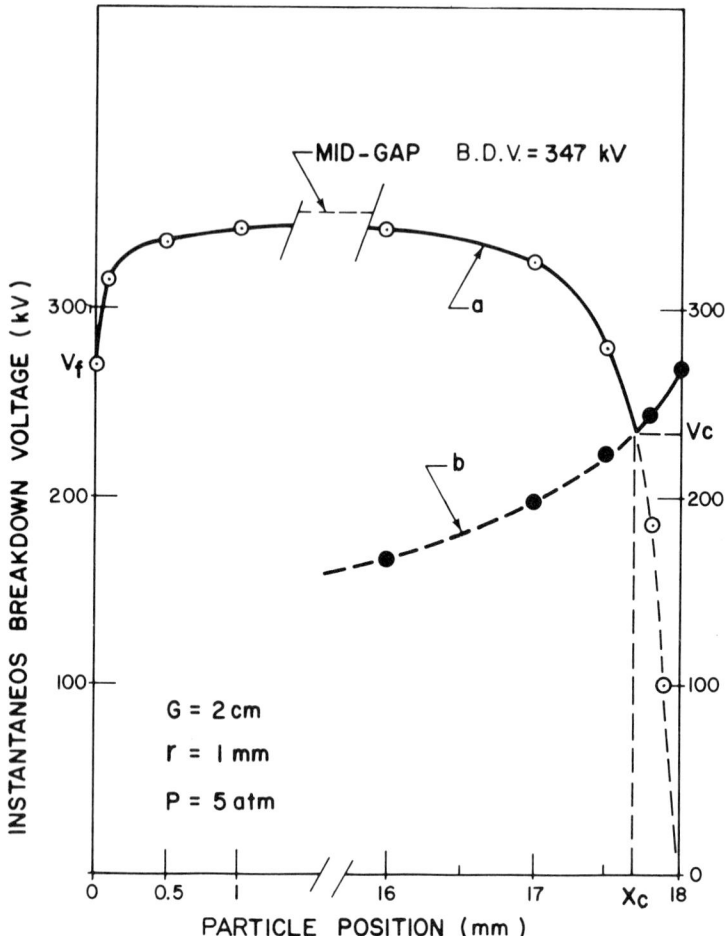

Fig. 10.12 Breakdown voltage profile of a particle-contaminated SF_6 2-cm gap, showing the effect of a 2-mm spherical particle position along the gap: (a) locus of breakdown voltage between a particle and the right-hand electrode, (b) locus of breakdown voltage between a particle-caused protrusion on the right-hand electrode and the left-hand electrode.

2. The high resistance of the coating dielectric impedes the development of pre-discharges in the gas, thus increasing the breakdown voltage.
3. In the presence of metallic contaminating particles, dielectric coating can be of further benefit in two main ways:
   a. The electric field necessary to lift a particle resting on the bottom of a GIS enclosure is much increased due to the coating (Parekh et al., 1979).
   b. Once a particle begins to move in the gas gap under the applied voltage, it may collide with either conductor. If the conductor is coated, the particle will acquire a drastically reduced charge, if any. Thus the risk of a breakdown initiated by a discharge is reduced significantly.

A particle resting on the bottom of a GIS duct acquires the charge necessary for lifting either by conduction through the dielectric coating or by means of microdischarges in the gas initiated at the particle surface.

Particle Charging Through Coating Layer Conductance

The dielectric coating layer has a finite nonzero, conductance, and therefore a conductive charging current will pass. The charge accumulated on the particle should reach a value at which the electrostatic lifting force on the particle overcomes the gravitational force (Anis and Ward, 1988). Using this mechanism, it can be shown that the lifting field E_1 and the corresponding lift-off charge q_1 of a filamentary particle in a coaxial GIS under ac configurations are

$$E_1 = \left[\frac{mg}{R \ln(R/r) A \alpha(\phi)}\right]^{1/2} \tag{10.6}$$

and

$$q_1 = \left[\frac{(1 + C_d/C_g)^2}{G^2} + \frac{1}{\omega^2 C_g^2}\right]^{-1/4} \frac{[mgR \ln(R/r)]^{1/2}}{\omega k} \Psi(\phi) \tag{10.7}$$

where

$$\Psi(\phi) = \left\{ \frac{\cos\phi - \cos[(2\pi + 2\phi)/3]}{\sin[(2\pi - \phi)/3]} \right\}^{1/2}$$

$$A = \frac{k/\omega}{\sqrt{(1 + C_d/C_g)^2/G^2 + 1/\omega^2 C_g^2}}$$

$$\alpha(\phi) = \left(\cos\phi - \cos\frac{2\pi + 2\phi}{3}\right) \sin\frac{2\pi - \phi}{3}$$

$$\phi = \cot^{-1}\left(\frac{\omega C_g (1 + C_d/C_g)}{G}\right)$$

where C_g and C_d are, respectively, the capacitance between the particle and the inner and outer electrodes; G the conductance of the particle-to-outer-electrode path; and ω is the angular frequency, k is a factor, less than unity, to account for the effect of image charges on the electrostatic force; k is approximately 0.82 for spherical and horizontal wire particles, and nearly unity for long vertical wire particles (Anis and Srivastava, 1981b). r and R are the inner and outer radii of the GIS system, and g is the gravitational acceleration.

Particle Charging Through Microdischarges

The other possible mechanism through which charges can be transferred to a contaminating particle from a coated conductor in a GIS is that of microdischarges. If the field near the particle is high enough, discharges may occur between the particle and the electrode through the dielectric coating, thus charging the particle. This mechanism was used to compute lifting fields of spherical particles (Parekh et al., 1979).

When the particle is in contact with the dielectric coating and the field is sufficiently high, electron avalanches will be initiated at a point on the particle surface and propagate along a field line toward the coating layer. The point on the particle surface at which avalanches will develop is that which ensures maximum avalanche size, which is a function of both field and space. The size of an electron avalanche started by one electron is given in Chapter 4 as

$$n_e = \exp\left[\int_{x_1}^{x_2} (\alpha - \eta)\, dx\right] \tag{10.8}$$

Gas-Insulated Switchgear

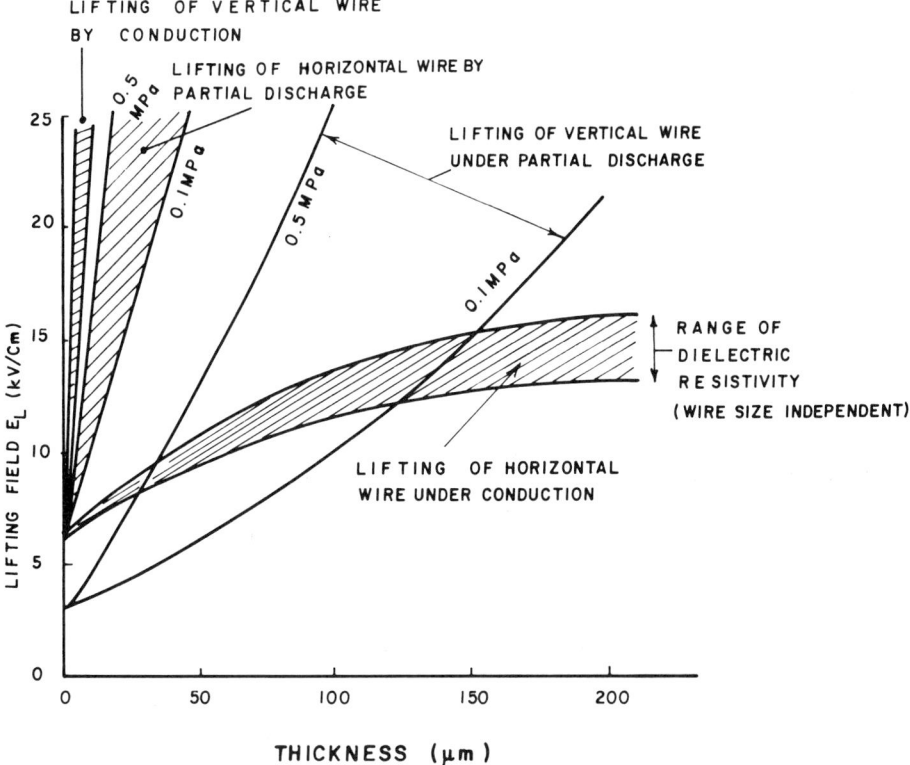

Fig. 10.13 Comparative lifting fields for conducting wire particles off coated electrodes in compressed-gas ducts.

where x_1 and x_2 are the avalanche limits in space, α the ionization coefficient, and η the electron attachment coefficient. Having n_e, the charge transferred from the conductor surface to the particle is en_e, e being the electronic charge. Equation (10.8) could be applied by first computing the field distribution in space, which is then related to $(\alpha - \eta)$ (Chapter 2).

The controlling mechanism causing the particle to lift off at a lower applied field will be either conduction or microdischarges. A comparison between the possible effectiveness of the two mechanisms for lift-off is shown in Figure 10.13. Two initial positions for wire particles are shown in each case, the horizontal and the vertical positions. It is clear from the figure that particle lift-off for all coating layer thicknesses beyond 100 μm is not possible through a microdischarge mechanism. Conduction through the coating or some other mechanism has to be responsible for lift-off. For thick coatings a particle will lift off from a horizontal position.

10.8 ELECTROMAGNETIC COMPATIBILITY IN GIS SUBSTATIONS

As the complexity of the electronics grows in a GIS substation, so does its susceptibility to increasing electromagnetic interference in the environment. To ensure high reliability for the installation, it is therefore necessary to make the electronic equipment compatible with its electromagnetic surroundings. The electromagnetic environment in GIS substations depends on the noise sources, which are essentially switching operations carried out with disconnect switches, earthing switches, and circuit breakers.

Other noise sources include lightning discharges, earth faults, and short circuits (Meppeling and Remde, 1986). Also to be taken into account are external sources such as local communication systems and power transmitter stations. The noise sources themselves are characterized by such variables as noise voltage, noise current, electromagnetic noise fields, and noise energy.

The disturbance variables are transmitted over a coupling path to the noise sink (electronic equipment or secondary units), where they are reduced by the coupling mechanism. Typical coupling paths are secondary cables, voltage and current transformers, earthing systems, and electromagnetic field coupling into control cubicles. Electromagnetic compatibility is achieved when the disturbance variables active at the noise sink are smaller than the immunity to disturbance of the sink.

Several means may be used to inhibit the coupling paths, thus approaching EMC compatibility. Some of those means are:

Meshed earthing system
Use of control cables and connectors with low transfer impedance
Cable shields earthed at both ends
Secondary circuits of instrument transformers earthed only once
Coaxial entry of secondary cable shields in cubicles
Filtering of the mains supply
Limitation of transient overvoltages
Use of fiber-optic cables for communication and data transmission

10.9 ON-SITE TESTING

In contrast to conventional switchgear installations, where the switchgear need not be installed according to any given sequence,

Gas-Insulated Switchgear

GIS installations require systematic assembly according to a plan laid down at the manufacturing stage. As soon as the components forming a common gas chamber are assembled, they are evacuated to about 100 Pa to remove any moisture that may have entered during the assembly phase. The high-voltage equipment is assembled at the works to form large units, tested, sealed in a gastight housing, evacuated and filled with dry nitrogen at 150 kPa, and then transported.

At the site of installation the following tests are performed during commissioning:

1. Every flanged joint made on site is tested for leaks. The maximum permissible values per joint are set so that the entire installation cannot lose more than 1% of its gas per year.
2. The moisture content of the SF_6 gas is measured between 2 and 3 weeks after the first filling. Experience shows that a final value is reached after this period.
3. Functional checks on the breakers, isolators, and earthing switches, and also in the settings of the supervisory instruments, are carried out as soon as the appropriate stage of completion is reached.
4. If the design of the voltage source and ambient conditions permit, measurement of partial discharges may be carried out.
5. The amount of air in the SF_6 is measured.
6. The earthing system is checked.
7. High-voltage tests are performed on the installation to detect damage or contamination that may have occurred in transit or during assembly.

10.9.1 High-Voltage On-Site Testing

Power-Frequency Voltage

Because it corresponds to the continuous stress met with in service, a power-frequency test should be carried out wherever possible. Of disadvantage in such cases is the high reactive power required for extensive installations and the high test voltages. The state of the insulation can be judged more accurately if frequent partial discharge tests are carried out together with the ac test. In most cases, however, the test results obtained on site cannot be conclusive, owing to the abnormally high ambient noise level. Certain drawbacks in conventional testing equipment can be avoided by the use of resonant circuits. In a few cases a power-frequency test can be carried out by energizing the voltage transformers on their low-voltage side. This distorts the shape of the wave, but although the distortion does not much affect the test result, it must be taken into consideration when measuring the voltage level. When the manufacturing tests have been carefully supervised, the test on the completed installation may be considered a repetition. The voltage amplitude can be limited to 80% of the rated test voltage without overlooking assembly errors.

Impulse Voltage

Applying impulse voltages to complete installations is not an entirely suitable way of checking the quality of erection because it indicates only shortcomings in the geometric arrangement, but contamination remains undetected. However, testing with oscillatory switching surges which correspond to the standard switching surge when flashover occurs are sometimes used.

DC Voltage

In cases where ac and impulse testing cannot be performed, it is possible to test the switchgear with dc voltage instead, although this stress rarely occurs in operation. A dc voltage test can still be carried out at 80% of the rms value of the rated power-frequency test voltage.

Since the switchgear contact gaps are already factory tested, testing of the installation can be limited to checking the dielectric strength between the live parts and the enclosure. It is, however, expedient to establish single testing sections in order to localize faults during testing. These sections are determined by the position of the isolators and breakers and are normally tested with ac or dc voltage for a period of 10 s, or with switching surges up to a maximum of three shots.

The final test, in which the largest possible number of installation sections are coupled together, is performed for a maximum of 1 minute for ac or dc voltage or up to five shots with oscillating switching surges. For stresses varying with polarity it may be sufficient for the test to be carried out with the more critical polarity.

10.10 MAINTENANCE

The GIS is virtually maintenance-free and is designed to avoid any opening of the enclosures during its lifetime of at least 30 years. The circuit breaker can normally withstand 20 interruptions at its rated short-circuit breaking current and 2000 interruptions at full-load current. Therefore, even the circuit breakers need not normally be opened for maintenance. The operating devices require minor maintenance a few times during the GIS lifetime.

10.10.1 Gas Handling

Charging the GIS with compressed SF_6 gas is effected by evacuating each gas compartment (Fig. 10.1b), and the section is filled with SF_6 gas to the required density. During this process the gas moisture content is checked. Nitrogen is used to dry out the internal parts of the GIS. Finally, moisture absorbers are mounted in circuit breakers and each section is filled with SF_6 gas to the working pressure. A special gas-handling chart is normally used for this purpose.

REFERENCES

Anis, H. and Srivastava, K. D. (1981a). *IEEE Trans.*, *PAS-100*: 3694–3702.
Anis, H. and Srivastava, K. D. (1981b). *IEEE Trans.*, *EI-16*: 327–338.
Anis, H. and Ward, S. (1988). *Proc. Conference on Electrical Insulation and Dielectric Phenomena*, Ottawa, pp. 312–317.
Cooke, C. M. (1978). *Proc. 1st Int. Symposium on Gaseous Dielectrics*, Knoxville, Tenn., pp. 162–189.
Cooke, C. M., Wootton, R. E., and Cookson, A. H. (1977). *Trans IEEE PAS-96*: 768–775.
Cookson, A. H. and Farish, O. (1973). *IEEE Trans.*, *PAS-92*: 871–876.
Endo, F., Ishikawa, T., Yamagiwa, T., and Ozawa, J. (1983). *Proc. 4th Int. Symposium on High Voltage Engineering*, Athens, Paper 32-05.
Maller, V. N. and Naidu, M. S. (1981). *Advances in High Voltage Insulation and Arc Interruption in SF_6 and Vacuum*, Pergamon Press Ltd., Oxford.
Masetti, C., Pigini, A., Bargigia, A., and Brambilla, R. (1982). *Proc. 4th BEAMA Int. Conference on Electrical Insulation*, Brighton, May, 119–123.
Meppeling, J. and Remde, H. (1986). *Brown Boveri Review*, 73(9): 498–502.
Parekh, H., Srivastava, K. D., and Van Heeswijk, R. G. (1979). *IEEE Trans.*, *PAS-98*: 748–755.
Reisinger, F. Muhr, M., Schenner, H., and Diessner, A. (1987). *Proc. CIGRE Symposium on New and Improved Materials for Electrotechnology*, Vienna, May, Report 1010-02.
Stone, G., Boggs, S., Braun, J., and Kurtz, M. (1987). *Proc. CIGRE Symposium on New and Improved Materials for Electrotechnology*, Vienna, May, Report 400.05.
Trinh, N. G., Mitchel, G. R., and Vincent, C. (1984). *Proc. 4th Int. Symposium on Gaseous Dielectrics*, Knoxville, Tenn., pp. 335–341.

11
Circuit Breaking

R. RADWAN *Cairo University, Giza, Egypt*

11.1 INTRODUCTION

Circuit breakers differ from switches in that they not only manually make and break the circuit and carry their normal currents, but are also capable of making and breaking the circuit under the severest system conditions. Breaking or making the circuit under load conditions represents no real problem for a circuit breaker since the interrupted current is relatively low and the power factor is high. Under short-circuit conditions however, the current may reach tens of thousands of amperes at a power factor as low as 0.1. It is the duty of a circuit breaker to interrupt such currents as soon as possible to avoid equipment damage. Loss of system stability is a consequence of slow fault clearance. Fault clearance time has been immensely reduced during the last 50 years due to the high technology adopted in circuit breaker design and the use of static relays. Fault clearing times on the order of two to three cycles have been achieved, of which the circuit breaker arcing occupies one-half to one cycle.

11.2 ARC INTERRUPTION

The circuit breaker contacts, fixed and moving, are usually made of highly conducting material with adequate contact area ensured by a suitable contact pressure. When the contacts part to break the circuit, the contact area decreases to a very small value just before contact separation. The heat produced within such small area causes its metal to melt and even evaporate (Section 6.2).

In air circuit breakers metal vapor and hot air bridge the gap between the contacts forming an electric arc. For oil circuit breakers the heat generated within the arc decomposes some oil and generates gases. The amount of gases generated is a function of the arc energy and is composed of about 66% hydrogen, 17% acetylene, 9% methane, and 8% others. At such extreme temperatures, these gases and the metal vapor will be highly ionized. They thus provide a conducting path between the circuit breaker's contacts and the circuit current will continue to flow through the arc.

All arc interruption methods are aimed at disturbing the energy balance of the arc. These methods are cooling the arc, increasing its length, and splitting it into a number of arcs in series. Before describing in detail the different types of circuit breakers, it may be well to define the basic rated quantities of circuit breakers and discuss the currents they are called upon to interrupt in ac circuits.

11.3 CIRCUIT BREAKER RATED QUANTITIES

In addition to the breaker's rated voltage and frequency, there are other rated quantities that are important for its operation and selection. These quantities are discussed briefly in the following paragraphs.

11.3.1 Rated Current

The rated circuit breaker current is the rms value of the current that it can carry continuously without the temperature rise of its components exceeding the specified limits.

11.3.2 Rated Breaking Current

The rated breaking current is the rms value of the current that it can break under specified conditions of recovery voltage. Figure 11.1 shows the short-circuit current flowing in an inductive circuit with negligible resistance containing an ac source and a circuit breaker with a fault applied to its terminals. The value of the rated breaking current may include the dc current component, in which case it is termed the "asymmetrical" breaking current. This value can be calculated as follows. The maximum value of the dc current component is

$$I_{dc} = \frac{\sqrt{2} \, V_p}{X}$$

where V_p is the source phase voltage and X is the total circuit inductive reactance per phase. The rms ac component

Circuit Breaking

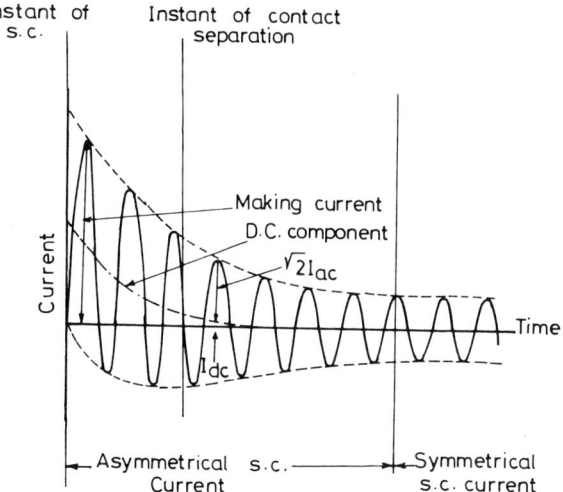

Fig. 11.1 Symmetrical and asymmetrical short-circuit currents.

$$I_{ac} = \frac{V_p}{X}$$

The rms asymmetrical breaking current is thus

$$I_b = \sqrt{2\left(\frac{V_p}{X}\right)^2 + \left(\frac{V_p}{X}\right)^2}$$

$$= \sqrt{3}\,\frac{V_p}{X} \tag{11.1}$$

The breaking current value above is the value of current at the instant of short circuit. However, the circuit breaker contacts part a few millisconds later and the interrupted current thus drops below this value (Fig. 11.1). Normally, the asymmetrical breaking current is taken by designers as 1.6 (V_p/X).

11.3.3 Rated Making Current

The rated making current is defined as the peak value of the current that a circuit breaker can make when closed onto a short

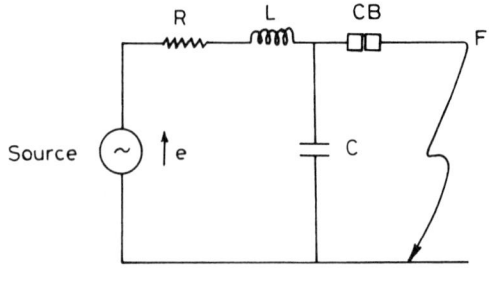

Fig. 11.2 (a) Single-phase equivalent circuit of a symmetrical three-phase-to-ground short circuit, (b) single-frequency transient recovery voltage.

circuit. Obviously, the duty imposed on the circuit breaker during this process is very severe, due to the high mechanical stresses produced within the circuit breaker. The rated making current is given by

I_m = maximum value of dc component + peak value of ac component

$$= \frac{\sqrt{2}\, V_p}{X} + \frac{\sqrt{2}\, V_p}{X}$$

$$= 2\sqrt{2}\, \frac{V_p}{X} \qquad (11.2)$$

The first current peak occurs after about one-fourth of a cycle from the instant of short circuit. To allow for the slight drop in current, the factor 2 (doubling-effect factor) is replaced by 1.8.

11.3.4 Rated Short-Time Current

The rated short-time current is the maximum current that the circuit breaker can carry for 1 s without damage to its conductors, insulation, operating mechanism, or tank.

11.3.5 Circuit Breaker Breaking Capacity

The breaking capacity of a three-phase circuit breaker (in MVA) is given by

$$MVA_b = \sqrt{3}\ VI_b \tag{11.3}$$

where V is the rated voltage in kV and I_b is the rated rms breaking current in kA. According to British practice, the rated breaking current is the rms value of the alternating component only at the instant of contact separation and in this case is termed the symmetrical breaking capacity. In American practice the rated breaking current includes the dc component, which increases the breaking capacity by a factor of 1.6 (Section 11.3.2) and in this case is termed the "asymmetrical" breaking capacity.

11.4 SWITCHED CURRENTS AND CIRCUITS

In addition to normal load currents, circuit breakers are called upon to interrupt short-circuit currents, small inductive currents, and capacitive currents. An example for small inductive and capacitive currents are the no-load current of a power transformer and the charging current of an extensive unloaded cable network, respectively. The duty imposed on a circuit breaker is strongly affected by the magnitude of the interrupted current and also by the circuit parameters and how far the fault is from the breaker terminals. The interruption of each of these currents and circuits is presented below.

11.4.1 Three-Phase Short Circuit

Symmetrical three-phase-to-ground faults at the circuit breaker terminals can be represented by an equivalent single-phase circuit as shown in Figure 11.2a, although interphase reactions cannot always be neglected (Guile and Paterson, 1980). The stray capacitance of the circuit breaker bushing and other connections, represented by C, determines the shape of the transient part of the recovery voltage (Fig. 11.2b). This is called the restriking voltage and builds up across the circuit breaker contacts. The severity of the circuit breaker duty is determined by the value of the short-circuit current together with the shape and amplitude of the restriking voltage.

In case of a fault at point F (Fig. 11.2) and the breaker interrupting it at a normal current zero, the restriking voltage can be obtained as follows. Let the supply voltage be in the form $e = E \cos \omega t$. The circuit voltage equation at the instant of contact separation is given by

$$Ri + L \frac{di}{dt} + \frac{1}{C} \int i \, dt = E \cos \omega t \qquad (11.4)$$

The solution of equation (11.4) gives an expression for the recovery voltage across the circuit breaker contacts which is in the form

$$e_r = E \left[\cos \omega t - \exp \frac{-Rt}{2L} \cos \omega_0 t \right] \qquad (11.5)$$

The transient recovery voltage will oscillate at a natural angular frequency $\omega_0 = 1/\sqrt{LC}$. If the effect of circuit resistance R is neglected and the natural frequency of oscillation is much greater than the supply frequency, the recovery voltage can be approximated to the form

$$e_r = E(1 - \cos \omega_0 t) \qquad (11.6)$$

The recovery voltage as expressed by equation (11.5) is a single frequency transient where a high-frequency oscillatory voltage is superimposed on the supply normal-frequency voltage during the transient period (Fig. 11.2b). Neglecting damping in the circuit, the maximum value of the transient recovery voltage will be 2E and it occurs after a time $\pi\sqrt{LC}$ from the instant of arc interruption. In actual circuits the effect of resistance and system losses are considered and the value of the maximum transient voltage is lower than 2E. The maximum rate of rise of the transient recovery voltage $(de_r/dt)_{max}$ can be easily derived from equation (11.6) and it is equal to $E/\sqrt{LC}$ at a time $(\pi/2)\sqrt{LC}$.

Since the currents in a three-phase system are displaced by 120° from each other, the arc in one of the phases will be extinguished, at a normal current zero, while the other two phases are still arcing. If the system neutral is isolated or the fault is not to ground, the recovery voltage across the first phase to clear would reach three times the maximum phase voltage provided that the three phases are balanced. The first-phase-to-clear factor is = 1.5. It may be less than this value for an earth fault in a system with earthed neutral where its value depends on the ratio between the zero- and positive-sequence impedances.

When a three-phase fault occurs away from the circuit breaker terminals, the transient recovery voltage will have more than one frequency component. The equivalent single-phase circuit representing this case is shown in Figure 11.3. This circuit comprises two parts: the source side (1) and the line side (2). Each part will produce an oscillatory voltage. The transient recovery voltage across the breaker terminals is the vector difference of both voltages and it is of a double-frequency nature (Fig. 11.4). The source-side frequency f_s is equal to $1/2\pi\sqrt{L_1 C_1}$ while the line-side frequency f_ℓ is $1/2\pi\sqrt{L_2 C_2}$.

11.4.2 Asymmetrical Short-Circuit Switching

Symmetrical short circuits on power systems are not very common. Single line-to-ground faults represent more than 90% of the total number of faults on high-voltage and extra-high-voltage systems. Double line faults are rare but they do occur on medium-voltage and low-voltage cables.

The power-frequency recovery voltage under asymmetrical faults to ground is equal to the maximum phase voltage when the power system is solidly earthed. Line-to-line faults produce relatively lower power-frequency voltages on the faulted phases ($\sqrt{3}/2$ × maximum phase voltage), since the two phases break the circuit simultaneously. Overvoltages produced by asymmetrical faults have been covered in the literature (e.g., Fakheri et al., 1983; Flurscheim, 1975).

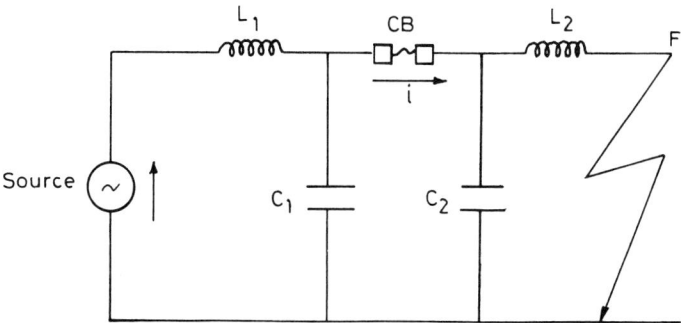

Fig. 11.3 Equivalent single-phase circuit representation of a three-phase fault away from the circuit breaker.

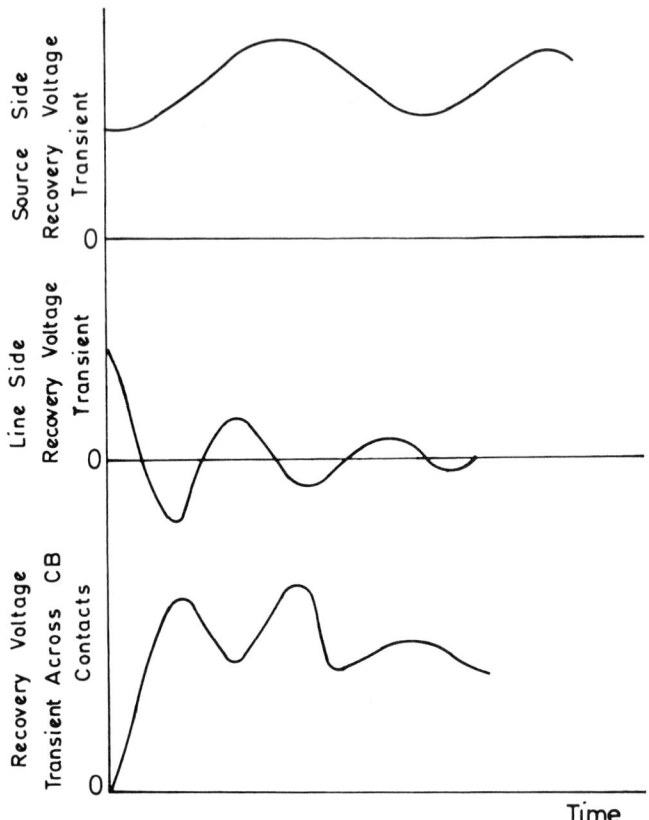

Fig. 11.4 Double-frequency transient recovery voltage.

11.4.3 Small Inductive Current

When the circuit breaker is interrupting a short-circuit current, the arc energy is adequate to keep the arc path highly ionized and conducting up to the instant of natural current zero when the arc gets interrupted. However, in the case of small inductive currents such as transformers' no-load currents, and powerful circuit breakers, the ionized gases get blown off violently and rapidly. The arc current thus gets forcibly extinguished before the natural current zero (Fig. 11.5). This "current chopping," with a very high rate of change, induces very high voltage transients L(di/dt). Having excessive rates of rise, these high voltages cause arc restrikes, and more important, they have serious effects on the system insulation, especially the terminal parts of the transformer winding. Protection schemes against such voltage transients are usually

Circuit Breaking

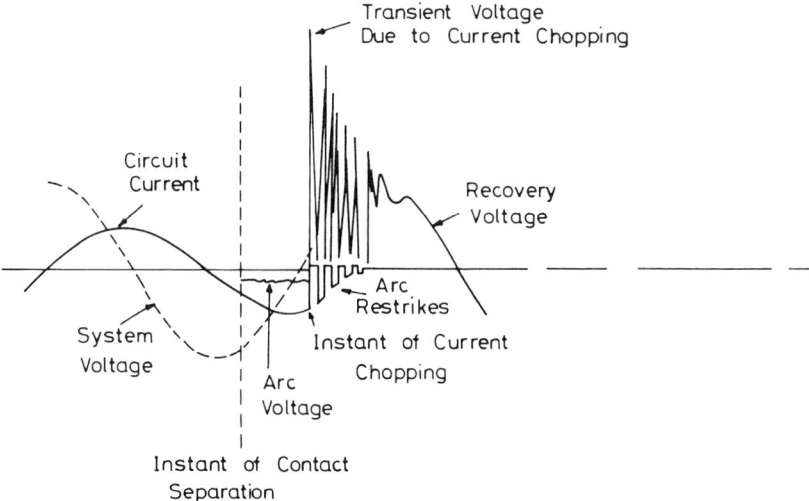

Fig. 11.5 Low inductive current chopping.

provided, which may be in the form of surge arresters, capacitive and resistive shunts, or the use of switching resistors.

The value of the voltage appearing across the circuit breaker terminals, when chopping inductive currents, can be estimated by considering the energy balance just before and after the instant of arc interruption. Referring to the circuit in Figure 11.3 with an arc current i flowing between the breaker contacts, the electromagnetic energy stored in the inductance L_2 is $(1/2)i^2 L_2$. When the arc is extinguished there will be a successive transfer of energy between the electromagnetic and electrostatic fields. Assuming that the arc current is chopped at a magnitude i, the energy balance equation will be in the form

$$\frac{1}{2} i^2 L_2 + \frac{1}{2} C_2 v_2^2 = \frac{1}{2} C_2 v_b^2 \tag{11.7}$$

The terms $(1/2)i^2 L_2$ and $(1/2)C_2 v_2^2$ are the energies stored in the inductance L_2 and capacitance C_2 at the instant just before arc interruption, respectively; v_2 is the voltage across C_2 just before the instant of arc interruption. The maximum voltage across the breaker contacts immediately after arc interruption is given by

$$v_b = v_2^2 + i^2 \frac{L_2}{C_2} \tag{11.8}$$

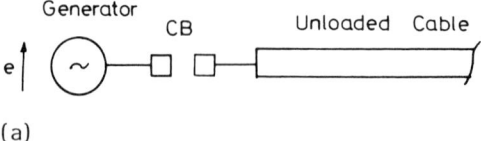

(a)

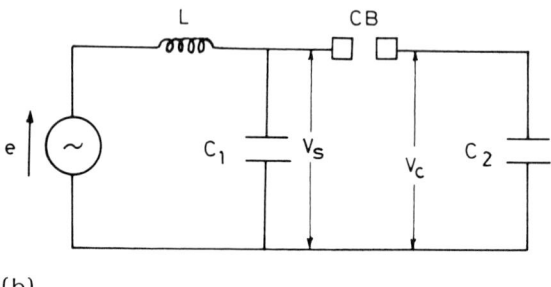

(b)

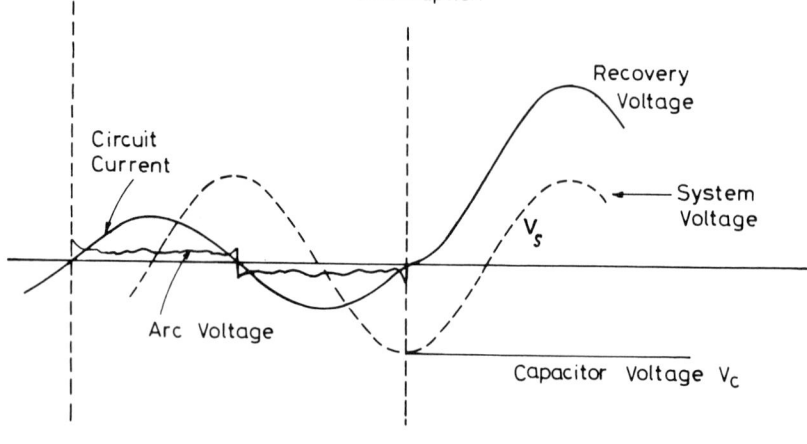

(c)

Fig. 11.6 (a) Representation of unloaded cable: (b) its equivalent circuit, (c) recovery voltage across circuit breaker contacts when interrupting its capacitive current.

Circuit Breaking

It is thus clear that the value of the restriking voltage is a function of the chopped current magnitude and L_2/C_2 The value of L L_2/C_2 depends on the type of load connected to the circuit breaker and may reach 10^5 Ω for power transformers.

11.4.4 Capacitive Current

When a circuit breaker interrupts a purely capacitive current such as that of a capacitor bank or a large network of unloaded cables, a high recovery voltage appears across its contacts. This can lead to restriking several times before complete arc interruption. The current flowing during restriking has a high frequency and its interruption may lead to voltage escalation.

The process of interrupting capacitive currents can be explained by considering the equivalent circuit shown in Figure 11.6b where C_1 represents the stray capacitance of the system on the source side, L its inductance, and C_2 is the cable network capacitance. When the breaker interrupts the capacitive current at its normal zero, the voltage on the source side V_s is at its peak value. The capacitance C_2 is charged to the same value and will remain constant if there is no leakage. The voltage across the circuit breaker contacts half a cycle later will thus reach twice the supply voltage peak value (Fig. 11.6c). The rate of rise of the recovery voltage is relatively low and the arc may be interrupted at the voltage peak. This is reasonably true for air-blast circuit breakers, where the gas blast powerfully deionizes the arc column within half a cycle. In other types, such as oil-circuit breakers, deionization of the arc column is relatively slow when interrupting small currents and the arc may restrike. At this instant an oscillatory current flows through the circuit with a frequency

$$f_n = \frac{1}{2\pi \sqrt{LC_2}} \quad \text{if } C_2 \gg C_1 \tag{11.9}$$

If this current is interrupted at its first zero, the voltage across the circuit breaker contacts will jump to three times the supply voltage. Again if restriking occurs once more with arc interruption at the first current zero, the recovery voltage will reach four times the supply voltage. Theoretically, this process may continue indefinitely and the recovery voltage gets escalated. In practical circuits, damping limits voltage escalation and the arc may get finally extinguished after the first restrike.

11.4.5 Short-Line Fault Switching

The fault current is inversely proportional to the impedance between the source and the fault. Consequently, terminal faults produce the

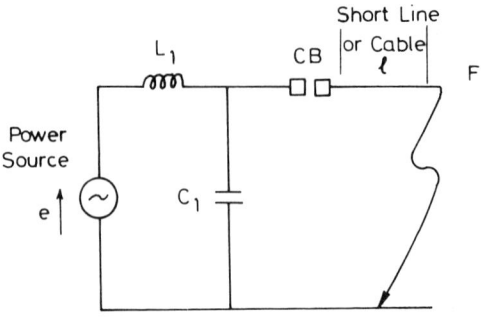

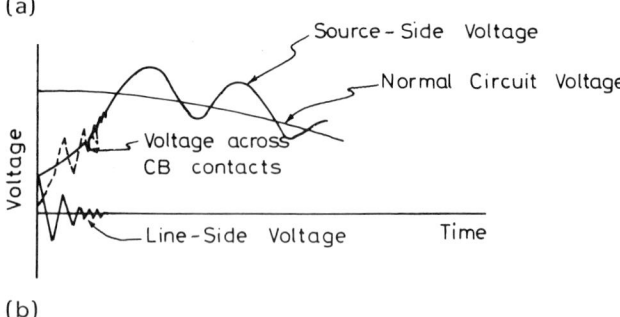

Fig. 11.7 Representation of (a) short-line and (b) voltage waveforms.

maximum fault currents that a breaker has to interrupt. However, the severity of this duty is less than if the fault is a few kilometers away. When a fault occurs at a distance of 1 to 8 km from the breaker terminals, the restriking voltage will be a double frequency transient (Fig. 11.7). The source-side voltage is a relatively slow rising voltage with a frequency in the range 500 to 5000 Hz. The line-side voltage is a high-frequency sawtoothed transient. The restriking transient voltage across the circuit breaker contacts is the vector difference between the two voltage components (Fig. 11.7). The frequency of the line-side voltage is in the range 30 to 100 kHz and is given approximately by

$$f_\ell = \frac{1}{2\pi\sqrt{L_2 C_2}} \tag{11.10}$$

where L_2 and C_2 are the equivalent lumped series inductance and shunt capacitance of the line from the breaker to the fault. If the inductance and capacitance of the line are considered as distributed parameters, the period of the line-side transient voltage is four

Circuit Breaking

times the time of travel between the breaker and the fault (Chapter 14) and is given by

$$T = \frac{4\ell}{v}$$

and

$$f_\ell = \frac{v}{4\ell} \qquad (11.11)$$

where v is the surge velocity and ℓ the length of the line or cable from the breaker to the fault.

11.5 RATE OF RISE OF RESTRIKING VOLTAGE

The previous analysis of the various currents and circuits has shown the nature of the transient recovery voltage across the breaker contacts immediately after arc interruption. Both the magnitude and rate of rise of this voltage will affect the arc interruption process. At the instant of arc interruption the dielectric medium starts to regain its strength while the voltage across the contacts builds up. If the voltage buildup is slow, there will be no risk for arc reignition. If the dielectric recovery is too slow, the arc will restrike. If, on the other hand, the dielectric regains its strength very quickly, such as in vacuum or air-blast interrupters, current chopping may occur with a consequent abnormal voltage rise, as mentioned in Section 11.4.3.

The rate of rise of restriking voltage (RRRV) depends on the interrupted current and its power factor, the circuit natural frequency, neutral grounding of the system, and the presence of any leakage across the interrupted circuit. In gas-blast breakers, the maximum RRRV is a function of the gas pressure and properties and the design features of the breaker. In HV and EHV air-blast circuit breakers, the high-frequency transient recovery voltage can be considerably damped by resistance switching. The basic idea of resistance switching is to interrupt the current on two steps. First a resistance is connected across the breaker contacts as soon as they part; thus the arc between them gets relatively easily interrupted and the circuit current drops to a magnitude decided by the resistance value. The second step is to interrupt the current completely. If the resistance is properly selected, the high-frequency voltage oscillation can be critically damped or overdamped. The value of the switching resistance for damping the voltage oscillations is given by the relation

$$R \leq \frac{1}{2}\sqrt{\frac{L}{C}} \qquad (11.12)$$

11.6 TYPES OF CIRCUIT BREAKERS

Circuit breakers are classified according to the switching medium in which their contacts part. The main switching media currently in common use are air, oil, sulfur hexafluoride (SF_6), and vacuum. With the rapid advancement in the field of solid-state devices, solid-state breakers have a promising future. The different types of circuit breakers are (a) air, (b) oil, (c) air-blast, (d) SF_6, (e) vacuum, and (f) solid-state. The construction and principle of operation of each type are outlined in the following sections.

11.6.1 Air Circuit Breakers

The arc established between the contacts of air breakers is interrupted in atmospheric air. They are commonly used in low-voltage systems with a normal current up to 3000 A. Due to their simple construction and maintenance, they may replace oil circuit breakers of the same ratings in systems where fire hazards exist. In heavy industries having large electric motors with frequent starting, air circuit breakers supersede oil breakers, due to oil contamination. They are also used extensively with electric furnaces.

To help interrupting the arc, deion champers, air chutes, magnetic blasts, and splitter plates are used. When air circuit breakers open, the arc is drawn between the main contacts and then moves upward, by thermal buoyancy and the magnetic effects, to the arcing contacts and then to the arc runners (Fig. 11.8). To speed the arc movement, special magnetic coils are fitted. These coils carry the arc current only during the arcing period.

11.6.2 Oil Circuit Breakers

Here oil provides an excellent medium for arc interruption. In oil breakers the arc can be described as "self-extinguishing." By its excessive heat it evaporates and decomposes some amount of its surrounding oil. The greater the arc energy, the higher will be the pressure of the gas bubble produced. The proper designs of oil breakers fully exploits these high pressures in fast arc interruption. Oil circuit breakers are classified as the bulk oil or minimum oil type. The methods of arc control and interruption are different from one type to the other.

Bulk Oil Circuit Breakers

Here oil serves both to insulate the live parts and to interrupt the arc. The contacts of bulk oil breakers may be of the plain-break type, where the arc is freely interrupted in oil, or enclosed within arc controllers. The latter have higher ratings.

Circuit Breaking

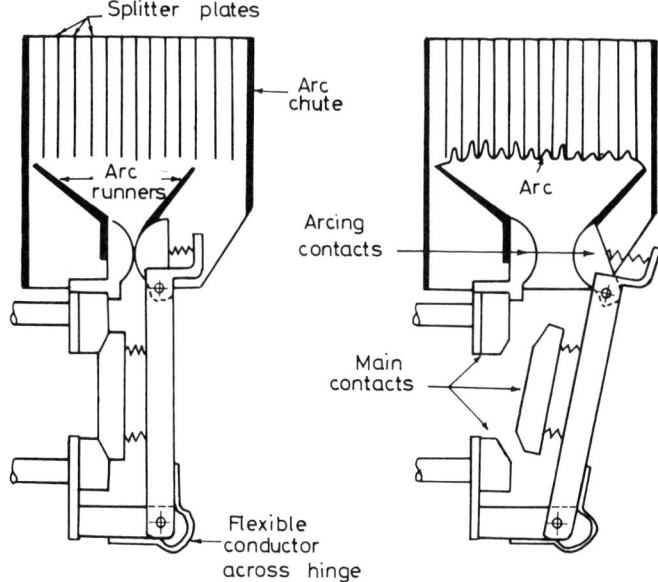

Fig. 11.8 Main parts of an air circuit breaker with closed and open contacts. [From Lythall (1986).]

Plain-break circuit breakers are evidently limited to the low-voltage range. They consist mainly of a large volume of oil contained in a metallic tank. Arc interruption here depends on the head of oil above the contacts and the speed of contact separation. The head of oil above the arc should be sufficient to cool the gases, mainly hydrogen, produced by oil decomposition.

Bulk-oil circuit breakers with arc control devices are used in the medium- and high-voltage ranges. In this type of breaker the arc is confined and interrupted in "explosion pots" rather than in the open volume of oil. The explosion pot encloses the fixed and moving contacts. When they part, the arc is drawn between them inside the pot. The gases so produced in the confined space will cause high turbulence of the oil and rush outside the pot. The rushing gases and oil will disturb the arc, and interruption may occur even before the moving contact has left the pot (Figs. 11.9, 11.10).

Minimum Oil Circuit Breakers

Bulk oil circuit breakers have the disadvantage of using large quantities of oil, with their associated handling and storing problems.

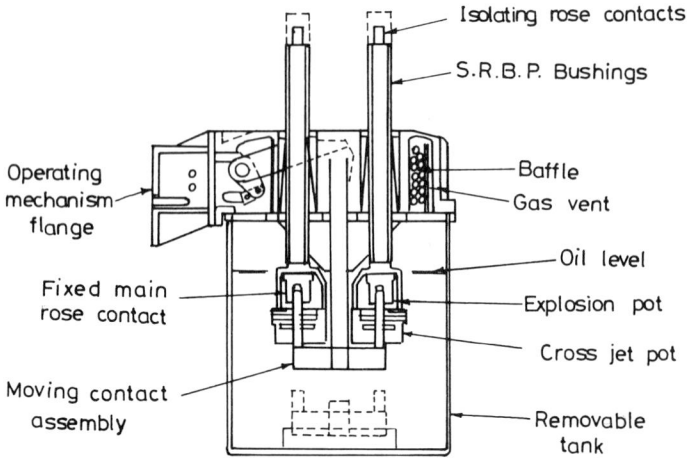

Fig. 11.9 Three-phase dead-tank bulk-oil 13.8-kV circuit breaker in its closed position. (Courtesy of South Wales Switchgear.)

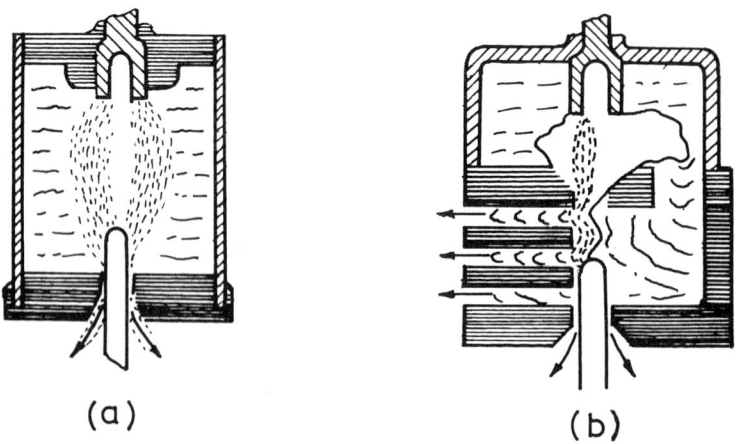

Fig. 11.10 Explosion pot arc controllers: (a) simple type, (b) cross jet.

Circuit Breaking

With frequent breaking and making heavy currents, the oil will deteriorate and may lead to circuit breaker failure. Minimum oil circuit breakers work on the same principles of arc control as those used in bulk-oil breakers. Their containers, however, are made of porcelain or other insulating material. In this type of breaker, arcing is confined to a much reduced volume of oil inside an explosion pot. Figure 11.11 shows one pole of a minimum oil circuit breaker. The lower chamber contains the operating mechanism and the upper chamber contains the moving and fixed contacts together with the arc-control device.

Single-break minimum oil breakers are available in the voltage range 33 to 132 kV with breaking capacities from 1500 to 5000 MVA. For higher voltages and ratings multibreak breakers are constructed from a number of modules in series. Equalizing resistors or capacitors are connected in shunt with each interrupter unit to ensure uniform voltage distribution across the breaks (Soderberg, 1978).

11.6.3 Air-Blast Circuit Breakers

The principle of arc interruption in air-blast circuit breakers is to direct a high-pressure blast of air longitudinally or perpendicularly to the arc. Fresh and dry air will thus rapidly replace the ionized hot gases within the arc zone and the arc length is considerably increased. Consequently, the arc may be interrupted at the first current zero.

The merits of air-blast breakers are (a) cheapness and availability of the interrupting medium, (b) chemical stability and inertness of air, (c) great reduction in the erosion of contacts from frequent switching operations, (d) high-speed operation, (e) short arcing time, (f) operation in fire hazard locations, (g) maintenance frequency reduced, and (h) consistent breaking time results from use of the interrupting medium pressure to open the contacts. However, the use of air compressors and high-pressure vessels increase the production cost. Upon arc interruption, air-blast breakers produce high-level noise when discharging to open atmosphere. In residential areas air-blast breakers should be equipped with silencers to reduce the noise to an acceptable level.

In many designs, contact separation is achieved by admitting high-pressure air from the air receiver. Prior to contact separation the air pressure in the interrupter head is atmospheric but rapidly builds up to separate the contacts by piston action. The contact separation should be enough to interrupt the arc and withstand the recovery voltage under high pressure. These designs with voltages up to 110 kV are equipped with series isolators to open the circuit while the interrupter head is pressurized. When the air supply is stopped, the contacts reclose under spring action.

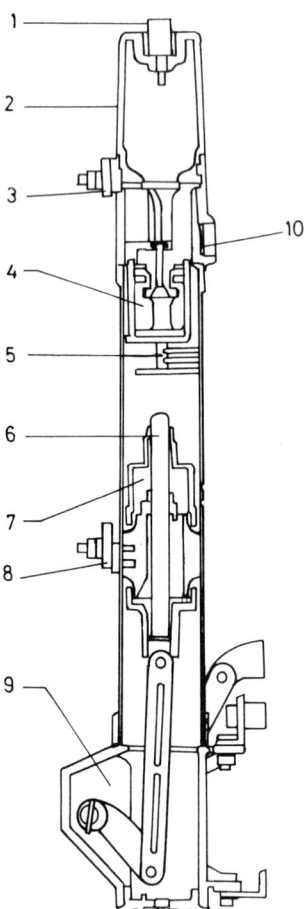

Fig. 11.11 One pole of a 12-kV minimum-oil circuit breaker. 1, Vent; 2, air chamber; 3, upper main terminal; 4, tulip contact; 5, arc control champer; 6, contact rod; 7, contact roller; 8, lower main terminal; 9, crank housing; 10, oil level observation glass. (Courtesy of Asea Brown Boveri AG, Switzerland.)

Circuit Breaking

Fig. 11.12 Air-blast circuit breaker rated 10,000 MVA at 300 kV having eight interrupter heads per phase and equipped with switching resistors and equalizing capacitors. (Courtesy of Asea Brown Boveri AG, Switzerland.)

Modern designs use permanently pressurized interrupter heads and airflow pipes (Fig. 11.12). The relative positions of the interrupter heads and the admission valve ensures simultaneous operation of the breaks. The number of breaks per pole or phase depends on the voltage level and it is customary to use four and eight breaks for the voltage levels 230 and 750 kV, respectively.

As with EHV minimum oil breakers, switching resistors and equalizing capacitors are usually fitted across the interrupters. The resistors help arc interruption and limit transient overvoltages. The capacitors are utilized to equalize the voltage across the open breaks.

The velocity and pressure of the air blast are independent of the interrupted current. Thus when it interrupts a small inductive current, there is a chance of its chopping, as explained in Section 11.4.3. The arcing time of air-blast breakers does not vary considerably with the interrupted current—different from the case of oil circuit breakers, where arc interruption relies on the turbulences produced in the oil.

11.6.4 SF_6 Circuit Breakers

Sulfur hexafluoride is an excellent insulating and arc-quenching medium. The physical, chemical, and electrical properties of SF_6 are superior to many of the other media. It has been used extensively during the past 30 years in circuit breakers, gas-insulated switchgear (GIS), high-voltage capacitors, bushings, and gas-insulated cables (Chapter 10).

Properties of SF_6

Sulfur hexafluoride has a high thermal conductivity and its thermal time constant is about 1000 times shorter than that of air, a great advantage in arc quenching. The velocity of sound in SF_6 is about 40% that in air, and therefore the required gas flow for arc interruption can be produced during an opening operation without the need of a gas compression plant (Schaumann and Evans, 1981).

SF_6 is chemically inert and does not attack metals or glass under normal conditions. However, at temperatures of the order of 1000°C, as an electric arcs, it decomposes to SF_4, SF_2, S_2, F_2, S, and F. The decomposition products recombine shortly after arc extinction (within about 1 μs). The highly toxic S_2F_{10} has never been traced in the arcing products. In SF_6 breakers the decomposition products attack the contacts, metal parts, and rubber sealings in the presence of moisture. The gas should therefore be meticulously dried. Most of the decomposition products can be absorbed by a mixture of soda lime (NaOH + CaO) and activated alumina placed in the arcing chamber. The high dielectric strength and good arc quenching properties of SF_6 are ascribed primarily to its high affinity for electron attachment (Chapter 3).

One of the major problems associated with the use of SF_6 is its condensation at high pressures and low temperatures. At a pressure as high as 14 bar, SF_6 liquefies at 0°C. SF_6 breakers to be used in areas where such low ambient temperatures are encountered may be equipped with special heaters for the gas. SF_6 is also highly sensitive to strong localized fields, moisture, and foreign solid particles (Chapter 10). SF_6 mixtures with other gases are used to overcome some of these problems.

Circuit Breaking

SF$_6$ circuit breakers cover voltage levels in the range 6.6 to 765 kV. The construction and principle of operation of each type are presented in the following sections.

Double-Pressure SF$_6$ Circuit Breakers

This is the early design of SF$_6$ circuit breakers. Its operation is similar in principle to that of air-blast circuit breakers. It comprises mainly a high-pressure metal reservoir, where most of the SF$_6$ is kept, and an interrupter compartment containing the breaker contacts. For the current breaking operation, the circuit breaker contacts part while the high-pressure gas is released from its reservoir to the interrupter compartment, where it blows out the arc. After the current interruption, the gas is pumped back to its reservoir.

In medium-voltage circuit breakers the interrupter compartment is an earthed metal "dead tank," whereas in EHV breakers the "live tank" type is more suitable. As with air-blast circuit breakers, EHV SF$_6$ breakers comprise a number of "interrupter compartment" modules in each phase. Because of its need for various auxilaries, such as gas compressors, filters, monitors, control devices, and their complicated design and construction, breakers of this type have been outmoded by simpler designs of the self-extinguishing and puffer types.

Self-Extinguishing SF$_6$ Circuit Breakers

The interrupting chamber is divided into two main compartments; one is the arc compartment. Both have the same gas pressure, about 5 atm, while the breaker is closed. When the breaker is being opened, the contacts separate and an arc is drawn between them. The heat generated in the arc heats the gas in the arc compartment and rapidly increases its pressure. The gas blasts from the arc compartment to the other compartment. This rapid expansion cools the arc column and extinguishes it at a current zero. A third compartment is incorporated to augment the gas pressure while interrupting smaller currents. By this arrangement the arcing time is independent of the current magnitude; and the currents, large and small get interrupted at their natural zero (Schaumann and Evans, 1981). Breakers of this type are normally used in medium-voltage (MV) networks and their breaking capacities range up to 500 MVA.

Puffer-Type SF$_6$ Circuit Breakers

These are also sometimes called "single-pressure" or "impulse"-type breakers. Their principle of arc interruption is by compressing the SF$_6$ gas during contact separation; the moving contact

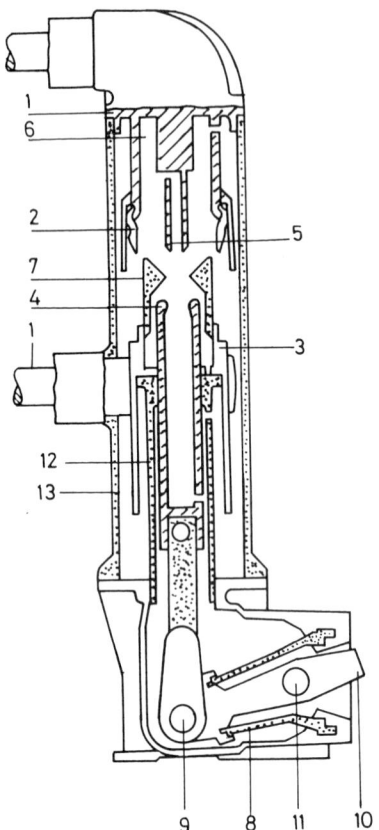

Fig. 11.13 Section through a puffer-type SF_6 circuit breaker.
1, Current terminals; 2, main contact, fixed; 3, main contact, moving as a piston; 4, arcing contact, moving; 5, arcing contact, fixed; 6, exhaust compartment; 7, insulating nozzle; 8, sheath seal; 9, transmission link; 10, operating lever; 11, lever fulcrum; 12, compression cylinder; 13, insulating cylindrical enclosure. (Courtesy of Asea Brown Boveri AG, Switzerland.)

Circuit Breaking

Fig. 11.14 362-kV two-break three-pole puffer SF_6 circuit breaker. (Courtesy of GEC Switchgear.)

acting as a piston. Thus the gas pressure in the interrupter compartment builds up rapidly to levels high above its steady value of about 3 to 5 atm. Its operation can easily be visualized by reference to Figure 11.13.

While the breaker is closed, the current flows between the current terminals (1) via the main contacts: the fixed (2) and the moving contact (3). When the breaker is being opened upon contact separation, the arc is drawn between the arcing contacts (4) and (5), through the insulating nozzle (7). During the contact travel, the moving cylinders of (3) and (4) act as pistons and pump compressed SF_6 gas, which flows axially and blows out the arc.

For the EHV range, up to 765 kV, a number of modules are arranged in series on insulating supports as shown in Figure 11.14, where two modules per phase are sufficient in a 362-kV breaker.

11.6.5 Vacuum Circuit Breakers

The advantages of vacuum as an insulant and arc interrupting medium has been known for many years. A hard vacuum, on the order of 10^{-4} Pa, has a dielectric strength and an arc interrupting ability superior to those of other media, including compressed gases and oil. In vacuum breakers a contact separation of about 1 cm is adequate. Being so compact the power needed to close or open it is much less than for other types of breakers. The rate of dielectric recovery of vacuum after arc interruption is about one order of magnitude faster than in air-blast breakers. Modern vacuum breakers successfully interrupt capacitive currents and small inductive currents and short line faults without producing excessive transient overvoltages. Being compact, with their simple mechanism, they do not need much maintenance. Their main shortcoming is that a failure in their hard vacuum cannot be easily detected while in service.

In a vacuum circuit breaker each phase consists of an evacuated interrupter compartment and an external operating mechanism (Fig. 11.15a). The contacts are of large surface areas, with spiral segments so that the arc current produces an axial magnetic field (Fig. 11.15b) to help move the arc over the contact surfaces and its rapid interruption (Yanabo et al., 1983). Moving the arc spots over the contacts minimizes metal evaporation and thus arc erosion (Section 6.11). The wall of the interrupter compartment is made either totally or partly of an insulating material (e.g., glass). The metal bellows (Fig. 11.15a) make possible the movement of the moving contact while maintaining the hard vacuum (Sunada et al., 1982).

In a vacuum breaker the process of arc extinction differs from that in other types of breakers. When its contacts part, their last points to separate get heated up to the metal's boiling point. The ionized metal vapor thus evolved provides the arc medium (Chapter 6). Here the arc is burning almost exclusively in metal vapor. Therefore, the metal or alloy of which the contacts are made directly affects the characteristics of the vacuum breaker.

The metal vapor swept radially outward from the arc column is bound to condense on the walls of the interrupter compartment. Therefore, insulating parts of the wall facing the arc region should be shielded from the metal vapor to keep their insulation strength. In the breaker illustrated in Figure 11.15a this part of the wall is already metallic; other parts take care of the insulation strength.

At low currents the arc is characterized by a diffuse appearance (Lafferty, 1980), while with large currents of kiloamperes, intense ionization confines the arc with an intense core, causing excessive heating and local melting of its spots on the contacts. Therefore, the arc is whirled magnetically (Fig. 11.15b) in order to minimize contact erosion.

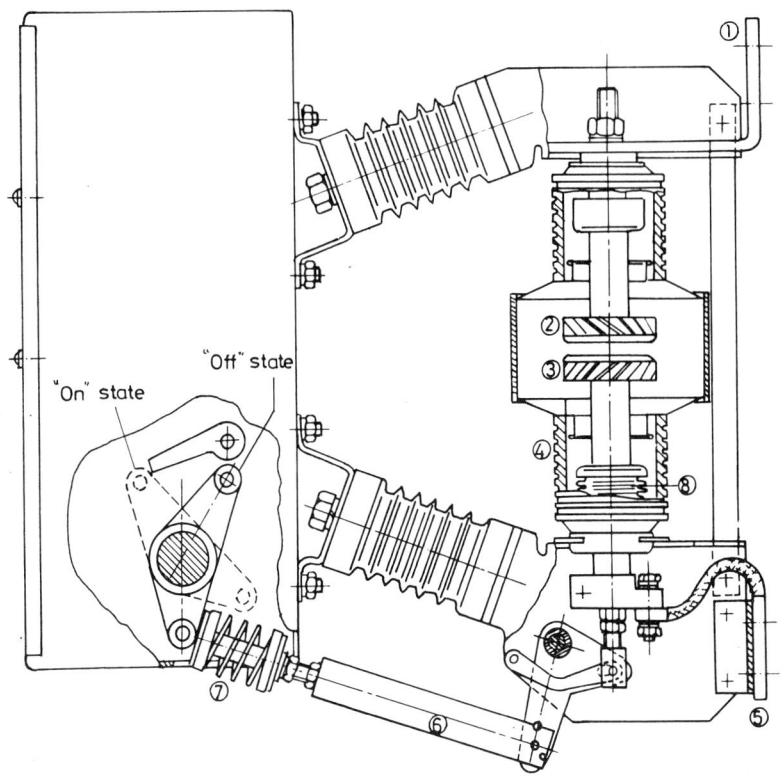

(a)

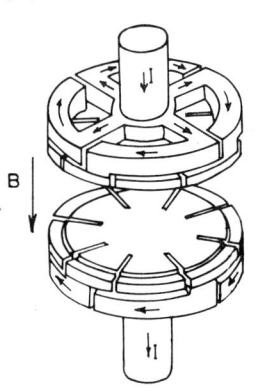

(b)

Fig. 11.15 (a) Construction of a 24-kV vacuum breaker. 1, Upper breaker terminal; 2, fixed contact; 3, moving contact; 4, interrupter body; 5, lower breaker terminal; 6, insulating coupler; 7, contact pressure spring; 8, metal bellows. (Courtesy of Siemens.) (b) Details of the contacts for the current, I, to produce an axial magnetic field B in the vacuum.

300   Chapter 11

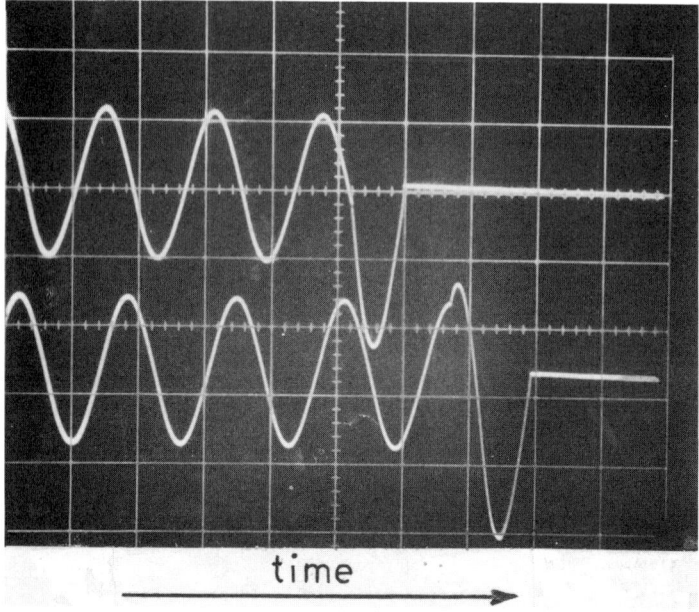

(a)

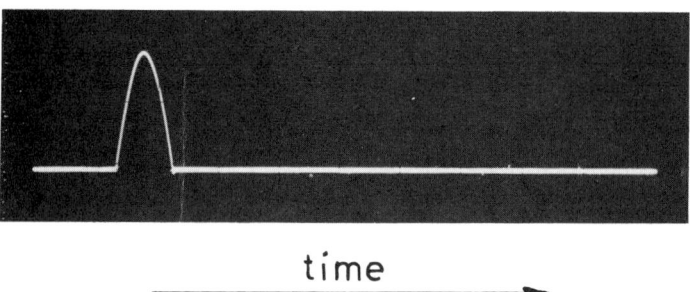

(b)

Fig. 11.16 Oscillograms of currents handled by the solid-state circuit breaker: (a, top waveform) current interrupted within about half a cycle from the instant it exceeds the overcurrent setting; (b) current flowing when the breaker closes a faulty circuit; (c) input current in a circuit where the current has a very pronounced dc component; (d) input current when the instant of circuit closing is properly synchronized.

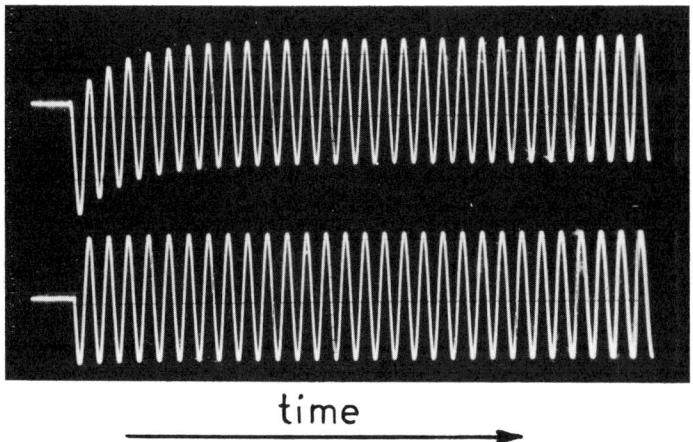

(c and d)

Fig. 11.16 (Continued)

11.6.6 Solid-State Circuit Breakers

This type of circuit breakers use solid-state devices such as thyristors, triacs, or power transistors. They do not suffer from the shortcomings of electromechanical devices. Solid-state breakers can clear a fault within only about one half-cycle (Fig. 11.16a). If it closes a circuit that happens to be faulty, it will still interrupt the fault current within only about one half-cycle (Fig. 11.16b) (Khalifa et al., 1979a,b). They have no moving contacts that get eroded by the electric arc and thus do not need to be maintained or replaced. The instant of making a circuit with a solid-state breaker can be synchronized such that the doubling of inrush currents into inductive circuits and its accompanying electromechanical overstresses on the circuit conductors and supports could be eliminated. Switching transient overvoltages can also be eliminated.

The principle of operation of solid-state breakers is illustrated by the simple circuit shown in Figure 11.17. In this figure two power thyristors are connected antiparallel. Each thyristor will allow the load current to flow during one complete half-cycle when properly triggered. As the instantaneous circuit current exceeds the set level, the current sensor and the gate control device will cut off the gate signals. This will turn off the thyristors and the

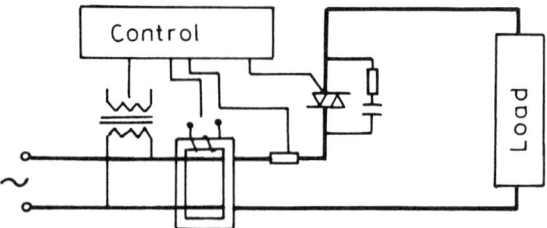

Fig. 11.17 Simplfied circuit with solid-state circuit breaker.

circuit current is broken at the ensuing zero crossing (Fig. 11.16a). In some low-voltage and low-current power systems applications, a triac may replace the power thyristors.

Up until now, solid-state breakers have been used in few power system applications in the low-voltage range. By series parallel connections of the thyristors solid-state breakers for higher voltages and currents can be constructed. The main disadvantage of solid-state breakers is their relatively high power losses. This situation would be much improved with the development of new semiconductor materials.

11.6.7 Direct-Current Circuit Breakers

High-voltage dc lines are now commonly used to transmit large bulks of electrical energy for long distances. They are point-to-point transmission systems. They cannot be tapped or paralleled due to the lack of HV dc breakers with high interrupting capacity.

Air-break circuit breakers are usually used to interrupt low dc currents at low voltages. They are installed in some traction systems, electrochemical plants, and similar applications.

All types of circuit breakers used in HV ac systems cannot be installed in HV dc systems unless equipped with additional circuits to bring the full dc current smoothly to zero for arc interruption (Yanabo et al., 1982). The basic principle of such circuits is shown in Figure 11.18. Before opening the main circuit breaker, switch S is closed to discharge the precharged capacitor C_1 in the opposite direction of the circuit current. This will force it to zero with a few oscillations and the arc will thus be interrupted at a current zero. The self-saturated reactor L_2 and capacitance C help to reduce the rate of decrease of the circuit current and the rate of increase of the transient recovery voltage, respectively.

Recently, Lee and his colleagues (1985) described a 500-kV dc circuit breaker to be used for switching load and fault currents up

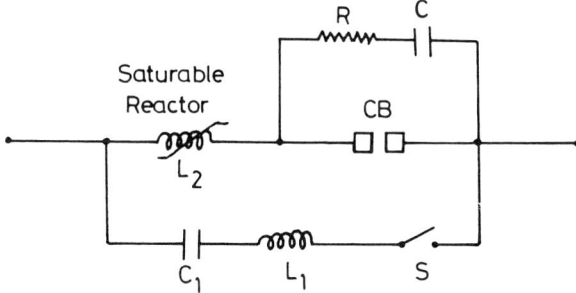

Fig. 11.18 Principle of commutation circuit for dc interruption.

to 2200 A. The breaker is a modified SF_6 puffer type. Four breaker modules are connected in series (Fig. 11.19). When the breaker contacts start to open to break the circuit current, an arc is drawn between them. As the contacts continue to open the arc voltage increases, due to arc lengthening and cooling. After the buildup of enough arc voltage, switch S_1 is closed and this causes a current diversion into C_S. The current diversion results in a current oscillation that grows. When its magnitude is large enough, a current zero is produced in the interrupter. As soon as the current zero in the interrupter is produced, the arc is interrupted and the full circuit current is diverted to C_S. When the voltage across C_S reaches the clipping voltage of the ZnO arrester, it will conduct

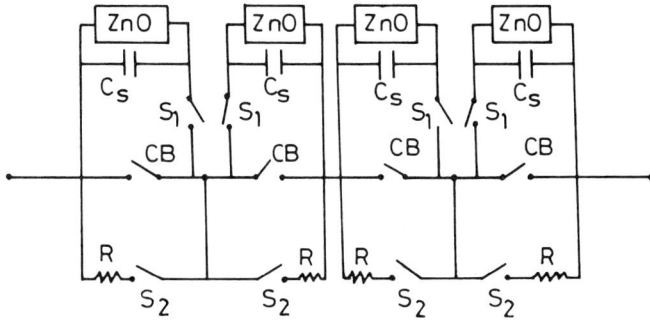

Fig. 11.19 Schematic of a 500-kV HV dc SF_6 puffer-type circuit breaker. [From Lee et al. (1985).]

and stop any further increase of voltage across the contacts. The circuit energy is absorbed by the arresters and the fault current is cleared in a few milliseconds.

Air-blast and SF_6 puffer 400-kV dc circuit breakers have been developed by Vithayathile and his co-workers (1985). Both breakers included commutating circuits and energy absorbing elements. The speed of operation of these breakers is comparable to that of ac breakers of similar types.

REFERENCES

Fakheri, A., Bhatt, N., Ware, B., Sybille, G., and Belanger, J. (1983). *IEEE Trans.*, *PAS-102*: 3315–3328.

Flurscheim, C. H. (1975). *Power Circuit Breakers Theory and Design*, IEE Monograph 17, Peter Peregrinus Ltd., Stevenage, Herts, England.

Guile, A. E. and Paterson, W. (1980). *Electric Power Systems*, Vol. 2, Pergamon Press Ltd., Oxford, p. 49.

Khalifa, M., Arifur-Rahman, S., and Enamul-Haque, S. (1979a). *Proc. IEE*, *126*(1): 75–76.

Khalifa, M., Arifur-Rahman, S., and Enamul-Haque, S. (1979b). *Proc. IEEE Industrial Commercial Power System Technical Conference*, Seattle, Wash., May, pp. 103–106.

Lafferty, J. M. (1980). *Vacuum Arcs, Theory and Applications*, John Wiley & Sons, Inc., New York.

Lee, A., Slade, P. G., Yoon, K. H., Porter, J., and Vithayathil, J. (1985). *IEEE Power Engineering Review* *10*: 32.

Lythall, R. T. (1986). *The J & P Switchgear Book*, Butterworth & Company (Publishers) Ltd., London, p. 32.

Schumann, R. and Evans, G. J. (1981). *Proc. Conference of the Electricity Supply Engineers Association of New South Wales*, Sydney, pp. 12-1–12-12.

Soderberg, G. (1978). *ASEA Journal*, *51*(1): 3–5.

Sunada, Y., Ito, N., Yanabo, S., Awaji, H., Okumura, H., and Kanai, Y. (1982). *Proc. CIGRE Conference on Large High Voltage Electric Systems*, Report 13-04.

Vithayathile, J. J., Courts, A. L., Peterson, W. G., Hingorani, N. G., Nilsson, S., and Poryer, J. W. (1985). *IEEE Power Engineering Review*, *10*: 29.

Yanabo, S., Tohru, T., Shoichi, I., Tsuneo, H., and Shozo, T. (1982). *IEEE Trans.*, *PAS-101*: 1958–1965.

Yanabo, S., Kaneko, H., Koike, T., and Tamagawa, T. (1983). *IEEE Trans.*, *PAS-102*: 1395–1402.

12
High-Voltage Cables

M. ABDEL-SALAM *Assiut University, Assiut, Egypt*

12.1 INTRODUCTION

Since Ferranti's first cable in 1881, underground cables have been used in power distribution networks in cities and densely populated areas. The last few decades have witnessed great development in cable technology. Underground cables can now supersede overhead lines for electric power transmission over short distances.

Cables have the following advantages compared with overhead lines:

1. No interruption of supply even under severe weather conditions such as thunderstorms (on overhead lines, insulator flashover, short circuits, etc., result in supply interruptions)
2. No liability of accident to the public
3. No way-leave troubles that may occur with short circuits of overhead lines crossing main roads
4. No objectionable effect on the aesthetics of the environment

Therefore, underground cables are installed in densely populated regions even when their cost is much higher than that of overhead lines. Cables are usually more expensive than overhead lines at all supply voltages, with a cost ratio of about 20, 8, and 2 at 400, 132, and 11 kV, respectively (King and Halfter, 1982).

Under certain circumstances, dc transmission has many advantages over ac. DC cables have their recommended applications as in case of long submarine transmission. In this chapter we present the many aspects of ac high-voltage cables, with particular emphasis on fundamentals. Readers interested in dc cables are invited to consult specialized books on cables (Weedy, 1988; King and Halfter, 1982).

12.2 CABLE CONSTRUCTION

Any cable used in the power industry usually comprises one, three, or four cores. Each core is a metallic conductor surrounded by insulation. The cable has an overall sheath. For underground cables, armoring is added for mechanical protection. In HV and EHV multicore cables, each core has a metallic screen.

12.2.1 Conductors

For all types of power cables, copper and aluminum are in common use. The metal purity is very important (>99.95%). Impurities seriously reduce the conductivity (Fig. 12.1). The conductors are usually stranded to secure flexibility. Aluminum was considerably cheaper than copper, but now the difference has diminished. The conductivity of aluminum is only 60% that of copper. Therefore, a larger aluminum cross-sectional area is required for the same current-carrying capacity. Both solid and stranded aluminum conductors are presently used in cable construction. When there is more than one layer of strands, alternative layers are spiraled in opposite directions. Flexibility obtained by spiraling of the strands is at the expense of a small increase in cable resistance, due to increased strand length. The increase in conductor resistance with stranding is significant only in case of aluminum, where the oxide layer prevents effective contact among the strands.

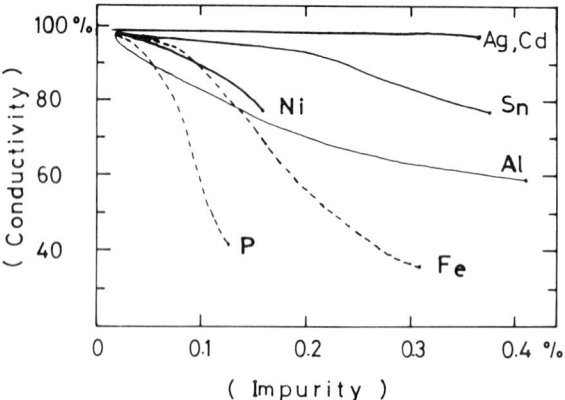

Fig. 12.1 Effect of impurities on the conductivity of copper.

High-Voltage Cables

Sodium has been proposed for use as a cable conductor (Humphrey et al., 1966). Sodium is abundant in seawater and is easily and cheaply manufactured by electrolysis of NaCl. Sodium conductors would be considerably larger in diameter than equivalent conventional ones, as the conductivity of sodium is about 30% of that for copper. Sodium possesses larger values of specific heat and latent heat of fusion. Thus sodium cables would have had good stable performance under short-circuit conditions. On the other hand, when sodium comes in contact with moisture a violent chemical reaction takes place, evolving hydrogen, which is highly inflammable. Also, sodium has virtually no mechanical strength.

12.2.2 Insulation

Insulating materials for power cables are generally classified as (a) impregnated paper, (b) synthetic materials, and (c) compressed gases. The main characteristics that cable insulating material should possess are:

1. High dielectric strength
2. High insulation resistance
3. Sufficiently low thermal resistivity
4. Reasonably long life
5. Low relative permittivity (ε_r) and loss tangent (tan δ) when used in ac cables
6. Chemical stability over a wide range of temperature
7. Easy handling, manufacture, and installation
8. Being economical

Cable insulation used to be oil-impregnated paper, with oil under high pressure in HV and EHV cables. Now polymers are in common use. Cross-linked polyethylene (XLPE) is used in LV and HV cables; poly(vinyl chloride) (PVC) and ethylene propylene rubber (EPR) are limited to LV cables. Attempts to use EPR in HV cables are being carried out (Ross, 1987). In Table 12.1 are given the ranges of voltage ratings in use for cables insulated with oil-impregnated paper and synthetic materials of the oil-filled, gas-filled, and compressed-gas types.

Impregnated Paper

Paper is well known to be hygroscopic. It is therefore impregnated with oil to improve its electrical properties. Typical values of 50-Hz breakdown stress are 6 to 8 MV peak per meter when

Table 12.1 Insulation and Voltage Rating of Various Cable Types

Cable type	Impregnated-paper cables	Plastic-insulated cables	Oil-filled cables	Gas-filled cables	Compressed-gas cables
Long-term electric strength (ac working stress)	2–4 MV/m (particularly for XLPE)	6–8 MV/m	15 MV/m	10–15 MV/m	20–25 MV/m
Voltage ratings	6.6, 11, and 22 kV (with screening each core in three-core cables)	PVC cables: 11, 22, and 33 kV PE cables: up to 22 kV XLPE cables: up to 345 kV EPR cables: up to 66 kV (132 kV under development)	33–525 kV (oil pressure up to 500 kPa)	Up to 220 kV (gas pressure = 1380 kPa)	400–500 kV (SF6 pressure = 300–500 kPa)

tested in thin layers before impregnation and 50 to 60 MV peak per meter after impregnation.

Synthetic Materials

Polymers such as PVC, XLPE, EPR, and polyethylene (PE) show some technical advantages over impregnated paper when used to insulate underground cables in domestic and power distribution networks (Olsson and Hjalmarsson, 1986). More detail is given in Chapter 8.

Compressed Gases

In some EHV cable installations, compressed gases such as sulfur hexafluorinde (SF_6), Freon (CCl_2F_2), and nitrogen (N_2) are used for insulation. As explained in Chapters 4 and 10, SF_6 has several advantages as an insulating medium. Cables insulated with compressed SF_6 are discussed in Chapter 10.

12.2.3 Screening

Experimental and theoretical investigations have revealed that conductor stranding can increase the maximum electric field in the cable insulation by about 20 to 30%. To mitigate this effect, conductor screens in the form of aluminum foil or semiconductor carbon paper tape are lapped over the stranded conductor. Screening the insulation around each core of a multicore cable confines the electric field and makes it symmetrically radial, thereby minimizing the possibility of surface discharges due to tangential field stresses. Laminated insulation usually has higher dielectric strength under normal than under tangential field stresses.

For paper-insulated cables, a metallic sheath is essential for mechanical protection against handling and to prevent ingress of moisture to cable insulation. In the past, metallic sheaths used to be manufactured from lead. High-purity lead is not only expensive but is mechanically weak and would crack under vibrations. Therefore, lead alloys containing small amounts of tin, antimony, or cadmium were commonly used for cable sheaths. Recently, aluminum sheaths have been introduced. The main advantages claimed for aluminum-sheathed cables are about a 50% saving in weight, improved mechanical properties, and better economy.

The drawbacks for aluminum sheaths are that they are less chemically inert and have a higher sheath loss than that of lead. The sheath losses are discussed in Section 12.5.

12.2.4 Armoring

For cables subject to mechanical stresses, armoring is required. A bedding of hessian is wound or a PVC sheath is extruded over the metal sheath to provide a mechanical cushion and chemical insulation between it and the armoring. The armoring consists of steel tapes or wires. Sometimes, nonmagnetic wires (from bronze) are used to increase the magnetic reluctance and thus minimize losses. An outer serving of hessian or PVC sheathing is then applied over the armoring to protect it against corrosion (Ross, 1987).

12.3 TYPES OF CABLES

Power cables can be classified into two distinct categories:

1. Solid-type cables in which the pressure within their insulation is atmospheric
2. Pressurized cables in which the pressure is always maintained above atmospheric either by oil in oil-filled cables or by gas in gas-filled and compressed-gas-insulated cables to help raise the insulation strength, as explained in Sections 12.3.4 and 12.3.5.

12.3.1 Paper-Insulated Cables

Three-core belted cables use impregnated paper wherein each of the conductors is insulated for half of the line voltage. Then extra insulation is applied as a circumferential belt over the three cores to provide sufficient insulation to withstand the phase voltage between each conductor and the sheath. A serious difficulty that occurs with belted cables is due to the electric field distribution throughout the insulation. The electric field is no longer radial to the conductor as in the case of single-core cables. The paper insulation is weaker under a tangential electric field than under a radial field. Therefore, in belted cables there is a tendency for leakage currents to flow along the layers under the tangential component of the field. Subsequently, heat is generated which may eventually lead to breakdown (Chapter 8). For this reason, the belted cables are restricted to voltages of less than 33 kV where the tangential-field component throughout the insulation is rather insignificant (Weedy, 1988; King and Halfter, 1982).

Tangential field components in belted cables can be eliminated by screening each core separately as in H-type cables. The conductor screens are in electrical contact with each other and with the overall metal sheath. The cable is therefore equivalent to three single-core cables with a radial field in each.

High-Voltage Cables

For voltages exceeding 220 kV, three-phase cable insulation is more efficiently employed if the cable design takes the form of three single cores rather than a three-core belted cable. Its size becomes less bulky and more economical.

Another serious problem with paper-insulated cables results from the inelastic property of the lead or aluminum sheath. With cyclic loading, the insulation and sheath expand with different rates and do not return to their original dimensions upon cooling. With time, voids are formed within the volume of insulation. These voids cause losses and ultimate breakdown of the cable insulation, as explained in Chapter 8.

Void formation seems to be a main cause for service failures of cables rated at 66 kV and above. For this reason the gas in the voids is replaced by oil under pressure in "Oil-filled cables," as explained in Section 12.3.3, or the gas is compressed by applying static pressure on the paper insulation in "gas-filled" cables, as explained in Section 12.3.4.

12.3.2 Synthetic-Insulation Cables

The reliability of paper-insulated cables has been so good that the introduction of thermoplastic materials in this field has not been rapid. Nevertheless, two advantages of plastic-insulated cables have accelerated their use: the resistance of plastic sheath to the ingress of corrosive moisture, and the omission of the lead or aluminum sheath which is essential for paper-insulated cables (Weedy, 1988; King and Halfter, 1982). Other advantages of plastic-insulated cables include lighter weight, reduced liability to damage during installation and colored cores for instant identification. Figure 12.2 shows the construction of a typical plastic-insulated cable.

Following the successful application of PVC cable designs and with the advent of thermosetting materials, XLPE-insulated single- and three-core cables were introduced for 11-kV and higher voltages (Brown, 1982).

Reports on the performance of PE- and XLPE-insulated cables installed in wet environments indicate the deleterious effects of moisture. This includes a decrease in breakdown strength and insulation resistance and an increase in dielectric loss due to water treeing (Bottger et al., 1987). This is discussed in more detail in Chapter 8.

12.3.3 Oil-Filled Cables

Oil-filled cables are paper insulated and always completely filled with oil under pressure (3 to 4 atm) with oil reservoirs along the route. The oil pressure inhibits the formation of voids. Their

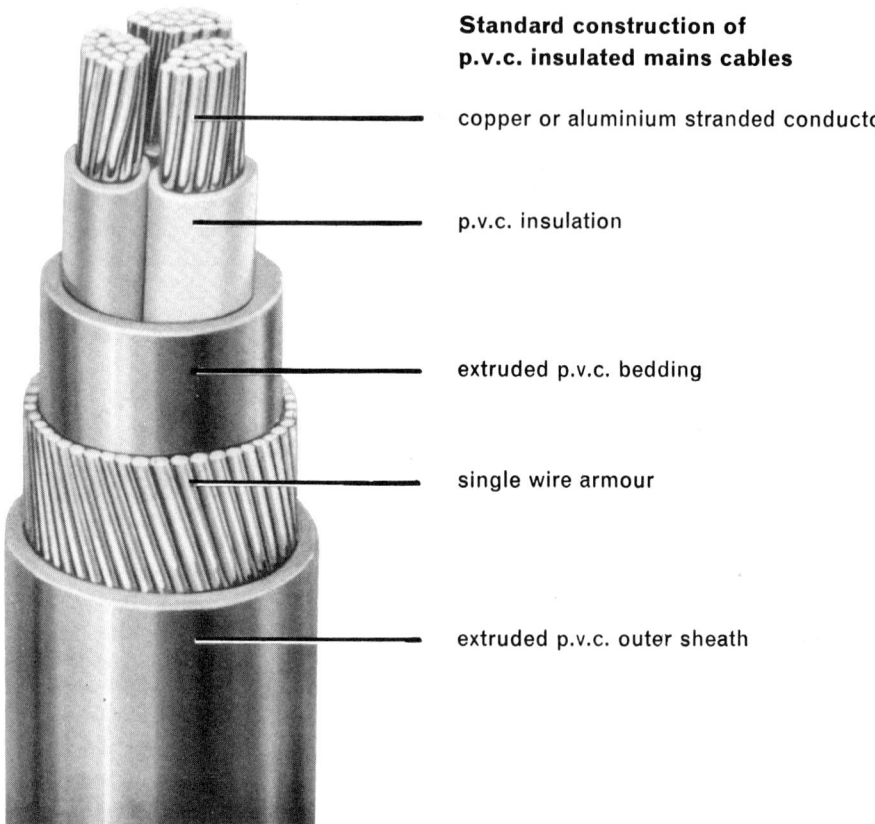

Fig. 12.2 Construction of a low-voltage PVC insulated cable.

contained gases being compressed would stand higher electric stresses. For this reason, the long-time breakdown strength of an oil-filled cable is approximately twice that of the equivalent paper-insulated solid cable. Moreover, the oil acts as a coolant (Fig. 12.3).

The design of the hydraulic system for oil-filled cables depends mainly on the cable route and its profile. A gradient in the profile will produce a rise in static oil pressure. Different oil-filling methods have been used in practice, their choice being dependent on the length of the cable.

A pressure reservoir at one cable end is applied to short lengths of cable routes. The reservoir consists of a number of

oil duct

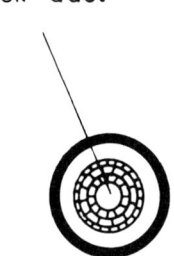

(a)

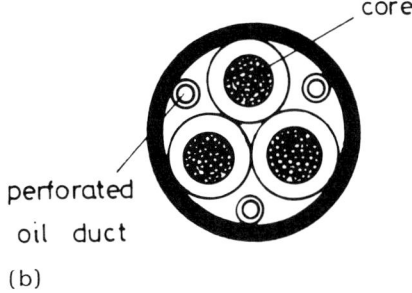

(b)

Fig. 12.3   Cross-sectional views of typical oil-filled cables.

atmospheric gas-filled sealed cells within a tank completely filled with cable oil and connected to the cable. During cable heating, the excess oil flows along the oil ducts in the cable to the pressure reservoir, where it compresses the sealed cells. When the cable cools at light loads, the oil in the cable contracts but the cells in the pressure reservoir reexpand and force oil back into the cable along the cable ducts and into the insulation. The formation of voids in the insulation is thus mitigated.

For comparatively long cable routes a feeding reservoir is added. In contrast to conventional pressure reservoirs, balanced pressure reservoirs are used to make it possible to adjust the pressure-volume characteristics of the reservoir to suit the profile of the cable. The disadvantages of oil filling are the higher initial cost and complication at joints. Oil leaks can be a considerable nuisance and maintenance costs are high (Ross, 1987).

12.3.4 Gas-Filled Cables

Gas-filled cables are paper-insulated cables but placed in steel pipes containing compressed gas, usually nitrogen, thus putting the insulation under pressure. The voids that may form in the cable insulation are filled with the gas at a pressure sufficient to suppress ionization. Gas-filling approximately doubles the working voltage of a cable of the same size. Moreover, the gas acts as a coolant and the steel pipe provides mechanical protection for the cable. There are in some EHV substations short lengths of cables insulated with compressed SF_6. These are described in Chapter 10.

12.4 CABLE INSULATION CHARACTERISTICS

12.4.1 Breakdown Stress

As explained in Chapter 8, the breakdown strength of the cable dielectric is greatly influenced by the time span during which the voltage is applied. The breakdown stress of impregnated paper reaches a value of about 50 MV/m if the stress duration is a fraction of a second, but drops to less than 20 MV/m for durations of 50 to 100 h. Therefore, short-term values of breakdown stresses have to exceed the long-term operating stress by a suitable safety margin.

12.4.2 Dielectric Loss

The losses occurring in a cable dielectric include leakage loss and hysteresis loss. The former occurs as the cable insulation has a finite volume resistivity. It therefore draws a leakage current, however small, under both dc and ac voltages. The additional dielectric loss under ac is that caused by the hysteresis involved in the process of dielectric polarization, as explained in Chapter 8. The total dielectric loss under ac is expressed by the loss angle δ.

Although the dielectric losses are usually insignificant in low-voltage and medium-voltage cables, they become much more significant in EHV cables. For instance, in a 400-kV oil-filled cable, the dielectric losses reach about 80% of the three-phase conductor losses (King and Halfter, 1982).

Several tests have shown a marked dependence on temperature of the loss factor (tan δ) of oil-impregnated paper insulation of cables, whether they are of the solid, oil-filled, or gas-filled type. Tan δ assumes values of 0.002 to 0.01, depending on the temperature and the type of cable insulation. A minimum value for the loss factor occurs at about 40 to 80°C. This characteristic results from the combined effects of temperature on both the resistivity of paper and the viscosity of oil. For cable insulation in general

High-Voltage Cables

the loss factor increases with applied electric stress. If the insulation contains gas voids, the rate of increase of the loss factor with applied voltage becomes rather high above the onset level for ionization of these gas voids. This is evident in the case of solid-type cables, much more than those with pressurized insulation. This point is explained in Chapter 8.

12.5 SHEATH PHENOMENA

The ac currents flowing along the cable conductors produce alternating magnetic fields, as is well known. These link the metallic sheaths of the cable and induce electromotive forces in them. The EMF induced along the sheath of every core of single-core cables drives a current if these sheaths are bonded. Their magnitudes depend on the core currents, the power frequency, the arrangement and spacing of the cores, and the sheath resistance (El-Kadi, 1984).

In addition to these circulating-current losses in single-core cables, eddy currents are induced by the alternating magnetic fields linking every part of the metallic sheaths. These eddy currents flow in both single-core and multi-core cables. Both types of sheath losses are discussed in the following paragraphs.

12.5.1 Circulating-Current Losses

For three-phase power transmission by three single-core cables, the cables are symmetrically disposed at the vertices of an equilateral triangle. Let $I_1 = I \underline{/0°}$, $I_2 = I \underline{/120°}$, and $I_3 = I \underline{/240°}$ be the line currents in phases 1, 2, and 3 and $I_{sh1} = I_{sh} \underline{/-\psi}$, $I_{sh2} = I_{sh} \underline{/120 - \psi}$, and $I_{sh3} = I_{sh} \underline{/240 - \psi}$ be the respective sheath currents, where ψ is the phase shift between the conductor and sheath currents. Let R_c and R_{sh} be the resistances of each cable conductor and sheath, respectively. Due to the symmetrical arrangement of the cables, the mutual inductances (M_{12}, M_{23}, M_{31}) between the conductor of a cable and the sheath of another cable are equal:

$$M_{12} = M_{23} = M_{31} = M = \frac{\mu}{2\pi} \ln \frac{D}{r_{sh}} \qquad (12.1)$$

where D is the cable-to-cable spacing, r_{sh} the mean sheath radius, μ the permeability of free space, and ω the angular frequency. The sheath voltage drop per meter of cable length E_{sh} is

$$E_{sh} = I_{sh1}R_{sh} - j\omega M(I_2 + I_{sh2}) - j\omega M(I_3 + I_{sh3})$$

$$= I_{sh1}R_{sh} + j\omega M(I_1 + I_{sh1}) \tag{12.2}$$

The leakage reactance that accounts for the flux between the sheath and conductor of any cable is so small that its corresponding voltage drop is neglected.

When all cable sheaths are open-circuited at one or both ends of the cable system,

$$I_{sh} = 0$$

and

$$E_{sh} = \omega M I = \frac{\mu \omega I}{2\pi} \ln \frac{D}{r_{sh}} \tag{12.3}$$

It is obvious that with long runs of heavily loaded cables, the sheath voltage to neutral may reach dangerous values, particularly under short-circuit conditions. This may result in arcing between sheaths, with considerable damage. Equation (12.3) shows that the voltage between sheaths increases with an increase in the spacing between cables.

When all cable sheaths are bonded at each end, equation (12.2) gives

$$I_{sh} = I \frac{\omega M}{\sqrt{R_{sh}^2 + \omega^2 M^2}} = I \frac{\omega M}{Z_{sh}} \tag{12.4}$$

and

$$\psi = 90° - \tan^{-1} \frac{\omega M}{R_{sh}} \tag{12.5}$$

It is obvious from equation (12.4) that cross-bonding of the sheaths along the route of the cable does not affect the value of the sheath current (I_{sh}), which is directly proportional to the conductor current (I). The sheath loss as a percentage of the core loss per phase could be easily evaluated, thus:

$$\frac{W_{sh}}{W_c} = \frac{R_{sh}}{R_c} \frac{\omega^2 M^2}{R_{sh}^2 + \omega^2 M^2} \tag{12.6}$$

Table 12.2 Dielectric and Sheath Losses Relative to the Conductor Losses in Three-Phase Cables of the Oil-Filled Type with Different Rated Voltages[a]

Voltage (kV)	Relative losses (%)		
	Conductor	Dielectric	Sheath
66	1.1	0.04	0.13
132	1.0	0.12	0.15
275	1.0	0.36	0.29
400	1.0	0.80	0.40

[a]Conductor area = 2000 mm^2. The cables were buried at a depth of 1 m in a soil of thermal resistivity = 1 K/mW.
Source: King and Halfter (1982).

Usually, the sheath loss is small with respect to the ohmic conductor loss (Table 12.2).

A three-phase cable system with sheaths bonded at both ends is equivalent to a three-phase 1:1 air-cored transformer. Its primary reactance per phase X_c is quite small. The secondary reactance per phase is practically equal to ωM.

It can easily be shown that bonding the three sheaths of the cable effectively changes the impedance per phase of the three-phase balanced cable from Z_c to Z_{ef}, where

$$Z_{ef} = Z_c + \frac{\omega^2 M^2}{Z_{sh}} \qquad (12.7)$$

Thus the phase conductor resistance is effectively increased while its inductance is reduced.

When the three single-core cables are disposed in a flat arrangement, each of the outer phases will suffer a higher induced EMF than that of the middle phase. In case of bonding, the outer phases will also have higher sheath losses.

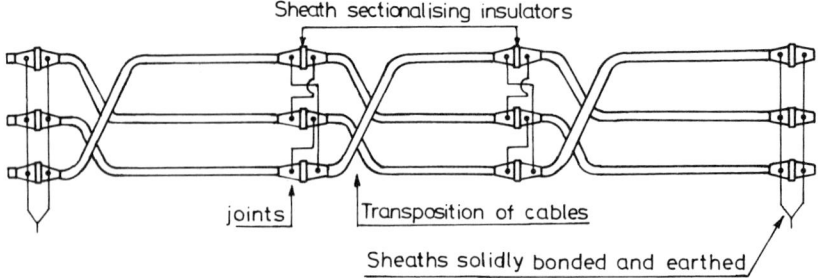

Fig. 12.4 Cross-bonding of cable sheaths. (Courtesy of Pirelli.)

12.5.2 Eddy-Current Losses

In three-phase cable systems, no point on any sheath is equidistant from all three current-carrying conductors, and hence the EMF induced in the sheath will vary from point to point on its circumference. Subsequently, an eddy current will flow circumferentially, resulting in a so-called sheath eddy loss. These seldom exceed 2% of the conductor loss and are usually neglected.

12.5.3 Cross Bonding in Three-Phase Single-Core Cable Systems

The simplest form of sheath bonding for three-phase single-core cables is to bond and earth the three sheaths only at one end or at an intermediate point and to accept the total induced voltages at the remote end. These voltages to ground, however, should be limited to a safe value (i.e., 100 V in protected positions and about 25 V in exposed positions). Therefore, with cable lengths exceeding a few hundred meters, "cross bonding" is resorted to for limiting sheath voltages (Kuffel and Poltz, 1981). This cross bonding of individual core sheaths is carried out in addition to their transposition (Fig. 12.4).

Consider a system consisting of three equal lengths of cable per phase, laid in trefoil, in which the continuity of the sheaths and insulation screens is broken at each joint. If the sheaths are cross bonded, the total voltage at the end of each sheath circuit will be zero for a balanced loading of the cables. At the ends, therefore, the three sheaths can be bonded and earthed without any sheath current flowing (Pirelli, 1986) (Fig. 12.4).

High-Voltage Cables

12.6 ARMOR LOSS

The armoring of a cable may be considered as a supplementary metallic sheath with corresponding losses. If the cable armoring is magnetic, additional losses due to magnetic hysteresis will take place. In single-core cables, the hysteresis losses are often of such a magnitude that they preclude the use of ferrous armoring materials. The derivation of armor losses has been dealt with thoroughly in the literature (e.g., Ametani, 1979).

In three-phase single-core cable systems, the armor losses are usually divided into:

1. Losses due to circulating and eddy currents in the armoring.
2. Losses due to the magnetic field produced by the cable conductor current itself and by currents in neighboring cables if unscreened. These combined magnetic fields could produce significant hysteresis losses.

12.7 CABLE CONSTANTS

12.7.1 Conductor Resistance

With dc the current gets uniformly distributed over the conductor cross section. With ac, however, the current's own magnetic field effectively drives it toward the skin of the conductor. This "skin effect" causes the ac resistance of a conductor to exceed its dc resistance. At 50 Hz, for example, the increase in resistance is 2.5% and 25% for conductor diameters of 2.5 and 3.8 cm, respectively. The effect is much more significant at higher frequencies. A similar effect is that of neighboring conductors carrying ac currents, known as the "proximity effect." The skin effect is evidently lower for tubular conductors than for solid cylindrical conductors; it is therefore a favored design for conductors in EHV cables. The duct inside the tubular conductor is very useful for cooling in oil-filled cables (Fig. 12.3).

12.7.2 Cable Capacitance

Consider the cross section of a single-core cable with a metallic sheath (Fig. 12.3). The capacitance per unit length of such a cable can easily be shown to have the value

$$C = \frac{2\pi \varepsilon_0 \varepsilon_r}{\ln(b/a)} \tag{12.8}$$

where a is the conductor radius, b the inner radius of the sheath, ε_o the permittivity of free space, and ε_r the relative permittivity of the cable insulation.

For three-core belted cables, accurate calculation of the capacitance per phase is very difficult. Resort is made to numerical computation (Malik and Al-Arainy, 1988). It is much easier to determine it by bridge measurements. There are capacitances between conductors and a capacitance of each conductor to the sheath.

12.7.3 Cable Inductance

Compared to overhead lines, the cable inductance is much smaller, and its accurate calculation is much more difficult. It is usually determined by bridge measurements. The inductance per phase of a cable is decided by the magnetic flux linkage, which depends on:

1. The amount of screening afforded by the metal sheaths
2. The presence of armoring and whether or not it is ferrous
3. The proximity of the cable to other conductors and ferrous objects

12.8 ELECTRIC FIELDS IN CABLE INSULATION

As explained in Chapter 2, the electric field intensity in single-core coaxial cables could be calculated. Its maximum value occurs at the surface of the inner conductor (Table 2.1) and is

$$E_{max} = \frac{V}{a \ln(b/a)} \quad (12.9)$$

where V is the applied voltage, a the conductor radius, and b the inner radius of the sheath. Thus for a given sheath size, the maximum electric stress varies with the conductor size and has a minimum value when $a = b/2.718$. In practice such a ratio is far from economical. However, an enlarged conductor diameter for the same net cross section is achieved with tubular conductors. This is a suitable design for oil-filled cables (Fig. 12.3).

In three-core cables with unscreened cores, the electric fields have tangential components at some points. The field distributions in such cases are evaluated by measurements on models or by numerical computations (Chapter 2). Some empirical formulas have been developed for estimating the field intensities at some point on the surfaces of the cable conductors (King and Halfter, 1982; Malik and Al-Arainy, 1987).

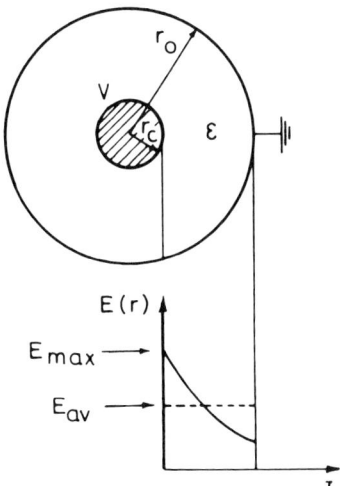

Fig. 12.5 Electric field distribution in a single-core cable with one uniform insulation.

The big difference between the maximum and average fields (i.e., E_{max} and E_{av}) in any cable means that the insulation is being inefficiently utilized; the inner parts are much more highly stressed than those near the sheath (Fig. 12.5). Some suggestions have been made for reducing the ratio E_{max}/E_{av}. One is to use multilayer insulation, the outer layer having a lower value of ε_r. Another suggestion is to subdivide the cable's homogeneous insulation by intermediate metallic sheaths to be energized at suitable fractions of the cable voltage. Both suggestions have been faced with numerous practical difficulties in both manufacture and operation.

12.9 CURRENT-CARRYING CAPACITY

The current-carrying capacity of a cable is defined as the maximum current it can carry continuously without the temperature at any point in its insulation exceeding the limit prescribed for it according to its thermal class (Section 8.3). Thus the rated current for a cable depends on the rates of heat generation within it and dissipation to its surrounding. This, in turn, depends on the thermal resistances of the different parts of the cable and of the

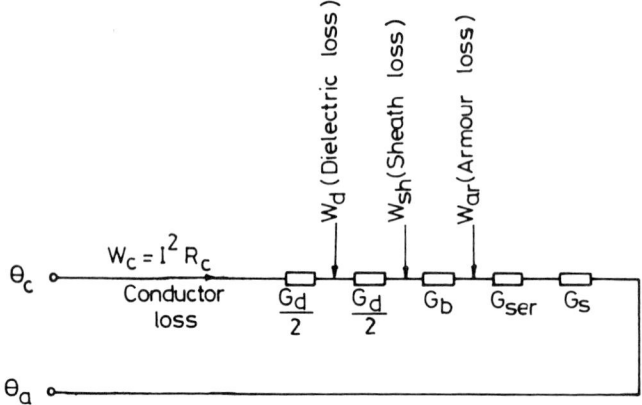

Fig. 12.6 Equivalent thermal circuit of a single-core buried cable.

soil in which it is buried. It depends also on whether the cable is buried directly in the soil or run through troughs or ducts; whether it is buried by itself, being one of a group of cables; or is laid near gas or water pipes. All these factors contribute to heating or cooling the cable under study, and should be considered (King and Halfter, 1982; El-Kadi et al., 1984).

The rated current of a cable could be evaluated by setting up the equivalent thermal circuit of the cable-soil system. Take the simple case of the single-core cable represented in Figure 12.6. There the maximum temperature of the insulation θ_c occurs at the conductor surface. The heat developed by the conductor power loss $W_c = I^2R$ flows through the thermal resistance G_d of the insulation, those of the bedding G_b, serving G_{ser}, and the soil G_s to the ambient, of which the temperature is θ_a. The dielectric loss W_d is assumed concentrated halfway through the insulation. The losses W_{sh} and W_{ar} in the sheath and armoring also contribute to the cable temperature rise, so they are included at the appropriate nodes in the equivalent circuit (Fig. 12.6).

Thus the thermal form of Ohm's law gives the relation

$$\theta_c - \theta_a = I^2 R_c (G_d + G_b + G_{ser} + G_s)$$
$$+ W_d \left(\frac{G_d}{2} + G_b + G_{ser} + G_s \right) + W_{sh}(G_b + G_{ser} + G_s)$$
$$+ W_{ar}(G_{ser} + G_s) \qquad (12.10)$$

High-Voltage Cables

As the dielectric loss flows through one-half of the internal thermal resistance of the cable, the temperature rise $\Delta\Theta_d$ above Θ_a is due only to the dielectric loss W_d and

$$\Delta\Theta_d = W_d \left(\frac{G_d}{2} + G_b + G_{ser} + G_s \right) \tag{12.11}$$

Substitution yields

$$W_{sh} = \lambda_{sh}(I^2 R) \quad \text{and} \quad W_{ar} = \lambda_{ar}(I^2 R)$$

where λ_{sh} and λ_{ar} are, respectively, the ratio of the sheath loss and armor loss to the conductor loss.

After simplification,

$$I = \left[\frac{\Theta_c - \Theta_a - \Delta\Theta_d}{R_c [G_d + G_b(1 + \lambda_{sh}) + (G_{ser} + G_s)(1 + \lambda_{sh} + \lambda_{ar})]} \right]^{1/2} \tag{12.12}$$

Equation (12.12) does not take into account correction factors for the skin and proximity effects (Section 12.7). Both effects, together with typical ratios λ_{sh} and λ_{ar}, could be evaluated (IEC, 1974).

What is left for calculating the current-carrying capacity of the cable is the determination of its own thermal resistance and the maximum permissible temperature. Calculation of the thermal resitance of a single-core cable is easy and is determined in terms of the cable radii and the thermal resistivity of the insulation. For a three-core belt-type cable, however, the calculation of thermal resistance is more complex, due to the arrangement of the thermal field, and computer programs are best used (Glicksman et al., 1978).

The external thermal resistance depends on the type of soil, the amount of moisture present, and how the cable cores are disposed in the earth. Theoretical and experimental studies for evaluating the internal and external thermal resistances and the temperature rise for different cooling systems of general buried cables have been reported in the literature (Abdel-Aziz and Riege, 1980; Burghardt et al., 1983; Prime et al., 1984).

12.10 JOINTING AND TERMINATING HIGH-VOLTAGE CABLES

Because of manufacturing, shipping, and installation limitations, all HV cables are produced and laid in a number of limited lengths that have to be joined together on site and terminated at required positions. Although cables are manufactured under carefully controlled conditions, the joints and terminations cannot always be made under similar circumstances. Nevertheless, when completed, they should be as reliable as the rest of the cable, to eliminate any risk of supply interruptions (King and Halfter, 1982; Pirelli, 1986).

In general, cable joints are classified as straight-through joints, branch or T-joints, trifurcating joints, stop joints of oil-filled cables, and outdoor sealing ends for terminating cables outdoors. The techniques of jointing and terminating cables depend on long experience with the specific type of cable, its conductors, and insulation materials. Extreme care is taken to ensure the high current-carrying capacity and high insulation strength of the joints and sealing ends.

12.11 LOCATING FAULTS IN CABLES

Faults in multiconductor cables may be divided according to their nature as follows:

1. High- or low-resistance earth faults involving one or more of the conductors
2. Open-circuit faults with low or high resistance at the break
3. Flashing (self-healing) faults

The faulty conductor and the type of fault could easily be identified by simple tests at both ends of the cable. These include measuring the insulation resistances by a megger of suitably high voltage. Normal practice is to identify the zone in which the fault has occurred by an approximate measurement of its distance X from either end of the cable, and then locating the fault more accurately. The distance X could be measured by either the "pulse" method or using a bridge. A dc bridge is suitable for low-resistance faults, while ac bridges are appropriate for faults of broken conductors, as noted below.

12.11.1 Pulse Method

A voltage pulse from a suitable source is applied between the faulty conductor and a sound conductor, or the metal sheath. This pulse will travel along the cable up to the fault, where it gets reflected. The amplitude and polarity of the reflected pulse depend on the resistance of the fault (Fig. 12.7; see also Chapter 14). Using such a trace of the pulses on a cathode-ray oscilloscope, the time t_x corresponds to the pulse traveling from the sending end of the

High-Voltage Cables

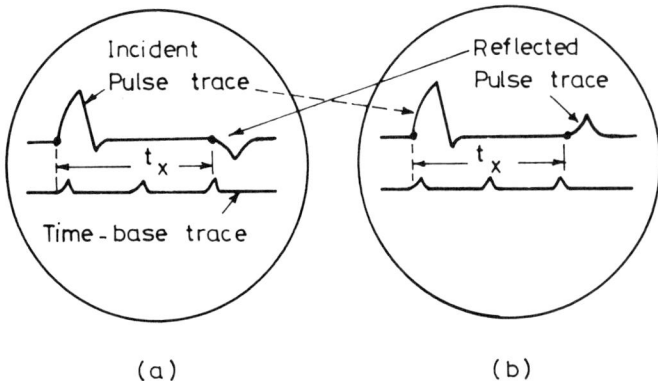

(a)   (b)

Fig. 12.7 Pulse method for locating the faulty zone of a cable;
(a) low-resistance fault; (b) high-resistance fault.

cable to the fault and back. Knowing v, the velocity of propagation of electromagnetic waves along the cable, the distance X to the fault can be calculated. The error in the result depends on errors in v and the measured t_x.

12.11.2 Bridge Method

Low-resistance faults could be located by connecting a dc bridge to the fault and a sound conductor at either end of the cable. These two conductors are connected at the far end by a jumper of sizable cross section forming a loop. Thus the balance conditions of the bridge are used in calculating the fault distance. Inaccuracy in measuring the distance X by this method results from errors such as insensitivity of the galvanometer, tolerances in the bridge elements, and the contact resistances at both ends and the jumper. Too high a resistance of the fault itself would no doubt affect the sensitivity of the bridge. A self-healing fault, for instance, would need to be burned further in order to be located by the bridge measurement with fair accuracy. The further burning of the fault is usually carried out using a special powerful energy source.

In case the fault is a break in one of the cable conductors, it could be located using an ac bridge (Fig. 12.8). The distance X of the fault is proportional to the capacitance C_x. The proportionality could be evaluated by measuring the entire length of the same cable by the bridge. Errors in measuring the distance X by such a bridge would result from the insensitivity of the detector

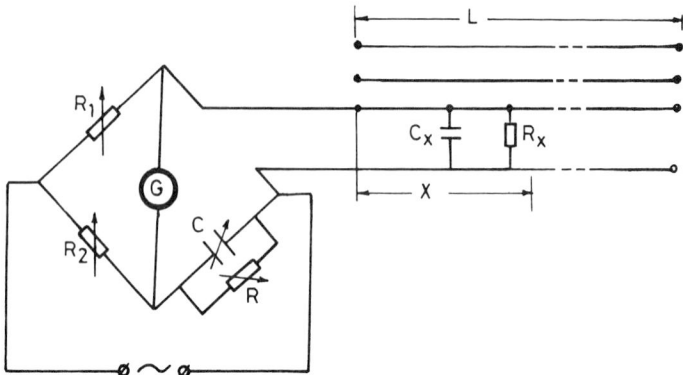

Fig. 12.8 AC bridge method for locating a break in one of the cable conductors. G is either earphones, meter, or scope.

G, tolerances in the bridge elements, and impedance between the two parts of the broken conductor.

For a more precise location of the fault, a train of high-frequency pulses is sent along the cable while their magnetic field is detected (e.g., by headphones and a search coil), along the cable route near the suspected fault location. There would be a detectable signal all along the cable up to the point of the fault, beyond which it ceases.

12.12 SUPERCONDUCTING CABLES

Superconductors are materials whose electrical resistivities become immeasurably small or actually zero below critical temperatures T_c while not being exposed to magnetic fields exceeding critical strengths H_c. Over the years intensive research has been carried out to discover more superconducting metals and alloys with the highest possible critical temperatures. Even as late as 1987 some ceramics were discovered, including oxides of copper, barium, yttrium, and lanthanum. These ceramic superconductors have critical temperatures approaching 90 K. For such materials the cost of refrigeration would be about four orders of magnitude lower than the case of conventional metallic superconductors (Anon., 1987).

If employed in cables, superconductors would mean loss-free transmission of very large powers. In machines, a drastic reduction in their sizes for large ratings would be possible. The

High-Voltage Cables

possibility of producing very strong magnetic fields by very high currents without resistance losses and no EMF for continuously driving it would make possible electromagnetic levitation or fast trains, magnetohydrodynamic propulsion of marine vessels, and many other applications. In electronics and computers it would mean an increase in switching rates and memory sizes.

12.12.1 Superconductivity

Experimental Evidence

A superconducting material manifests zero electrical resistivity and zero magnetic permeability as long as it is kept at temperatures lower than T_c. Its zero permeability is its repulsion to magnetic fields, which accounts for the great possibility of magnetic levitation.

The critical temperature T_c and the critical magnetic fields of strengths lower than H_c are correlated as follows:

$$H_c(T) = H_0 \left(1 - g \frac{T^2}{T_c^2}\right) \qquad (12.13)$$

where g is a constant; g was considered equal to unity by King and Halfter (1982). Equation (12.13) states that the critical field strength $H_c(T)$ at the absolute temperature T is lower than the critical field strength H_0 above which the material's superconductivity is destroyed at absolute zero temperature. This relation is diagramatically shown in Figure 12.9 for different metals.

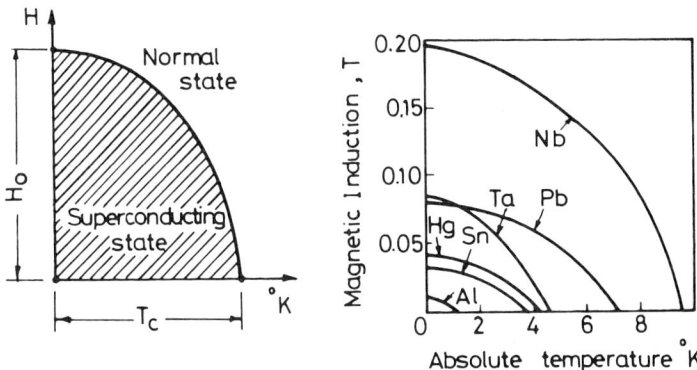

Fig. 12.9   Graphs illustrating the states of superconducting metals.

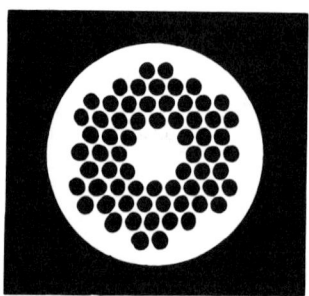

Fig. 12.10 Superconducting filamentary wires clad in a copper wire. Filament diameter = 57 μm, overall wire diameter = 0.7 mm. (Courtesy of BBC.)

Superconducting coils have actually been manufactured for producing very strong magnetic fields in cycletrons and other particle accelerators (Maix et al., 1987). To ensure stability of superconductor performance, it is drawn in the form of several filaments of the superconductor alloy 10 to 60 μm in diameter. The group of filaments is embedded inside a conducting wire of copper or copper-nickel with an overall diameter of about 1 mm. The stable rated current of such a wire (Fig. 12.10) under magnetic fields of about 2 T is about 300 A.

Theory

A theory to explain superconductivity was proposed in 1957 by Bardeen, Cooper, and Schrieffer, based on the concept of a perfect pair of free electrons (Hummel, 1985). Different from normal conductors, in which the current is carried by a stream of individual electrons, the current is carried through a superconductor by pairs of electrons, as explained very briefly below.

According to this theory, positive ions located in the crystal lattice along the path of the electrons through the superconductor help to cancel out some of the repulsive forces between pairs of electrons having mutually suitable energies. They have equal and opposite movements. Their energy levels lie near free states. Thus, as each electron pair drifts through the superconductor, one of the two electrons perturbs its atomic lattice. A displaced ion reverts back into its original position with a momentary oscillation. A phonon acts as an elastic coupling between the two electrons as they pass together through the atomic lattice with no net perturbation. The perturbation of the atomic lattice of normal conductors caused by the flowing electrons accounts for their electrical resistivities.

Cryogenic cooling of a suitable metal or alloy to near 0 K damps out vibrations of its lattice atoms and furnishes the condition suitable for free electron pairs. This theory no doubt would need modification to accommodate superconducting ceramics discovered recently.

12.12.2 Prospects for Superconducting Cables

The dielectric loss in superconducting cables should be minimum as they impose an additional load on the refrigeration system. Insulating materials, including paper, cannot be used because of the unacceptable loss tangent. Synthetic materials are not permitted for use because of their differential contraction, which causes cracking. In the early days of superconducting cables, a relatively low-loss dielectric such as helium or vacuum was used for cable installation, together with solid dielectric spacers. Later, liquid nitrogen and liquid hydrogen have been introduced.

Until 1987, active studies have been progressing for developing the least expensive design for superconducting cables with suitable core material and size. Some researchers have also studied the suitability of liquid air-impregnated paper as an insulating material. Other materials, such as glass—epoxy tapes and enamels, have been considered (Kofler and Telser, 1987; Bobo et al., 1987).

Now, with the discovery of ceramics that exhibit superconductivity at temperatures approaching normal room temperatures, many problems have vanished. On the other hand, the ceramic superconductors available so far have unsuitable mechanical and physical properties. They are brittle and chemically unstable (Bogner et al., 1987). However, the coming decade or two may well witness the solution of these various technical problems and the commercial availability of normal-temperature superconducting cables and wires for numerous power and electronics applications.

12.13 CABLE TESTING

Detailed descriptions of the procedures for the various type, routine, and acceptance tests on cables are laid down in the appropriate standards (Chapter 18).

REFERENCES

Abdel-Aziz, M. and Riege, H. (1980). *IEEE Trans.*, PAS-99: 2386–2392.

Ametani, A. (1979). *IEEE Trans.*, *PAS-98*:902–910.
Anon. (1987). *High-Tech Materials*, 4(9):3.
Bobo, J., Poitevin, J., and Nithart, H. (1987). *Proc. CIGRE Symposium on New and Improved Materials for Electrotechnology*, Vienna, May, Report 100-04.
Bogner, G., Lambrecht, D., and Sabrié, J. (1987). *Electra*, *114* (October):96–107.
Bottger, O., Golz, W., and Saure, M. (1987). *CIGRE Symposium on New and Improved Materials for Electrotechnology*, Vienna, May, Report 620-07.
Brown, M. (1983). *IEEE Trans.*, *PAS-102*:373–381.
Burghardt, R., Mathews, H., Purnhagen, D., and Engelhardt, J. (1983). *IEEE Trans.*, *PAS-102*:2133–2144.
El-Kadi, M. (1984). *IEEE Trans.*, *PAS-103*:2043–2050.
El-Kadi, M., Chu, F., and Radhakrishna, H. (1984). *IEEE Trans.*, *PAS-103*:2735–2740.
Glicksman, L., Sanders, J., and Robsenow, W. (1978). *IEEE Trans.*, *PAS-97*:134–139.
Hummel, R. (1985). *Electronic Properties of Materials*, Springer-Verlag, Berlin.
Humphrey, L., Hess, R., Addis, G., Ruprecht, A., Ware, P., Steeve, E., Schneider, J., Matthysse, I., and Scoran, E. (1966), *IEEE Spectrum*, *11* (November):73.
IEC (1974). *Calculation of the Continuous Current Rating of Cables (100% Load Factor)*, Publication 287-2, International Electrotechnical Commission, Geneva.
King, S. Y. and Halfter, N. A. (1982). *Underground Power Cables*, Longman Group Ltd., Harlow, Essex, England.
Kofler, H. and Telser, E. (1987). *Proc. CIGRE Symposium on New and Improved Materials for Electrotechnology*, Vienna, May, Report 100-02.
Kuffel, E. and Poltz, J. (1981). *IEEE Trans.*, *PAS-100*:369–374.
Maix, R., Rauch, J., Benz, H., Vecsey, G., Jakob, B., and Zichy, J. (1987). *Brown Boveri Review*, *74*:40–48.
Malik, N. and Al-Arainy, A. (1987), *IEEE Trans.*, *PWRD-2*, 589–595.
Malik, N. and Al-Arainy, A. (1988), *International Journal of Electrical Engineering Education*, *25*(1):27–32.
Olsson, J. and Hjalmarsson, G. (1986). *ASEA Journal*, *59*:8.
Pirelli (1986). *Self-Contained Oil-Filled Cables*, Pirelli, Milan, p. 55.
Prime, J., Valdes, J., Macias, C., and Porven, A. (1984). *IEEE Trans.*, *PAS-103*:2794–2798.
Ross, A. (1987). *IEE Power Engineering Journal*, *1*:51.
Weedy, B. M. (1988). *Thermal Design of Underground System*, John Wiley & Sons, Inc., New York.

13
Grounding Systems

A. EL-MORSHEDY *Cairo Uiversity, Giza, Egypt*

13.1 INTRODUCTION

The use of electricity brings with it an electric shock hazard for humans and animals, particularly in the case of defective electrical apparatus. In electricity supply systems it is therefore a common practice to connect the system to ground at suitable points. Thus in the event of fault, sufficient current will flow through and operate the protective system, which rapidly isolates the faulty circuit. It is therefore required that the connection to ground be of sufficiently low resistance.

It is essential to mention here that the terms "ground" and "earth" cannot quite be used interchangeably. The proper reference ground may sometimes be the earth itself, but most often, with small apparatus, it is its metallic frame or grounding conductor. The potential of this ground conductor may be quite different from zero (i.e., that of earth itself) (IEEE, 1972). The term "ground" is therefore used throughout this chapter.

Grounding is of major importance in our efforts to increase the reliability of the supply service, as it helps to provide stability of voltage conditions, preventing excessive voltage peaks during disturbances. Grounding is also a measure of protection against lightning. For protection of power and substations from lightning strokes, surge arresters are often used which are provided with a low earth resistance connection to enable the large currents encountered to be effectively discharged to the general mass of the earth. Depending on its main purpose, the ground is termed either a power system ground or a safety ground.

13.2 RESISTANCE OF GROUNDING SYSTEMS

The value of resistance to ground of an electrode system is the resistance between the electrode system and another "infinitely large" electrode in the ground at infinite spacing. The soil resistivity is a deterministic factor in evaluating the ground resistance. It is an electrophysical property. The soil resistivity depends on the type of soil (Table 13.1), its moisture content, and dissolved salts. There are effects of grain size and its distribution and effects of temperature and pressure.

Homogeneous soil is seldom met, particularly when large areas are involved. In most cases there are several layers of different soils. For nonhomogeneous soils, an apparent resistivity is defined for an equivalent homogeneous soil, representing the prevailing resistivity values from a certain depth downward (Sverak et al., 1982; Nahman and Salamon, 1984; Komaragiri and Mukhedkar, 1981; Sunde, 1968).

The moisture content of the soil reduces its resistivity (Fig. 13.1). As the moisture content varies with the seasons, the resitivity varies accordingly. The grounding system should therefore be installed nearest to the permanent water level, if possible, to minimize the effect of seasonal variations on soil resistivity. As water resistivity has a large temperature coefficient, the soil resistivity increases as the temperature is decreased, with a discontinuity at the freezing point.

The resistivity of soil depends on the amount of salts dissolved in its moisture. A small quantity of dissolved salts can reduce the resistivity very remarkably. Different salts have different effects on the soil resistivity, which explains why the resistivities of apparently similar soils from different locations vary considerably (Fig. 13.2).

Table 13.1 Typical Values of Resistivity of Some Soils

Type of soil	Resistivity ($\Omega \cdot m$)
Loam, garden soil	5 – 50
Clay	8 – 50
Sand and gravel	60 – 100
Sandstone	10 – 500
Rocks	200 – 10,000

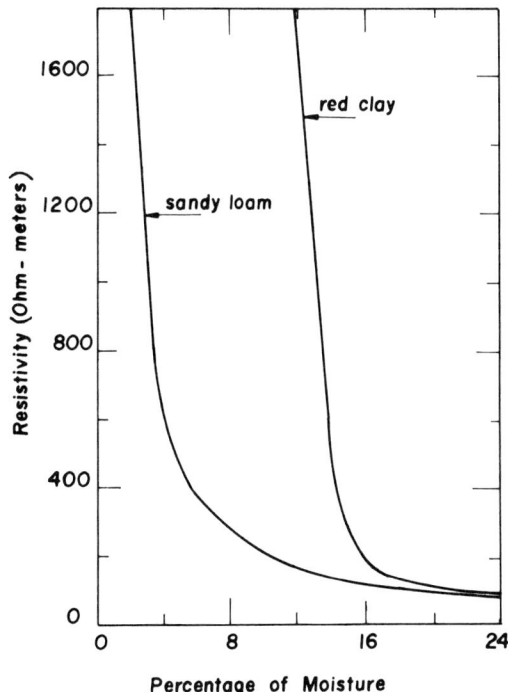

Fig. 13.1 Variation of soil resistivity with its moisture content.

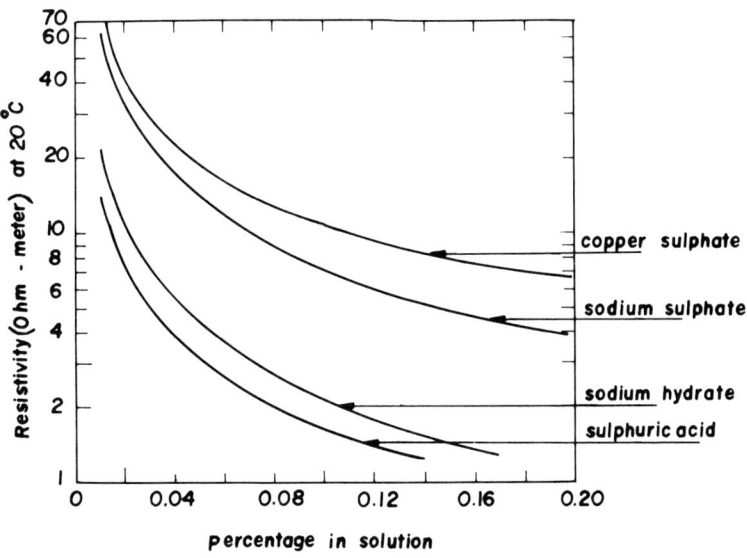

Fig. 13.2 Resistivities of different solutions.

The distribution of grain size has an effect on the manner in which the moisture is held. The finer the grading, the lower will thus be the resistivity. There is not much experimental work on the effect of pressure, but it is reasonable to assume that higher pressures resulting in a more compact body of earth will result in lower values of resistivity.

13.2.1 Resistance of a Grounding Point Electrode

The simplest possible electrode is the hemisphere (Fig. 13.3). The ground resistance of this electrode is made up of the sum of the resistances of an infinite number of thin hemispherical shells of soil. If a current I flows into the ground through this hemispherical electrode, it will flow away uniformly in all directions, through a series of concentric hemispherical shells. Considering each individual shell with a radius x and a thickness dx, the total resistance R up to a large radius r_1 would be

$$R = \int_r^{r_1} \frac{\rho \, dx}{2\pi x^2} = \frac{\rho}{2\pi}\left(\frac{1}{r} - \frac{1}{r_1}\right) \tag{13.1}$$

where ρ is the earth resistivity. As $r_1 = \infty$

$$R_\infty = \frac{\rho}{2\pi r} \tag{13.2}$$

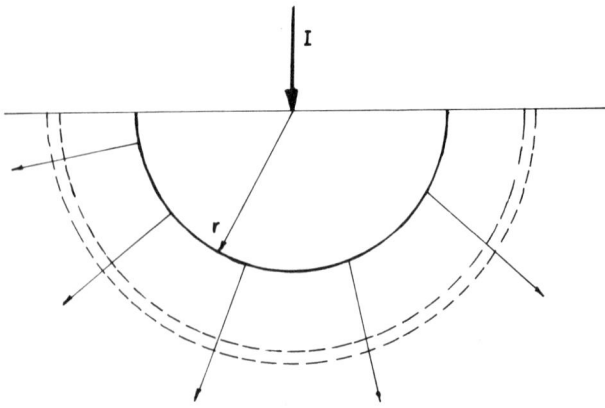

Fig. 13.3 Current entering ground through a hemispherical electrode.

Grounding Systems

The general equation for electrode resistance is

$$R = \frac{\rho}{2\pi c} \tag{13.3}$$

where c is the electrostatic capacitance of the electrode combined with its image above the surface of the earth. This relation is applicable to any shape of electrode.

13.2.2 Resistance of Driven Rods

The driven rod is one of the simplest and most economical form of electrodes. Its ground resistance could be calculated if its shape is approximated to that of an ellipsoid of revolution having a major axis equal to twice the rod's length and a minor axis equal to its diameter d; then

$$R = \frac{\rho}{2\pi \ell} \ln \frac{4\ell}{d} \tag{13.4}$$

If the rod is taken as cylindrical with a hemispherical end, the analytical relation for R takes the form

$$R = \frac{\rho}{2\pi \ell} \ln \frac{2\ell}{d} \tag{13.5}$$

If, however, the rod is assumed as carrying current uniformly along its length, the formula becomes

$$R = \frac{\rho}{2\pi \ell} \left[\ln \left(\frac{8\ell}{d} \right) - 1 \right] \tag{13.6}$$

Table 13.2 gives approximate formulas for the resistances of electrodes with various shapes.

The resistance of a single rod is in general not sufficiently low, and it is necessary to use a number of rods connected in parallel. They should be driven as far apart as possible so as to minimize the overlap among their areas of influence. In practice, this is very difficult, so it becomes necessary to determine the net reduction in the total resistance by connecting rods in parallel. One of the approximate methods is to replace a rod by a hemispherical electrode having the same resistance. The method assumes that each equivalent hemisphere carries the same charge. Evaluating their average potential and total charge, the capacitance and hence the resistance of the system can be calculated.

Table 13.2 Approximate Formulas for Resistance of Various Electrodes

Electrode	Formula
Ground rod	$R = \dfrac{\rho}{2\pi\ell}\left[\ln\left(\dfrac{8\ell}{d}\right) - 1\right]$
Two ground rods $S > \ell$	$R = \dfrac{\rho}{4\pi\ell}\left[\ln\left(\dfrac{8\ell}{d}\right) - 1\right] + \dfrac{\rho}{4\pi S}\left(1 - \dfrac{\ell^2}{3S^2}\right)$
$S < \ell$	$R = \dfrac{\rho}{4\pi\ell}\left(\ln\dfrac{32\ell^2}{dS} - 2 + \dfrac{S}{2\ell} - \dfrac{S^2}{16\ell^2}\right)$
Horizontal wire	$R = \dfrac{\rho}{4\pi\ell}\left(\ln\dfrac{16\ell^2}{dh} - 2 + \dfrac{h}{\ell} - \dfrac{h^2}{4\ell^2}\right)$
Horizontal strip (section a by b)	$R = \dfrac{\rho}{4\pi\ell}\left[\ln\dfrac{8\ell^2}{ah} + \dfrac{a^2 - \pi ab}{2(a+b)^2} - 1 + \dfrac{h}{\ell} - \dfrac{h^2}{4\ell^2}\right]$

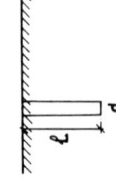

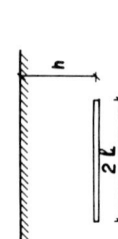

Grounding Systems

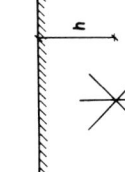

Four-point star

$$R = \frac{\rho}{8\pi\ell}\left(\ln\frac{4\ell^2}{dh} + 2.9 - 2.14\frac{h}{\ell} + 2.6\frac{h^2}{\ell^2}\right)$$

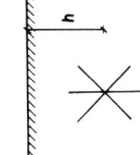

Six-point star

$$R = \frac{\rho}{12\pi\ell}\left(\ln\frac{4\ell^2}{dh} + 6.85 - 6.26\frac{h}{\ell} + 7\frac{h^2}{\ell^2}\right)$$

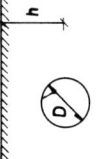

Ring of wire

$$R = \frac{\rho}{2\pi^2 D}\ln\frac{16D^2}{dh}$$

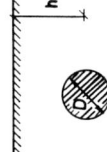

Horizontal round plate

$$R = \frac{\rho}{4D} + \frac{\rho}{8\pi h}\left(1 - 0.036\frac{D^2}{h^2}\right)$$

Vertical round plate

$$R = \frac{\rho}{4D} + \frac{\rho}{8\pi h}\left(1 + 0.018\frac{D^2}{h^2}\right)$$

The resistance of n rods in parallel is thus found to exceed (1/n) of that of a single rod because of their mutual screening. The screening coefficient η for n electrodes in parallel is defined as

$$\eta = \frac{\text{resistance of one electrode}}{n \text{ (reistance of n electrodes in parallel)}}$$

It is difficult to determine the value of η for complicated systems and usually it is listed in tables obtained by calculations and measurements (Dawalibi and Blattner, 1984).

13.2.3 Grounding Grids

A common method for obtaining a low ground resistance at high-voltage substations is to use interconnected ground grids. A typical grid system for a substation would comprise 4/0 bare solid copper conductors buried at a depth of from 30 to 60 cm, spaced in a grid pattern of about 3 to 10 m. At each junction, the conductors are securely bonded together.

The size of grid conductors required to avoid fusing under the fault current I is estimated as (IEEE, 1987)

$$a = I \sqrt{\frac{76t}{\ln\{(234 + T_m)/(234 + T_a)\}}} \qquad (13.7)$$

where a is copper cross section (circular mils), t is the fault duration (seconds), T_m is maximum allowable temperature, and T_a is the ambient temperature. Such a grid not only effectively grounds the equipment, but has the added advantage of controlling the voltage gradients at the surface of the earth to values safe for human contact. Ground rods may be connected to the grid for further reduction in the ground resistance when the upper layer of soil is of much higher resistivity than that of the soil underneath (Garrett and Holley, 1980).

Resistance to Ground and Mesh Voltages of Grounding Grids

The resistance to ground determines the maximum potential rise of the grounding system during a ground fault. The following equation for grid resistance could be used:

$$R = \frac{\rho}{L}\left(\ln\frac{2L}{\sqrt{dh}} + K_1 \frac{L}{\sqrt{A}} - K_2\right) \qquad (13.8)$$

where L is the total length of all conductors, A the total area of the grid, d the grid conductor diameter, and K_1 and K_2 the factors presented graphically as functions of length-to-width ratio of the area. For practical design purposes various approximate formulas based on the similarity of a grid and a round plate of equal area have been proposed (Table 13.2).

The mesh voltage represents the maximum touch voltage to which a person can be exposed at the substation. It is the potential difference between the grid conductor and a point at the ground surface above the center of the grid mesh. Mesh voltages of ground-grounding grids are calculated using a relation of the form (IEEE Committee, 1982)

$$E_{mesh} = K_m K_i \rho \frac{I}{L} \qquad (13.9)$$

where I is the current flowing into the ground and K_m is a coefficient that takes into account the effect of number n of the grid conductors their spacing S, diameter d, and depth of burial h,

$$K_m = \frac{1}{2\pi} \ln \frac{S^2}{16hd} + \frac{1}{\pi} \ln \left(\prod_{j=3}^{j=n} \frac{2j-3}{2j-2} \right) \qquad (13.10)$$

K_i is an irregularity correction factor, to allow for nonuniformity of ground current flow from different parts of the grid:

$$K_i = 0.65 + 0.172n \qquad (13.11)$$

Scale Models of Grounding Grids

Scale model tests with an electrolytic tank are very useful for determining the ground resistance and surface potential distributions during ground faults in complex grounding arrangements where accurate analytical calculations are hardly possible (El-Morshedy et al., 1986). By measuring the voltage applied to the model and the current flowing through the electrolyte between the model grid and the return electrode, the effective grid resistance could be evaluated.

Surface potential profiles for a 4 × 4 mesh grid 10 cm × 10 cm, 1 mm conductor diameter, at a depth of 1 cm are shown in Figure 13.4 for the normal profile. The surface potential is given as a percentage of applied grid voltage and the horizontal axis is in centimeters measured from the center of the grid. The maximum and minimum potentials throughout the grid could be determined. Using such a model, the effect of changing the grounding mesh parameters and adding grounding rods of different depths at different locations could easily be estimated. Sample results are given in Figure 13.5 (El-Morshedy et al., 1986).

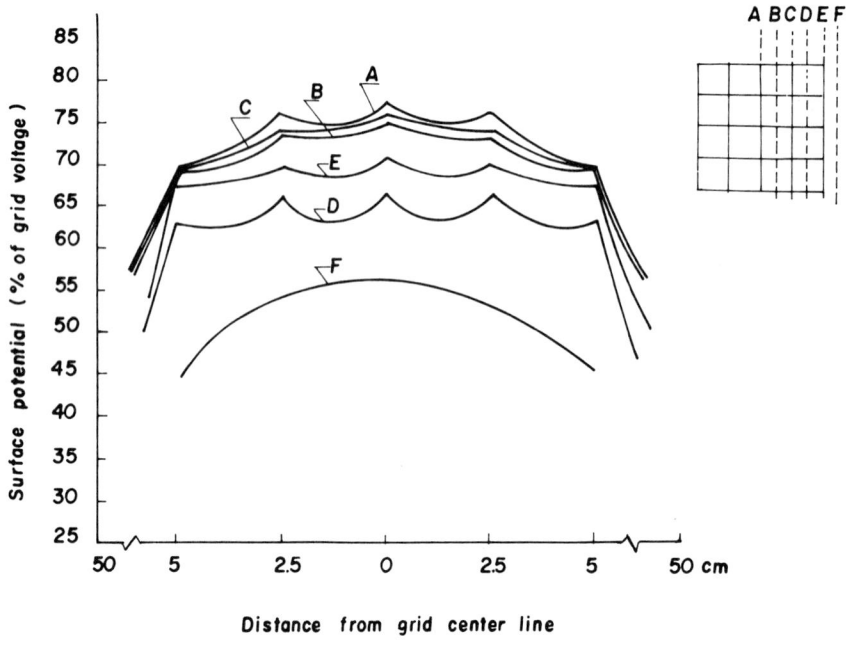
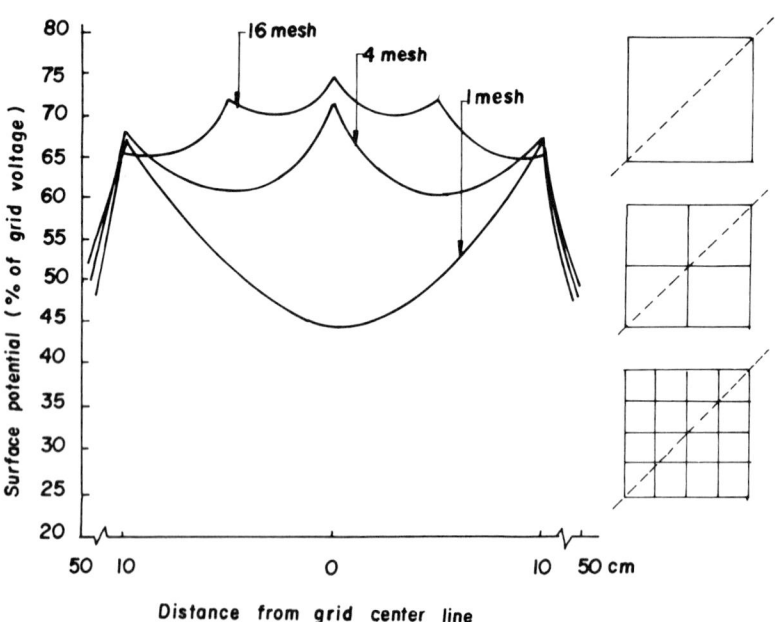

Fig. 13.4 Profiles of surface potential for a 4 × 4 mesh, 10 cm × 10 cm grid. Conductor diameter = 1 mm, its depth below electrolyte surface = 1 cm.

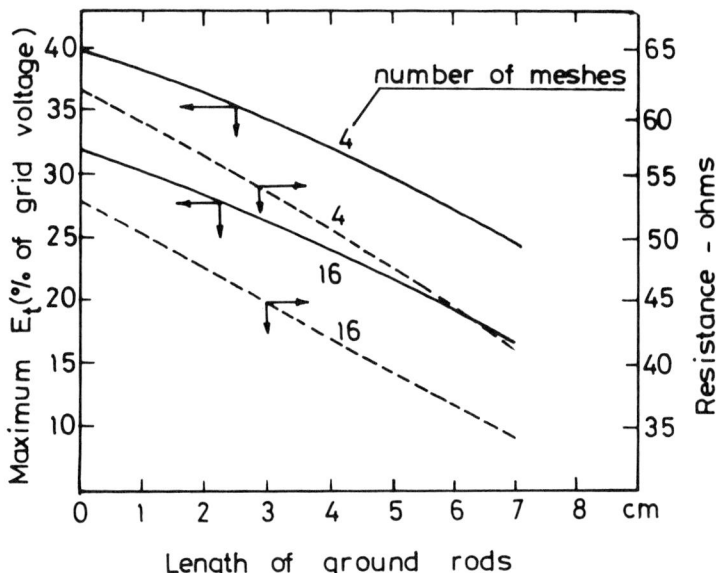

Fig. 13.5 Maximum touch voltage and ground resistance for a model of grounding grids of different mesh as functions of the length of ground rods. Grid size = 20 × 20 cm, conductor diameter = 1 mm, depth below electrolyte surface = 1 cm.

13.3 IMPULSE IMPEDANCE OF GROUNDING SYSTEMS

The impulse impedance of a grounding system is necessary for determining its performance while discharging impulse currents to ground, as in the case of lightning and transient ground faults.

13.3.1 Performance of Driven Rods

In Figure 13.6a the current I gets into the rod electrode and from there diffuses into the ground. In addition to its resistivity, the soil has a dielectric constant ε_r. When the electrode voltage changes with time, there will be a conductive current in addition to a capacitive current. The capacitance of the ground electrode is (IEEE, 1987)

$$C = \frac{\varepsilon_r \ell}{18 \ln(4\ell/d)} \times 10^{-9} \quad F \tag{13.12}$$

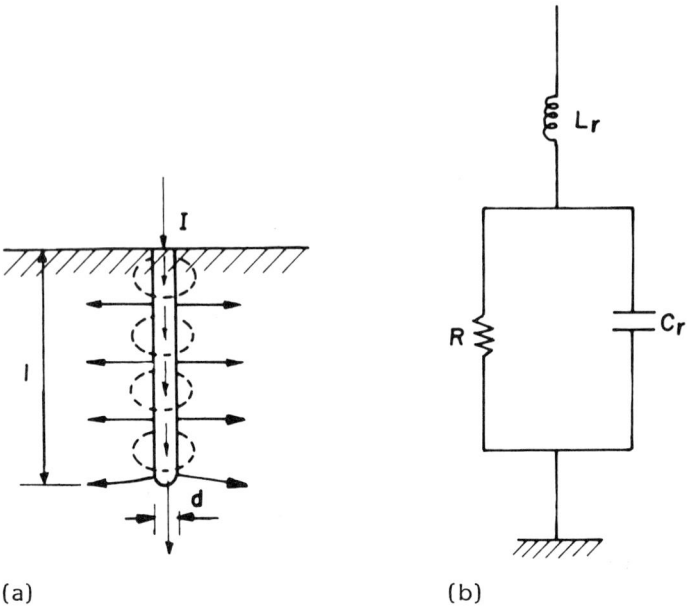

Fig. 13.6 (a) Impulse current spreading from a driven rod. Its magnetic field is also represented. (b) Equivalent circuit for a driven rod under impulse.

The current in the electrode and ground forms a magnetic field. It is highest where the current is most concentrated (i.e., at the top). The inductance of such a rod is

$$L = 2\ell \ln \frac{4\ell}{d} \times 10^{-7} \qquad H \qquad (13.13)$$

At high frequencies of about 1 MHz the ground will draw a considerable capacitive current in addition to its conductive current. The total current flows through the self-inductance of the electrode. Figure 13.6b represents the equivalent circuit (Gupta and Thapar, 1980). At low frequencies the inductance and capacitance can be safely neglected, whereas at extremely high frequencies the ground would be represented by a distributed network.

13.3.2 Performance of Grounding Grids

Ground wires of considerable length are usually buried as counterpoises along transmission lines or near high-voltage substations for lightning protection. A buried wire can be represented by the usual transmission-line circuit of distributed constants. Its distributed resistance to ground is:

For a deeply buried wire:

$$R = \frac{\rho}{2\pi} \ln \frac{4\ell}{d} \qquad \Omega \cdot m \qquad (13.14a)$$

For a wire buried very near the surface:

$$R = \frac{\rho}{\pi} \ln \frac{4\ell}{d} \qquad \Omega \cdot m \qquad (13.14b)$$

The impulse impedance of a buried ground wire in operational form (Verma and Mukhedkar, 1981) is expressed as

$$Z = \sqrt{R(r + SL_d)} \qquad (13.15)$$

where r is the metallic resistance of the wire per meter (Ω) and L_d is the distributed inductance per meter (H). In case of a grounding grid its equivalent circuit is analyzed in response to the applied impulse current. The circuit parameters could be estimated knowing the dimensions of the grid. To account for the impulse current, however, flowing off each conductor along its length, its effective inductance is reduced to one-third the steady-state value.

Naturally, the impulse impedance is initially higher than the power-frequency impedance, but decreases with time to the steady-state value at a rate depending on the circuit and wave parameters. The inductance of the grid is the governing factor contributing to its impulse impedance.

13.4 PRINCIPLES OF DESIGN OF SUBSTATION GROUNDING SYSTEM

The ground of a substation is very important, as it provides the ground connection for the system neutral, the discharge path for surge arresters, and ensures safety to operating personnel. It also provides a low-resistance path to ground to minimize the rise in ground potential. The ground-potential rise depends on fault-current magnitude and the resistance of the grounding system.

Low-resistance substation grounds are difficult to obtain in desert and rocky areas. In such cases, the use of grids will provide the most convenient means of obtaining a suitable ground connection. Many utilities add ground rods for further reduction of the resistance. The size of the grid and the number and length of driven rods depend on the substation size, the nature of the soil, and the ground resistance desired.

The practical design of a grid requires inspecting the layout plan of equipment and structures. The grid system usually extends over the entire substation yard and sometimes several meters beyond. To equalize all ground potentials around the station, the various ground cables or buses in the yard and in the substation building should be bonded together by heavy multiple connections and tied into the main station ground. It is also necessary to adjust the total length of buried conductors, including cross connections and rods, to be at least equal to those required to keep local potential differences within acceptable limits.

13.4.1 Ground Conductor Size

The ground conductor should have low impedance and should carry prospective fault currents without fusing or getting damaged, taking into account future expansion of the connected power system. The size of ground conductor is given by equation (13.7).

13.4.2 Conductor Length Required for Gradient Control

Equation (13.9) gives the value of the mesh voltage; the value of the step voltage, E_{step}, is given by the formula (Fig. 13.7) (IEEE, 1987)

$$E_{step} = K_s K_i \rho \frac{I}{L} \qquad (13.16)$$

where K_s is a coefficient that takes into account the effect of number, spacing S, and depth h of burial of the ground conductors.

$$K_s = \frac{1}{\pi}\left(\frac{1}{2h} + \frac{1}{S+h} + \frac{1}{2S} + \frac{1}{3S} + \cdots\right) \qquad (13.17)$$

The number of terms within the parentheses is equal to the number of parallel conductors in the basic grid, excluding cross connections.

Grounding Systems

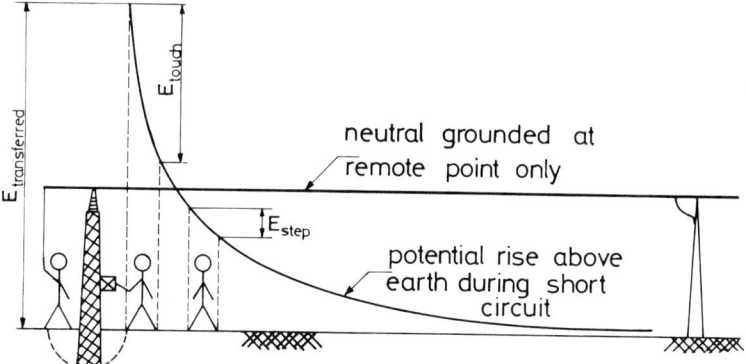

Fig. 13.7 Step, touch, and transferred voltages near a grounded structure.

The tolerable step voltage with duration t_s, E_{step}, which is the voltage between any two points on the ground surface that can be touched simultaneously by the feet is (IEEE, 1987),

$$E_{step} = \frac{165 + \rho_s}{\sqrt{t}} \tag{13.18}$$

where ρ_s is the resistivity of ground beneath the feet, in ohm-meters, taking its surface treatment into account.

The tolerable touch voltage, E_{touch}, which is the voltage between any point on the ground where a person may stand and any point that can be touched simultaneously by either hand is (IEEE, 1987),

$$E_{touch} = \frac{165 + 0.25\rho_s}{\sqrt{t}} \tag{13.19}$$

Equating the value of E_{mesh} to the maximum value of E_{touch} yields

$$\frac{K_m K_i \rho I}{L} = \frac{165 + 0.25\rho_s}{\sqrt{t}} \tag{13.20}$$

The approximate length of buried conductor required to keep voltage within safe limits is thus

$$L = \frac{K_m K_i \rho I \sqrt{t}}{165 + 0.25\rho_s} \tag{13.21}$$

13.5 NEUTRAL GROUNDING

Grounding of the neutral points of generators, transformers, and transmission schemes is an important item in the design of power systems, as it has a considerable bearing on the levels of transient and dynamic overvoltages stressing the equipment insulation. It also directly affects the levels of short-circuit currents in the power network and accordingly, the ratings of switchgear needed to cope with them.

The methods of system neutral grounding include resistance and low reactance for effective grounding. They also include tuned reactance, solid grounding, and grounding through a high-impedance such as that of a potential transformer (Fig. 13.8) (Brown et al., 1978; Gulachenski and Courville, 1984). Each of these methods has advantages and limitations. For example, with isolated or high-impedance grounding, excessive overvoltages appear on the system in the case of line-to-ground (L-G) faults. The "healthy" phases acquire transient overvoltages several times higher than the normal peak phase voltage (Chapter 14). Also, some contingencies

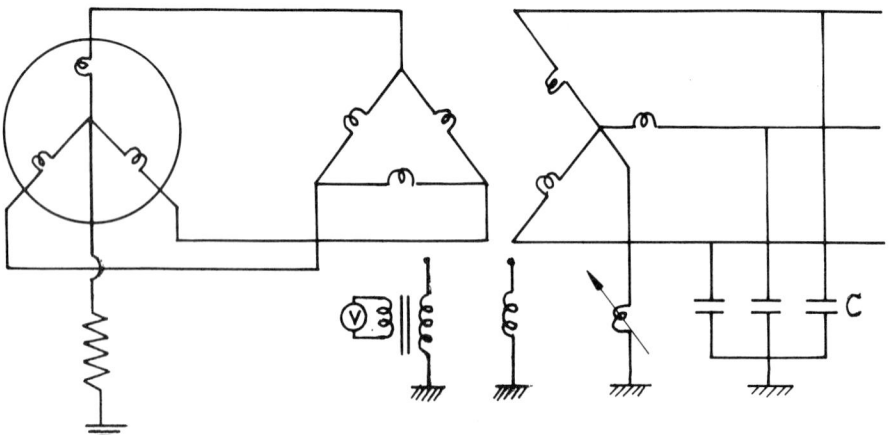

Fig. 13.8 Examples of power system neutral grounding.

may develop ferroresonance, causing high power-frequency overvoltages. In both cases, the equipment insulation, if not suitably designed, would be very vulnerable. Further, with the same method of neutral grounding, the magnitudes of L-G fault currents would be so low that only special protective gear could detect them. This type of neutral grounding, however, is favored in only some LV and MV isolated networks where the need of supply continuity is extremely pressing and the equipment insulation is adequate.

Neutral grounding through a reactor tuned to match the system capacitance C to ground (= $3\omega C$) neutralizes the system L-G fault currents almost completely. Thus the fault arc becomes unstable and easily gets extinguished. This method has been in common use in some high-voltage networks in Europe. It helps maintain the continuity of the supply without endangering the system insulation.

For resistance grounding, a resistance of a suitable design is connected between the system neutral point and the grounding electrode group as an addition to the system ground resistance. This technique is suitable for generators, as it helps maintaining their stability by the power consumed during L-G faults; otherwise, the generators might race out of step (Ershevich et al., 1982).

By grounding solidly or through a small reactance the system overvoltages are limited to their possible minimum (Chapter 14). On the other hand, the L-G short circuit currents would be excessive unless the grounding reactor is designed with a suitable magnitude.

13.6 GROUNDING OF GAS-INSULATED SUBSTATIONS

The main difference between gas-insulated substations (GIS) and conventional substations is their metallic enclosures. Under faults, these enclosures carry induced currents of significant magnitude. The currents must be confined to specific paths (Chapter 10).

13.6.1 Grounding of Enclosures

The following requirements should be met to minimize the undesirable effects caused by circulating currents:

1. All metallic enclosures should normally operate at ground voltage level.
2. No significant voltage differences should exist between individual enclosure sections.
3. The supporting structures and any part of the grounding system must not be influenced by the flow of induced currents.

4. Precautions should be taken to prevent excessive currents being induced into adjacent frames and structures.
5. As GIS substations have limited space, reinforced-concrete foundations may cause irregularities in a current discharge path. The use of simple monolithic slab reinforced by steel serves as an auxiliary grounding device. The reinforcing bars in the foundations can act as additional ground electrodes.

13.6.2 Touch Voltages for GIS

The enclosures of GIS should be properly designed and adequately grounded so as to limit the potential difference between individual sections within the allowable limit of 65 to 130 V during faults (Sverak et al., 1982). The analysis of GIS grounding includes estimation of the permissible touch voltage. For this, equation (13.19) could be used with $\rho_s = 0$ for a metal-to-metal contact. Dangerous touch and step voltages within the GIS area are drastically reduced by complete bonding and grounding of the GIS enclosures, and by using grounded conductive platforms connected to the GIS structures.

13.7 TRANSMISSION-LINE TOWER GROUNDING

To design and operate a transmission system with a low-outage rating and safety to the maintenance personnel as well as the public, it is necessary to have a suitable grounding system. This can be furnished by overhead ground wires, counterpoises, and by earthing tower bodies and foundations.

The degree to which low tower footing resistance can be met depends on local soil conditions. The method used to reduce the equivalent footing resistance and the degree to which this is carried out is a matter of economics. Experience indicates that some means of reducing the footing resistance to an equivalent of 10 Ω, as measured with the ground wires removed, is more economical than adding extra insulation.

Unfortunately, improved grounding cannot be economically effective in the event of direct lightning strokes to phase conductors. The importance of having low tower footing resistance in this case is in avoiding a high rate of back flashovers and thereby improving the conditions for successful fault suppression by ground-fault neutralizers. Ground wires are so located as to shield the line conductors adequately from direct lightning strokes. Their design is discussed in detail elsewhere (IEEE Committee, 1975; 1978).

Grounding Systems

With underground cables the situation is different. Dangerous voltages to earth may result from insulation failure, charges due to electrostatic induction, flow of currents through the sheath, or from the voltage rise during faults discharging to the station ground system to which the sheaths are connected.

13.8 SAFETY GROUND

Safety ground is meant to ensure that persons working with electrical equipment will not be exposed to the danger of electric shocks. Safety of operating personnel requires grounding of all exposed metal parts of power equipment. There is no simple relation between the resistance of the ground system as a whole and the maximum shock current to which a person might be exposed. Thus a low station ground resistance is not in itself a guarantee of safety (Sverak et al., 1981, El-Kady and Vainberg, 1983).

13.8.1 Range of Tolerable Currents

The effects of an electric current passing through the vital parts of a human body depend on its duration, magnitude, and frequency. Human beings are most vulnerable to currents with frequencies of 50 to 60 Hz. The human body can tolerate slightly larger currents at 25 Hz and approximately five times larger direct currents and still larger currents at 3 to 10 kHz (IEEE Committee, 1982).

Currents of 1 to 6 mA, often termed "let-go currents," although unpleasant to sustain, do not impair a person's ability to control his or her muscles. Currents of 9 to 25 mA may be quite painful and can make it hard to release energized objects grasped by the hand. For still higher currents, uncontrollable muscular contractions could make breathing difficult. In the range of 100 mA ventricular fibrillation, stoppage of the heart, or inhibition of respiration might occur and cause severe injury or death. It has been observed (Sverak et al., 1981) that for the same effect the current magnitude I varies with its duration t according to a relation of the form $I \sim t^{-1/2}$. High-speed clearing of ground faults is evidently crucial, as the risk of electric shock would thus be greatly mitigated.

To avoid danger from electric shock, one's body should under no circumstances be a part of an electric circuit. There are several means of reducing the hazard of electric shock, including grounding, isolation, guarding, insulation and double insulation, shock limitation, isolation transformers, and employing high-frequency and direct current. The first four methods are normally adopted in high-voltage systems.

13.8.2 Tolerable Step, Touch, and Transferred Voltages

Using the magnitude of the tolerable body current and the appropriate circuit constants, it is possible to calculate the tolerable potential difference between possible points of contact. The step voltage (Fig. 13.7) increases the closer one gets to the site of a ground fault or to the grounding point. If the enclosure of grounded equipment in which an earth fault has occurred is touched by a person, he or she will be subjected to a potential termed a "touch voltage" (Fig. 13.7).

When a person standing within the station area touches a conductor grounded at a remote point or when a person standing at a remote point touches a conductor connected to the station ground grid, he or she is subjected to a transferred potential. The shock voltage in this case may be equal to the full voltage rise of the ground grid under fault conditions (El-Kady and Vainberg, 1983). The tolerable step and touch voltages with durations t_s for persons weighing 50 to 70 kg could be estimated as

$$E_{step} = [1000 + 6C(\rho_s)] \frac{0.116}{\sqrt{t}} \qquad (13.22)$$

$$E_{touch} = [1000 + 1.5C(\rho_s)] \frac{0.16}{\sqrt{t}} \qquad (13.23)$$

where C is a factor depending on the soil homogeneity and equals 1 for uniform soil.

13.8.3 Electric Shock Hazard in Hospitals

Increased application of electric instrumentation has greatly increased the risk of electric shock, especially when a patient in a hospital is actually connected into the circuit. The hazard increases when probes or needles are inserted into the body. Sometimes these are placed directly on or inside a heart chamber. Some medical authorities believe that a current as small as 20 µA at 50 to 60 Hz applied directly to the heart could produce ventricular fibrillation. The hazard is increased if pacemakers are used.

13.9 MEASUREMENT OF GROUND RESISTANCE

Ground resistance measurement consists of measuring the resistance of a body of earth surrounding a grounding electrode system. One

Grounding Systems

end of the resistance is definitely available at the grounding system itself, while the other end is not practically available and is called "remote soil." In practice, however, about 98% of the total resistance is contained within a finite distance from the grounding system. The power frequency reactances of the grounding system can safely be neglected unless its ohmic resistance is extremely low. The resistance is measured using alternating current to avoid possible polarization effects under direct current (IEEE, 1983; Dawalibi and Blattner, 1984). The practical and reliable method for measuring the resistance of a ground is that of "fall of potential."

13.9.1 Fall-of-Potential Method

This method has several variations and is applicable to all types of ground resistance measurements. It involves passing through the grounding system G a current I to return from another electrode C (Fig. 13.9a). The passage of this current produces at a distance x from G a voltage drop V_x in the soil. V_x is measured by a potential probe P. The simplest form of the fall of potential method is obtained when G, P, and C are on a straight line and P is located between G and C. When V_x/I is plotted as a function of the potential probe distance x, curves similar to those shown in Figure 13.9b are produced. If the distance D is large enough with respect to the grounding system dimensions, the center part of the fall-of-

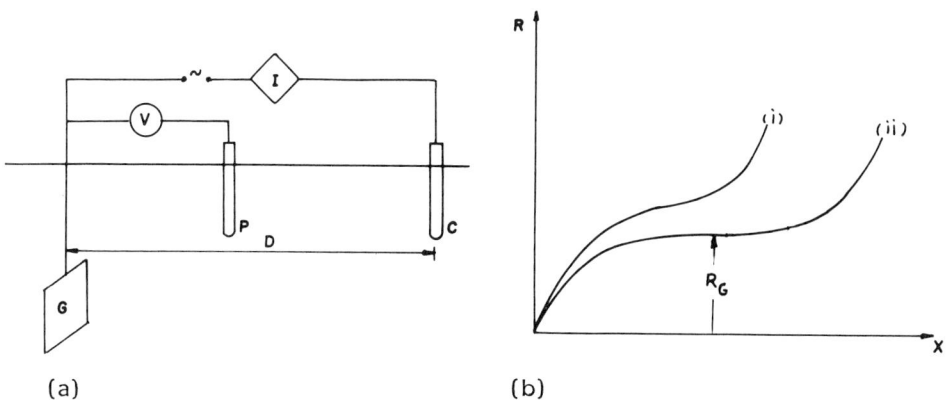

Fig. 13.9 Ground resistance measurement by the fall-of-potential method; (a) circuit connection; (b) resistance plots for the electrode separation, D, being too short (i) and adequate (ii).

potential curve tends to be flat. It is usually accepted that the flat section of the curve gives the correct magnitude of the resistance R_G.

13.10 MEASUREMENT OF EARTH RESISTIVITY

Soil resistivity tests should be carried out before designing grounding systems. Since the performance and cost of any grounding electrode or ground grid is directly related to soil resistivity, accurate evaluation is a matter of considerable importance.

13.10.1 Four-Point Method

The most common method of measuring soil resistivity employs four electrodes driven into the soil. The theoretical basis of this method has been known. Small electrodes are inserted into four small holes in the earth all to a depth of b meters and spaced along a straight line at intervals of a meters, and making electrical contact with the earth only at the bottom. A test current I is injected through the earth between the two outer electrodes and the potential E between the two inner electrodes is measured with a potentiometer or high-impedance voltmeter. The ratio between the observed potential and the injected current is referred to as the apparent resistance, which is a function of soil resistivity and the electrode geometry. The apparent soil resistivity is computed by multiplying the apparent resistance by a geometric multiplier. The apparent soil resistivity is

$$\rho = \frac{4\pi a R}{1 + 2a/\sqrt{a^2 + 4b^2} - a/\sqrt{a^2 + b^2}} \qquad (13.24)$$

where R is the apparent resistance (Ω) and ρ is soil resistivity ($\Omega \cdot m$). For spacing a, much greater than depth b, it can be easily shown that equation (13.24) reduces to

$$\rho = 2\pi a R \qquad (13.25)$$

13.10.2 Four-Long Cylindrical Electrodes

It is important to note that in the relation (13.25) the electrodes are considered as "points" buried at a depth b. The following new formula based on the field theory relating soil resistivity to measured resistance was developed (Baishiki et al., 1987). It

Grounding Systems

relaxes the requirement that rod electrodes have a large spacing compared to their buried length.

$$\rho = \frac{4\pi R b}{2 \ln[(2+E)/(1+F)] + 2F - E - a/b} \quad (13.26)$$

with

$$E = \sqrt{4 + \left(\frac{a}{b}\right)^2}$$

$$F = \sqrt{1 + \left(\frac{a}{b}\right)^2}$$

where b is the electrode driven length in meters. When $a \gg b$, equations (13.24) and (13.26) reduce to (13.25).

13.11 GROUND RESISTANCE METERS

Any instrument for measuring the resistance of a grounding point and the resistivity of soil mainly comprises an ac source of an adjustable frequency and a measuring circuit. DC current would not be suitable for such tests because of the entailed electrolysis and polarization, introducing serious errors. The frequency to be used should differ from that of any neighboring network; otherwise, any pickup would cause errors in the measurements.

Figure 13.10 shows a schematic of the circuit in common use, where a hand-driven dc generator G is coupled to a mechanical inverter CR and to a mechanical rectifier PR. The inverter inverts the dc current I to ac (of a frequency adjusted by the speed of cranking the generator) before it is fed into the ground. The rectifier PR converts the ac potential picked up between the electrodes P1 and P2 back to dc before it is fed to its respective coil in the ohmmeter CC. This "ohmmeter" compares the measured potential with the main circuit current I.

The circuit shown in Figure 13.10 is for measuring the soil resistivity according to the four-point method. For measuring the resistance of a grounding electrode already installed it would be connected to the ground resistance meter at the two electrodes C1 and P1 (Fig. 13.10), combined. Some variations of the ground resistance meter use a battery instead of the generator G, and include electromechanical vibrators or solid-state devices for the current inversion and rectification.

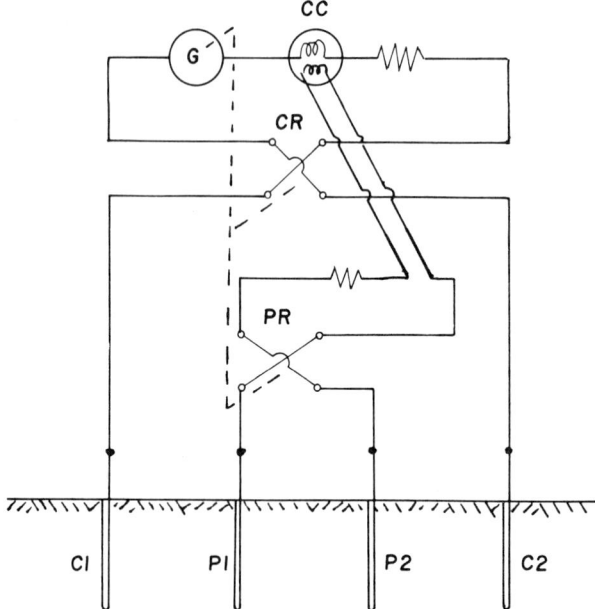

Fig. 13.10 Schematic connection of a ground resistance meter used for measuring the ground resistivity with the four-electrode method.

REFERENCES

Baishiki, R. S., Osterberg, C. K., and Dawalibi, F. (1987). *IEEE Trans.*, PWRD-2:64–71.

Brown, P., Johnson, I., and Stevenson, J. (1978). *IEEE Trans.*, PAS-97:683–694.

Dawalibi, F. and Blattner, C. (1984). *IEEE Trans.*, PAS-103:374–382.

El-Kady, M. and Vainberg, M. (1983). *IEEE Trans.*, PAS-102:3080–3087.

El-Morshedy, A., Zeitoun, A. G., and Ghourab, M. M. (1986). *Proc. IEE*, 133C:287–292.

Ershevich, V. V., Krivwshkin, L. F., Neklepaev, B. N., Sheimovich, V. D., and Slavin, G. A. (1982). *CIGRE paper 32-09*.

Garrett, D. and Holley, H. (1980). *IEEE Trans.*, PAS-99:2008–2011.

Gupta, B. and Thapar, B. (1980). *IEEE Trans.*, *PAS-99*: 2357 – 2362.
IEEE (1972). *IEEE Recommended Practice for Grounding of Industrial and Commercial Power Systems*, Publication 142, Institute of Electrical & Electronics Engineers, New York.
IEEE (1983). *IEEE Guide for Measuring Ground Resistance and Potential Gradients in the Earth*, Publication 81, Institute of Electrical & Electronics Engineers, New York.
IEEE (1987). *IEEE Guide for Safety in Alternating Current Substation Grounding*, Publication 80, Institute of Electrical & Electronics Engineers, New York.
IEEE Committee (1975). *IEEE Trans.*, *PAS-94*: 1241 – 1247.
IEEE Committee (1978). *IEEE Trans.*, *PAS-97*: 2243 – 2252.
IEEE Committee (1982). *IEEE Trans.*, *PAS-101*: 4006 – 4023.
Komaragiri, K. and Mukhedkar, D. (1981). *IEEE Trans.*, *PAS-100*: 2993 – 3001.
Nahman, J. and Salamon, D. (1984). *IEEE Trans.*, *PAS-103*: 880 – 885.
Sunde, E. D. (1968). *Earth Conduction Effects in Transmission Systems*, Macmillan Publishing Co., Inc., New York.
Sverak, J. G., Dick, W. K., Dodds, T. H., and Heppe, R. H. (1981). *IEEE Trans.*, *PAS-100*: 4281 – 4290.
Sverak, J. G., Benson, R. U., Dick, W. K., Dodds, T. H., Garret, D. L., Idzkowski, J. E., Keil, R. P., Patel, S. G., Ragan, M. E., Smith, G. E., Verma, R., and Zukerman, L. G. (1982). *IEEE Trans.*, *PAS-101*: 4006 – 4023.
Verma, R. and Mukhedkar, D. (1981). *IEEE Trans.*, *PAS-100*: 1023 – 1030.

14
Overvoltages on Power Systems

H. ANIS *Cairo University, Giza, Egypt*

14.1 INTRODUCTION

The examination of overvoltages on the power system includes a study of their magnitudes, shapes, durations, and frequency of occurrence. This study should be performed at the point not only where an overvoltage originates but also at all other points along the transmission network to which the surges may travel.

14.2 TYPES OF OVERVOLTAGES

The voltage stresses on transmission network insulation are found to have a variety of origins. Normal operation ac (or dc) voltages do not stress the insulation severely. However, they remain the initial factor that determines its dimensions. Overvoltages stressing a power system can generally be classified into two main types:

1. *External overvoltages*: generated by atmospheric disturbances. Of these disturbances, lightning is the most common and the most severe.
2. *Internal overvoltages*: generated by changes in the operating conditions of the network. Internal overvoltages can be divided into (a) switching overvoltages and (b) temporary overvoltages.

14.3 LIGHTNING OVERVOLTAGES

According to theories generally accepted, lightning is produced in an attempt by nature to maintain a dynamic balance between the positively charged ionosphere and the negatively charged earth

(Marshall, 1973). Over fair-weather areas there is a downward transfer of positive charges through the global air–earth current. This is then counteracted by thunderstorms, during which positive charges are transferred upward in the form of lightning (Lewis, 1965).

During thunderstorms, positive and negative charges are separated by the movements of air currents forming ice crystals in the upper layer of a cloud and rain in the lower part. The cloud becomes negatively charged and has a larger layer of positive charge at its top. As the separation of charge proceeds in the cloud, the potential difference between the concentrations of charges increases and the vertical electric field along the cloud also increases. The total potential difference between the two main charge centers may vary from 100 to 1000 MV. Only a part of the total charge — several hundred coulombs — is released to earth by lightning; the rest is consumed in intercloud discharges. The height of the thundercloud dipole above earth may reach 5 km in tropical regions.

14.3.1 The Lightning Discharge

The lightning discharge through air occurs as one of the forms of streamer breakdown of long air gaps explained in Chapter 4. The channel to earth is first established by a stepped discharge called a leader stroke. The leader is generally initiated by a breakdown between polarized water droplets at the cloud base by the high electric field or a discharge between the negative charge mass in the lower cloud and the positive charge pocket below it. In Figure 14.1 the development stages of a lightning flash are depicted.

As the downward leader approaches the earth, an upward leader begins to proceed from earth before the former reaches earth. The upward leader joins the downward one at a point referred to as the striking point. This is the start of the return stroke, which progresses upward like a traveling wave on a transmission line. At the earthing point a heavy impulse current reaching the order of tens of kiloamperes occurs, which is responsible for the known damage of lightning. The velocity of progression of the return stoke is very high and may reach half the speed of light. The corresponding current heats its path to temperatures up to 20,000°C, causing the explosive air expansion that is heard as thunder. The current pulse rises to its crest in a few microseconds and decays over a period of tens or hundreds of microseconds (Ragaller, 1980).

14.3.2 Lightning Voltage Surges

The most severe lightning stroke is that which strikes a phase conductor on the transmission line as it produces the highest

Overvoltages on Power Systems 359

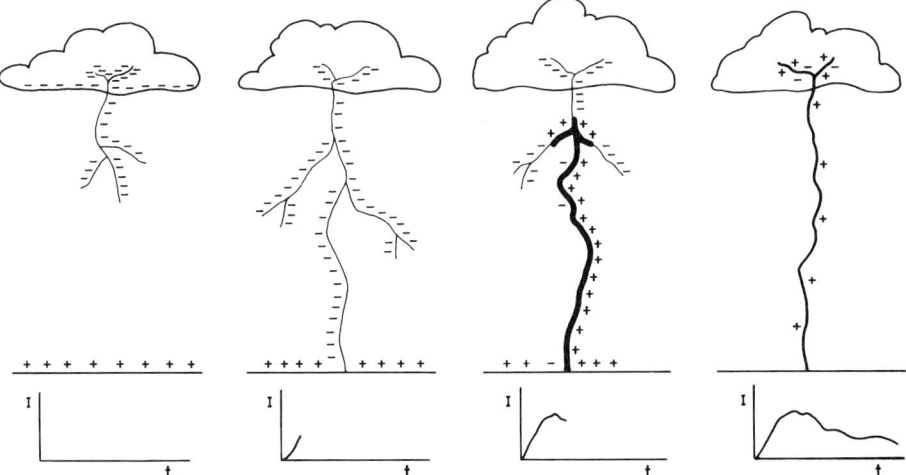

Fig. 14.1 Developmental stages of a lightning flash and the corresponding current surge.

overvoltage for a given stroke current. The lightning stroke injects its current into a termination impedance Z, which in this case is half the line surge impedance Z_0 since the current will flow in both directions as in Figure 14.2. Therefore, the voltage surge magnitude at the striking point is

$$V = \tfrac{1}{2} I Z_0 \tag{14.1}$$

The lightning current magnitude is rarely less than 10 kA (Berger, Anderson, and Kroninger, 1975) and thus for a typical overhead line surge impedance Z_0 of 300 Ω, the lightning surge voltage will probably have a magnitude in excess of 1500 kV. Equation (14.1) assumes that the impedance of the lightning channel itself is much larger than $\tfrac{1}{2} Z_0$; indeed, it is believed to range from 100 to 3000 Ω. Equation (14.1) also indicates that the lightning voltage surge will have approximately the same shape characteristics. In practice, however, the shapes and magnitudes of lightning surge waves get modified by their reflections at points of discontinuities as they travel along transmission lines (Section 14.6).

Lightning strokes represent true danger to life, structures, power systems, and communication networks. Lightning is always a major source of damage to power systems where equipment insulation

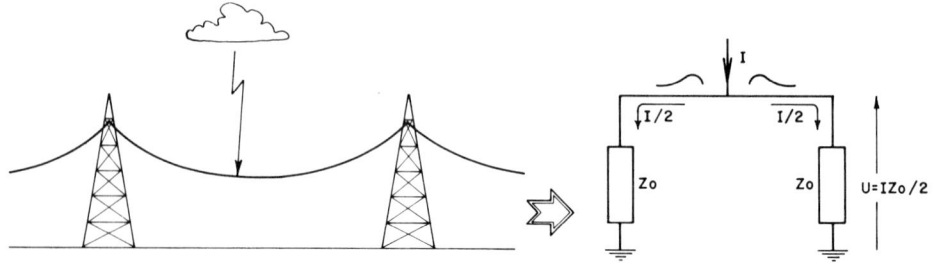

Fig. 14.2   Development of lightning overvoltage.

may break down under the resulting overvoltage and the subsequent high-energy discharge.

14.4   SWITCHING OVERVOLTAGES

With the steady increase in transmission voltages needed to fulfill required increase in transmitted powers, switching surges have become the governing factor in the design of insulation for EHV and UHV systems. In the meantime, lightning overvoltages come as a secondary factor in these networks. There are two fundamental reasons for this shift in relative importance from lightning to switching surges as higher transmission voltages are called for:

1. Overvoltages produced on transmission lines by lightning strokes are only slightly dependent on the power system voltages. As a result, their magnitudes relative to the system peak voltage decrease as the latter is increased.
2. As shown in Chapter 15, external insulation has its lowest breakdown strength under surges whose fronts fall in the range 50 to 500 µs, which is typical for switching surges.

According to the International Electrotechnical Commission (IEC) recommendations, all equipment designed for operating voltages above 300 kV should be tested under switching impulses (i.e., laboratory-simulated switching surges) (IEC, 1973).

14.4.1   Origin of Switching Overvoltages

There is a great variety of events that would initiate a switching surge in a power network. The switching operations of greatest relevance to insulation design can be classified as follows:

Overvoltages on Power Systems 361

1. *Energization of transmission lines and cables.* The following specific switching operations are some of the most common in this category:
    a. Energization of a line that is open circuited at the far end
    b. Energization of a line that is terminated by an unloaded transformer
    c. Energization of a line through the low-voltage side of a transformer
2. *Reenergization of a line.* This means the energization of a transmission line carrying charges trapped by previous line interruptions when high-speed reclosures are used.
3. *Load rejection.* This is effected by a circuit breaker opening at the far end of the line. This may also be followed by opening the line at the sending end in what is called a line dropping operation.
4. *Switching on and off of equipment.* All switching operations involving an element of the transmission network will produce a switching surge. Of particular importance, however, are the following operations:
    a. Switching of high-voltage reactors
    b. Switching of transformers that are loaded by a reactor on their tertiary winding
    c. Switching of a transformer at no load
5. *Fault initiation and clearing.*

14.4.2 Energization of an Unloaded Transmission Line

When an unloaded transmission line is switched on, the sinusoidal supply voltage is suddenly applied to it as represented by the single-phase circuit of Figure 14.3. The transformer is represented here by its leakage inductance and the line by its inductance and capacitance to ground. The switching operation is effected at an instant T seconds beyond that of zero voltage. The voltage across the capacitor C is the one under study here, as it represents the voltage at the open-circuit end of the line. The resistance R includes all series resistances of the line and transformer.

The circuit performance after switching may be expressed by the following differential equation:

$$v_s(t) = Ri(t) + L\frac{di(t)}{dt} + \frac{1}{C} \int i(t)\, dt \tag{14.2}$$

The supply voltage $v_s(t)$ beyond the switching instant is given by

$$v_s(t) = V_s \sin(\omega t + \omega T) \tag{14.3}$$

Equations (14.2) and (14.3) could easily be manipulated (e.g., using operational calculus) and the expression for the voltage across the line capacitance takes the form

$$v_c(t) = V_c \sin(\omega t + \omega T - \theta) + A e^{-\alpha t} \sin(\omega_1 t + \beta) \qquad (14.4)$$

where

$$\theta = \tan^{-1} \frac{-R}{\omega L - 1/\omega C}$$

$$V_c = \frac{V_s}{\omega C \sqrt{R^2 + (\omega L - 1/\omega C)^2}}$$

$$A = -V_c \frac{\sin(\omega T - \theta)}{\sin \beta}$$

$$\beta = \tan^{-1} \frac{\omega_1 \sin(\omega T - \theta)}{\omega \cos(\omega T - \theta) + \alpha \sin(\omega T - \theta)} \qquad (14.5)$$

$$\alpha = \frac{R}{2L}$$

$$\omega_0 = \frac{1}{\sqrt{LC}}$$

$$\omega_1 = \sqrt{\omega_0^2 - \alpha^2} = 2\pi f_1$$

Example

A typical 132/500-kV, 200-MVA transformer would have an inductance (referred to its 500-kV side) of about 400 mH; a 500-kV transmission line may have a series inductance of about 1 mH/km and a capacitance to ground of about 0.015 µF/km. For a transmission line of 100 km, the total series inductance L of Figure 14.3 would be 500 mH and the capacitance C would be 1.5 µF. Neglecting all series resistances (i.e., letting $\alpha = 0$), the frequency of oscillation of the transient voltage component as given in equation (14.4) would be

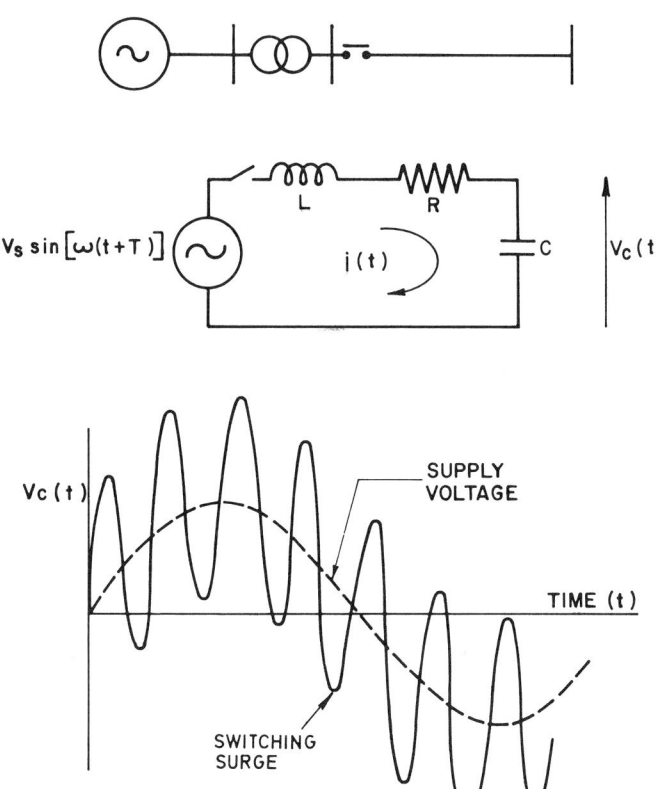

Fig. 14.3 Energization switching transient.

$$f_1 = \frac{1}{2\pi \sqrt{LC}}$$

$$= 184 \text{ Hz}$$

This, in effect, means that the time to crest of the switching surge, which is about $1/4f_1$, is

$$t_{crest} = \frac{10^6}{4 \times 400}$$

$$= 1360 \text{ μs}$$

In Figure 14.3 the expected switching surge is sketched for the previous case when energization is made at T = 0.

14.5 TEMPORARY OVERVOLTAGES

Temporary overvoltages (i.e., sustained overvoltages) differ from transient switching overvoltages in that they last for longer durations, typically from a few cycles to a few seconds. They take the form of undamped or slightly damped oscillation at a frequency equal or close to the power frequency. The classification of temporary overvoltages as distinct from transient switching overvoltages is due mainly to the fact that the responses of power network insulation and surge arresters to their wave shapes are different. Some of the most important events leading to the generation of temporary overvoltages are discussed briefly below.

14.5.1 Load Rejection

When a transmission line or a large inductive load that is fed from a power station is suddenly switched off, the generator will speed up and the busbar voltage will rise. The amplitude of the overvoltage can be evaluated approximately as illustrated in Figure 14.4 by

$$V = E \frac{X_c}{X_c - X_s} \qquad (14.6)$$

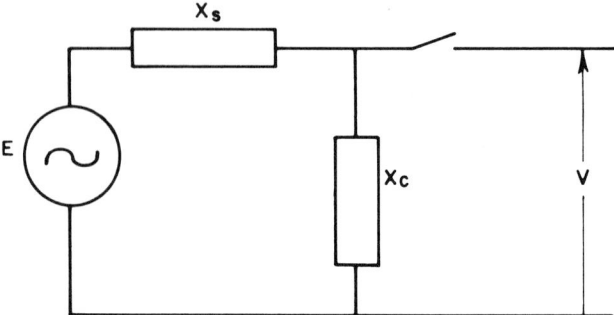

Fig. 14.4 Equivalent circuit during load rejection.

Overvoltages on Power Systems

where E is the voltage behind the transient reactance, which is assumed to be constant over the subtransient period and equal to its value before the incident, X_S the transient reactance of the generator in series with the transformer reactance, and X_c the equivalent capacitive input reactance of the system.

14.5.2 Ferranti Effect

The Ferranti effect of an uncompensated transmission line is given by

$$\frac{V_r}{V_s} = \frac{1}{\cos \beta_0 \ell} \qquad (14.7)$$

where V_r and V_s are the receiving-end and sending-end voltages, respectively; and ℓ is the line length in (km). β_0 is the phase shift constant of the line per unit length. It is equal to the imaginary part of $\sqrt{ZY}$, where Z and Y are the impedance and admittance of the line per unit length. For a lossless line $\beta_0 = \omega \sqrt{LC}$, where L and C are the inductance and capacitance of the line per unit length. β_0 has a value of about 6° per 100 km at normal power frequency (Diesendorf, 1974).

14.5.3 Ground Fault

A single line-to-ground fault will cause the voltages to ground of the healthy phases to rise. In the case of a line-to-ground fault, systems with neutrals isolated or grounded through a high impedance may develop overvoltages on healthy phases higher than normal line-to-line voltages. Solidly grounded systems, on the other hand, will only permit phase-to-ground overvoltages well below the line-to-line value. An earth fault factor is defined as the ratio of the higher of the two sound phase voltages to the line-to-neutral voltage at the same point in the system with the fault removed.

14.5.4 Harmonic Overvoltages Due to Magnetic Saturation

Harmonic oscillations in power systems are initiated by system non-linearities whose primary source is that of the saturated magnetizing characteristics of transformers and shunt reactors. The magnetizing current of these components increases rapidly and contains a high percentage of harmonics for voltages above the rated voltage. Therefore, saturated transformers inject large harmonic currents into the system.

14.6 TRAVELING WAVES

Any overvoltage surge appearing on a transmission line due to internal or external disturbances will propagate in the form of a traveling wave toward the ends of the line. During its travel the overvoltage surge will, in general, experience attenuation and distortion inflected upon it by earth resistance, line losses, and corona discharges.

To be able to develop the time equation of the traveling wave, a transmission line may be looked upon as being made up of a chain of lumped components such as that shown in Figure 14.5, in which

R = line resistance (Ω/m)

L = line inductance (H/m)

C = line capacitance-to-ground (F/m)

G = line leakage conductance (S/m)

dx = length of a line element (m)

The magnetic flux linkage produced by the wave's current is given by

$$\phi = iL \, dx \tag{14.8}$$

Consequently, the total voltage drop over an element dx is given by

$$-dv = \left(R + L \frac{\partial}{\partial t}\right) i \, dx \tag{14.9}$$

The total change in wave current is caused by the leakage component vG dx and the current used to charge the line capacitance to ground. Therefore, the total current change takes the form

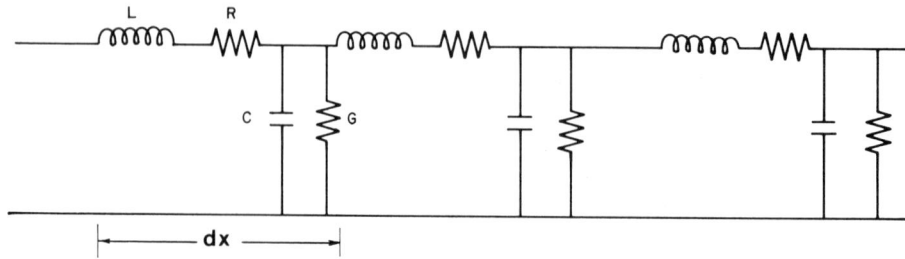

Fig. 14.5 Transmission-line equivalent circuit.

Overvoltages on Power Systems 367

$$-di = \left(G + C\frac{\partial}{\partial t}\right) v\, dx \tag{14.10}$$

By differentiation we can easily get

$$\frac{\partial^2 v}{\partial x^2} = [(R + Ls)(G + Cs)]v \tag{14.11}$$

and

$$\frac{\partial^2 i}{\partial x^2} = [(R + Ls)(G + Cs)]i \tag{14.12}$$

s being the differential operator.

Equations (14.11) and (14.12) are the known transmission-line traveling-wave equations, in which the term $(R + Ls)$ is the impedance operator Z, and $(G + Cs)$ is the admittance operator Y. The solution to equations (14.11) and (14.12) has the general form

$$v = e^{x\Gamma} F_1(t) + e^{-x\Gamma} F_2(t) \tag{14.13}$$

and

$$i = -\sqrt{\frac{Y}{Z}}\, [e^{x\Gamma} F_1(t) - e^{-x\Gamma} F_2(t)] \tag{14.14}$$

where the operator Γ is defined as

$$\Gamma = \sqrt{ZY} \tag{14.15}$$

The time functions $F_1(t)$ and $F_2(t)$ are integration constants with respect to x only. The exact solution to equations (14.11) and (14.12) depends on the nature of Γ (i.e., on the line constants) and on the boundary conditions of the wave. The operator Γ can easily be shown to take the form

$$\Gamma = \frac{1}{\gamma}\sqrt{(s + \zeta)^2 - \eta^2} \tag{14.16}$$

in which γ has the dimensions of velocity (m/s) and

$\zeta = \tfrac{1}{2}\left(\dfrac{R}{L} + \dfrac{G}{C}\right)$ is the attenuation constant

$\eta = \tfrac{1}{2}\left(\dfrac{R}{L} - \dfrac{G}{C}\right)$ is the wavelength constant

14.6.1 Lossless Line Equation

An ideal transmission line with no losses should have both $R = 0$ and $G = 0$. Then the operator Γ reduces to

$$\Gamma = s\sqrt{LC} = \frac{s}{\gamma} \tag{14.17}$$

while both the attenuation and wavelength constants vanish. Remembering that according to Taylor's expansion

$$e^{\pm xs/\gamma}F(t) = F\left(t \pm \frac{x}{\gamma}\right)$$

the voltage and current waves along a lossless line take the general forms

$$v = F_1\left(t + \frac{x}{\gamma}\right) + F_2\left(t - \frac{x}{\gamma}\right) \tag{14.18}$$

$$i = -\sqrt{\frac{C}{L}}\left[F_1\left(t + \frac{x}{\gamma}\right) - F_2\left(t - \frac{x}{\gamma}\right)\right] \tag{14.19}$$

This solution, as it is expressed in terms of $(t \pm x/\gamma)$, indicates that the wave is traveling along x at a velocity γ while maintaining its shape such that for a given value of $(t \pm x/\gamma)$ the wave's instantaneous value is unchanged.

The voltage (or current) wave is also seen in equations (14.18) and (14.19) to be made up of a component F_2 traveling forward along x and a component F_1 traveling backward (Fig. 14.6). The magnitudes of the voltage and current waves are related as follows:

For the forward component:

$$\frac{v}{i} = \sqrt{\frac{L}{C}}$$

For the backward component:

$$\frac{v}{i} = -\sqrt{\frac{L}{C}}$$

This means that the voltage and current waves traveling along the line have a fixed ratio termed the line's surge impedance Z_0.

$$Z_0 = \sqrt{\frac{L}{C}} \tag{14.20}$$

Overvoltages on Power Systems 369

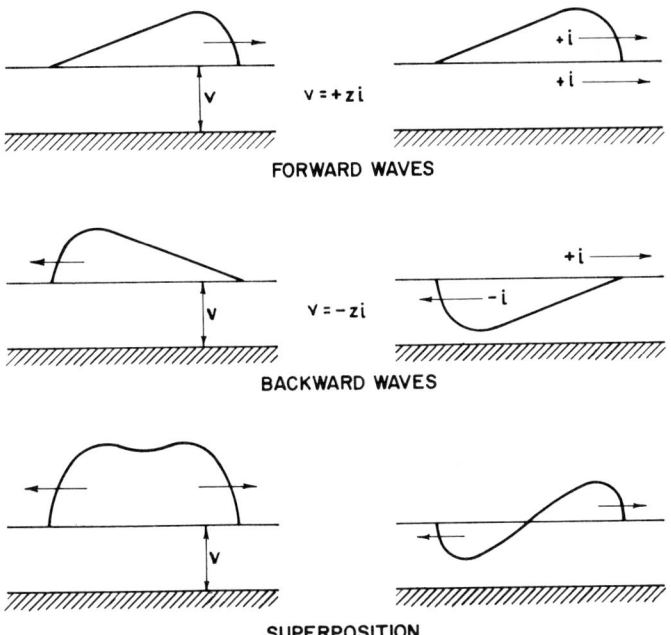

Fig. 14.6 Voltage and current traveling-wave components.

The surge impedance is clearly independent of the line length. In practice, it is about 300 to 400 Ω for overhead transmission lines and about 30 to 80 Ω for underground cables.

14.6.2 Velocity of Wave Propagation

For a lossless transmission line, it can easily be shown from equation (14.9) that

$$\frac{\partial v}{\partial x} = -L \frac{\partial i}{\partial t} \qquad (14.21)$$

Similarly,

$$\frac{\partial i}{\partial x} = -C \frac{\partial v}{\partial t} \qquad (14.22)$$

Realizing that dx/dt is the velocity at which the surge propagates, an expression for that velocity may easily be deduced from equations (14.21) and (14.22) as

$$\gamma = \frac{1}{\sqrt{LC}} \qquad (14.23)$$

As we remember, the inductance L per unit length of a conductor with radius r and height h above ground is approximately given by

$$L = \frac{\mu_0}{\pi} \ln \frac{2h}{r} \quad H/m \qquad (14.24)$$

Also, the capacitance C per unit length of the same conductor arrangement is given by

$$C = \frac{\pi \varepsilon_0}{\ln(2h/r)} \quad F/m \qquad (14.25)$$

where μ_0 and ε_0 are the permeability and permittivity of free space, respectively. Therefore, the velocity of wave propagation is given by

$$\gamma = \frac{1}{\sqrt{\mu_0 \varepsilon_0}} \qquad (14.26)$$

which is equal to the speed of light and is independent of line geometry. At that speed a surge will travel 1 km in about 3.3 μs.

In case of cables, the dielectric constant ε_r of the insulation is larger than unity while the permeability remains very much equal to μ_0. Therefore, the velocity of wave propagation should be expected to be smaller than the speed of light by a factor of $\sqrt{\varepsilon_r}$.

14.6.3 Reflection and Refraction of Traveling Waves

When the traveling wave on a transmission line reaches a point beyond which the line constants are different (e.g., a cable connected to an overhead line or where equipment or more lines are connected), a part of the wave is reflected back along the line (the reflected component) and another part is passed on to the new section (the refracted component). Consider a junction between

Overvoltages on Power Systems 371

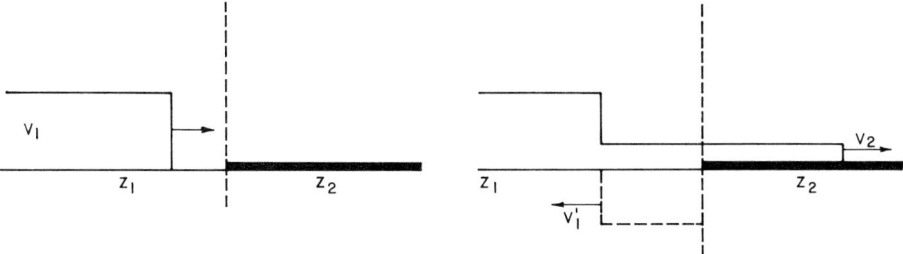

Fig. 14.7 Wave reflection at a junction.

two lines whose surge impedances are Z_1 and Z_2 such as that shown in Figure 14.7. A surge that is traveling along line 1 has its voltage and current related by

$$\frac{v_1}{i_1} = Z_1 \tag{14.27}$$

At the junction, the reflected and refracted waves will have voltages v_1' and v_2 and currents i_1' and i_2, respectively. These voltages and currents are related by

$$i_1' = \frac{-v_1'}{Z_1} \tag{14.28}$$

and

$$i_2 = \frac{v_2}{Z_2} \tag{14.29}$$

Simply, applying Kirchoff's laws at the junction yields

$$v_1 + v_1' = v_2 \tag{14.30}$$

and

$$i_1 + i_1' = i_2 \tag{14.31}$$

Therefore,

$$\frac{v_1}{Z_1} - \frac{v'_1}{Z_1} = \frac{v_2}{Z_2} \tag{14.32}$$

The reflected and refracted wave voltages can finally be put in the form

$$v'_1 = \frac{Z_2 - Z_1}{Z_2 + Z_1} v_1 \tag{14.33}$$

$$v_2 = \frac{2Z_2}{Z_2 + Z_1} v_1 \tag{14.34}$$

The quantity $(Z_2 - Z_1)/(Z_2 + Z_1)$ is termed the reflection coefficient, and the quantity $2Z_2/(Z_2 + Z_1)$ is the refraction (transmission) coefficient.

14.6.4 Lattice Diagram

The successive reflection of a traveling wave at the sending and receiving ends of a transmission line may lead to a buildup in the voltage or current at some points along the line. A method—normally called the lattice diagram (Bewley, 1963)—is applied in Fig. 14.8 to the simple case of a lossless transmission line with surge impedance Z_0 connected at the sending end to an infinite bus (i.e., one that has a zero internal impedance) and terminated at the receiving end by a load impedance Z. The time-space diagram following the issuing of a voltage surge at the sending end of 1 p.u. is illustrated. The time taken by surge to cross the entire line is constant and denoted by T. Incident waves at the receiving end are reflected with a reflection coefficient of

$$\alpha = \frac{Z - Z_0}{Z + Z_0}$$

Subsequent reflections at the sending end occur with a reflection coefficient of -1.

The voltage buildup at any chosen point along the line can be estimated by the algebraic summation of the voltage surge magnitudes after each surge reflection. For example, the voltage at

Overvoltages on Power Systems 373

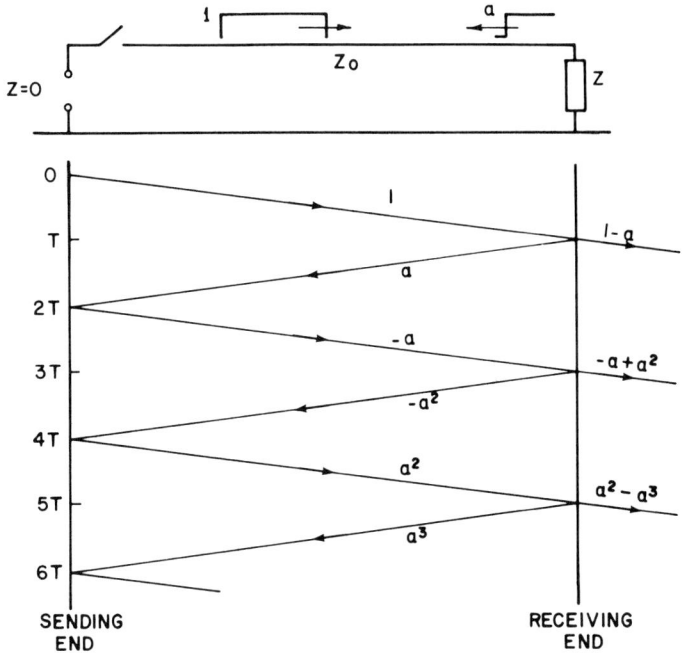

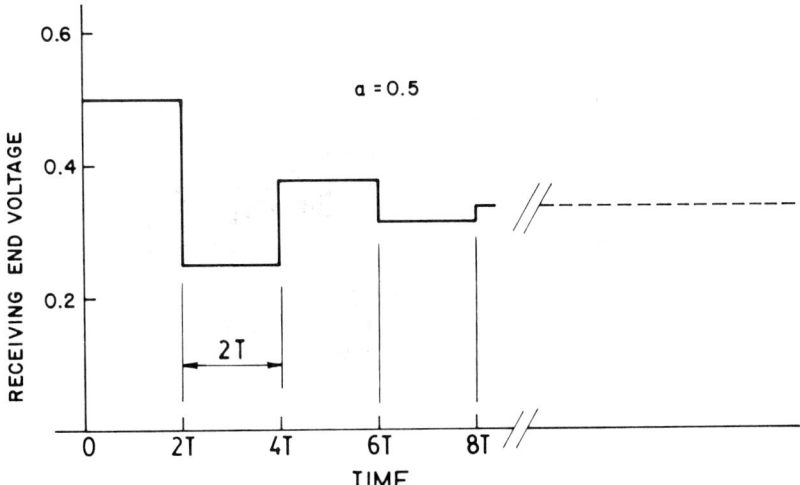

Fig. 14.8 Lattice diagram.

the receiving end (R.E.) of the system shown in Figure 14.8 can be evaluated to take the form indicated for the case:

$$\text{R.E. voltage } [rT < t < (r+2)T] =$$

$$(1-a)(1-a+a^2-a^3+\cdots) \qquad (14.35)$$

After a sufficiently long time the receiving-end voltage in this case settles at a final value of

$$\text{R.E. voltage (final)} = \frac{1-a}{1+a} \quad \text{p.u.} \qquad (14.36)$$

as shown in Figure 14.8.

14.6.5 Approximations

The discussion above assumed that the line is a lossless single conductor passing over a perfectly conducting ground. In reality, a traveling wave suffers from attenuation and distortion, due to energy loss and corona effects. For steep wavefronts, skin effect in conductors and ground return paths produce some attenuation and distortion (Semlyen, 1981). Corona discharge has the effect of retarding the portion of the wavefront above corona inception voltage, as discussed in Chapter 5 (Abdel-Salam and Stanek, 1987).

Multiconductor lines further distort a traveling wave. The voltage can be looked upon as made up of line components and ground components each having different velocities of propagation and different rates of attenuation. For ground components, the electrostatic charges induced in the ground are near the surface, while the return current is well below the surface. The depth of the current return path depends on frequency and soil resistivity. For a perfectly conducting earth the return current flows at the ground surface at a velocity equal to the speed of light. With finite earth conductivity the conductor's image goes well below the surface and the velocity of propagation is reduced. Distortion in this case results from the different velocities of the two voltage components (Ametani and Schinzinger, 1974; Nakagawa, 1981; Cristina and d'Amore, 1983).

14.7 CONTROL OF OVERVOLTAGES

The adverse effects of overvoltages on power networks can be reduced in two ways: by using protective devices—chiefly surge arresters—or by reducing their magnitudes wherever the surge

Overvoltages on Power Systems 375

originates. The latter way is commonly known as overvoltage control. The techniques employed to control switching surges and temporary overvoltages are outlined briefly below.

14.7.1 Control of Switching Surges

Following are some of the techniques currently in use to control the magnitudes of switching surges.

Resistor Switching

This is one of the most common methods for reducing energization overvoltages. It is effected by initially applying the supply voltage to the line through a resistor. After a suitable period of time, normally between one-third and one-half of a cycle, the preinserted resistor is short circuited, allowing the full supply voltage to be applied to the line. By the end of the preinsertion period, the magnitude of the energization surge is usually much reduced by the effect of system damping. This effect is evident from equation (14.4), which describes the voltage surge waveform. The initial amplitude of the energization surge when a preinsertion resistor of value R is used would be only $Z_0/(R + Z_0)$ of that reached in the absence of the resistor, where Z_0 is the surge impedance of the line.

When the resistor is shorted at the end of the preinsertion period, another surge will develop. If R is too small, control of the first surge becomes ineffective; if it is too large, the second surge becomes dangerous. An optimal value of R would normally be a fraction of Z_0, and depends on transmission-line length.

Phase-Controlled Closure

Referring to equation (14.4), it can be seen that the amplitude of the energization surge depends on the switching phase angle ωT. By properly timing of the closing of the circuit breaker poles, the resulting switching overvoltage can be greatly reduced. Phase-controlled switching should be carried out successively for the three poles to accomplish a reduction in the initial voltages on all three phases. This is extremely difficult with conventional circuit breakers but is quite possible with solid-state circuit breakers, as explained in Chapter 11.

Use of Shunt Reactors

Shunt reactors are used on many high-voltage transmission lines as a means of shunt compensation to improve the performance of the line, which would otherwise draw large capacitive currents from the supply. They have the additional advantage of reducing energization surge magnitudes. This is accomplished mainly by the reduction in temporary overvoltages, as will be seen in the next section.

Drainage of Trapped Charges

Charges are trapped on the capacitance to ground of transmission lines after their sudden reenergization. If the line is reenergized soon after, usually by means of automatic reclosures, these charges may cause an increase in the resulting surge. If in the simple system of Figure 14.3 the capacitance C has an initial voltage $V_c(0) = V_0$ caused by trapped charges, the surge voltage will include an extra component V_0, which if of the same polarity as the surge's peak voltage will increase the overvoltage on the line (Bickford and Doepel, 1967).

In practice, trapped charges may be partially drained through the switching resistors incorporated in circuit breakers, as mentioned earlier. Magnetic-type potential transformers also drain trapped charges via a low-frequency oscillation which is highly damped by the effect of magnetic saturation.

14.7.2 Control of Temporary Overvoltages

By referring to Figure 14.4 and its related overvoltage equation (14.6), it is evident that by increasing the capacitive input reactance of the transmission line X_c the magnitude of the temporary overvoltage V is reduced. If a shunt reactor of reactance X_r is added to the transmission line, the equivalent input reactance of that line will be increased from X_c to

$$X'_c = \frac{X_c}{1 - X_c/X_r} \qquad (14.37)$$

In equation (14.37), the goal of increasing X_c is obviously achieved by means of decreasing X_r (down from its infinite value in the absence of the reactor), thus reducing the overvoltage magnitude according to equation (14.6).

Furthermore, the second harmonic component of temporary overvoltages can be successfully suppressed, or even eliminated, by the use of surge arresters with nonlinear resistor (varistor) characteristics (Ragaller, 1980). A properly designed varistor would conduct in such a way as to provide large losses at the frequency in question.

Overvoltages on Power Systems 377

14.8 STATISTICAL CHARACTERISTICS OF OVERVOLTAGES

All types of overvoltages on a power system are subject to statistical fluctuation in their magnitudes, shapes, and durations. The sources of randomness can generally be grouped into two main categories:

1. Randomness produced simply by the existence of a large variety of overvoltage-producing events
2. Randomness created inherently in the overvoltage-producing process

Examples of the latter category are outlined in the following sections.

14.8.1 Statistical Variations in Lightning Surges

Many factors contribute to fluctuations in the lightning stroke current, which will, in turn, reflect on the consequent overvoltages. The distribution of thunder clouds in the area, the size and charge distribution within each cloud, the weather conditions (e.g., wind, temperature, pressure, etc.), and the distribution of projected objects in the area are among those factors. Figure 14.9 depicts a cumulative frequency distribution of lightning current magnitudes. It has been reported that, in general, higher currents are associated with longer fronts. The marked statistical fluctuation in lightning surges emphasizes the importance of taking this fluctuation into account when dimensioning the insulation and evaluating the risk of insulation failure.

14.8.2 Statistical Variations in Switching Surges

Since all switching surges are products of circuit breaker actions, the random variation in switching surges is attributed primarily to the performance of those devices, as well as the network circumstances at the time of switching. Random angles of switching are mainly responsible for this variation. The following factors contribute to the randomness in switching angles:

1. The mechanical movement of circuit breaker contacts produces fluctuations about the aiming angle of interruption or closure.

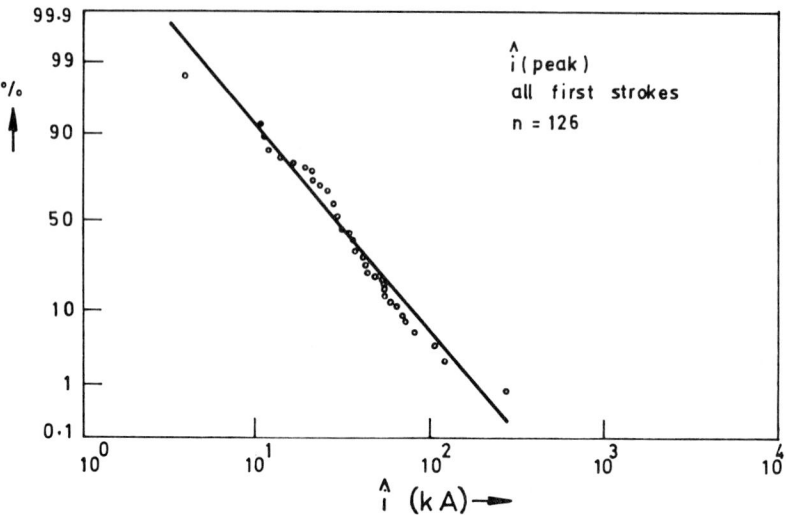

Fig. 14.9 Probability distribution of lightning peak current. [From Berger et al. (1975).]

2. Circuit closure may be prematurely effected following a breakdown between the breaker's contacts.
3. Arc interruption, whose timing is relevant to the production of recovery transient voltages, involves a number of physical processes, many of which are inherently random, as described in Chapter 6.

Figure 14.10 shows typical distributions of switching surge magnitudes. Uncontrolled surges may fluctuate in magnitude at a coefficient of variation (standard deviation relative to the mean) of about 25%, while proper control measures can reduce this fluctuation to only about 12%.

14.8.3 Statistical Variation in Temporary Overvoltages

Temporary overvoltages are not subject to randomness in the same sense that switching surges are. Any change in temporary overvoltage magnitude would have to be related to changes in network parameters. If these changes are considered, however, over a long period of time, a statistical distribution of overvoltage magnitudes may be produced. The distributions are highly skewed

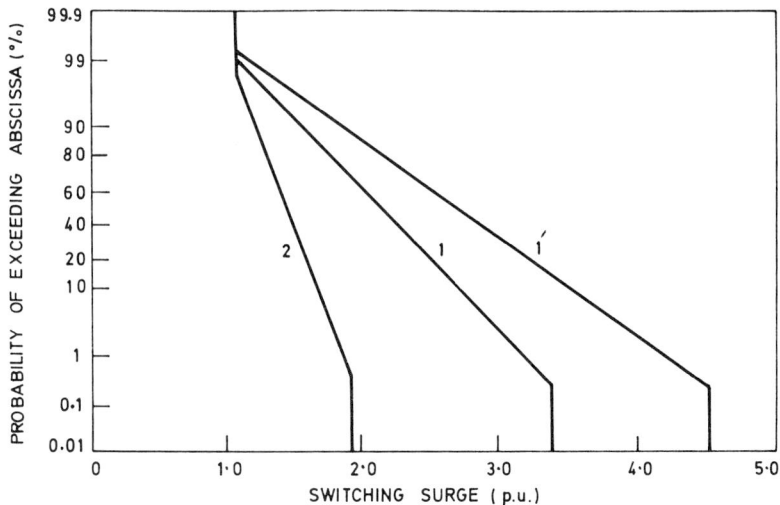

Fig. 14.10 Switching surge magnitudes. 1 and 1', Uncontrolled surges; 2, controlled surges. [From Diesendorf (1974).]

with very restricted variations — almost deterministic — at lower voltage magnitudes within 1.25 p.u. and a fairly wide variation, with more than a 15% coefficient of variation, at higher voltage magnitudes (Ragaller, 1980).

REFERENCES

Abdel-Salam, M. and Stanek, E. (1987). *IEEE Trans.*, *IA-23*: 481–489.
Ametani, A. and Schinzinger, R. (1976). *IEEE Trans.*, *PAS-95*: 773–781.
Berger, K., Anderson, R. B., and Kroninger, H. (1975). *Electra*, *41*: 23–37.
Bewley, L. V. (1963). *Traveling Waves on Transmission Systems*, Dover Publications, Inc., New York.
Bickford, J. P. and Doepel, P. S. (1967). *Proc. IEE*, *114*: 465–477.
Cristina, S. and d'Amore, M. (1983). *IEEE Trans.*, *PAS-102*: 1685–1693.
Diesendorf, W. (1974). *Insulation Coordination in High Voltage Electric Power Systems*, Butterworth & Company (Publishers), Ltd., London.

IEC (1973). *High Voltage Test Techniques*, Publication 60-1,2,3, International Electrotechnical Commission, Geneva.
Lewis, W. W. (1965). *The Protection of Transmission Systems Against Lightning*, Dover Publications, Inc., New York.
Marshall, J. L. (1973). *Lightning Protection*, Wiley-Interscience, New York.
Nakagawa, M. (1981). *IEEE Trans.*, *PAS-100*: 3626–3633.
Ragaller, K., ed. (1980). *Surges in High-Voltage Networks*, Plenum Press, New York.
Semlyen, A. (1981). *IEEE Trans.*, *PAS-100*: 848–856.

15
Insulation Coordination

H. ANIS   *Cairo University, Giza, Egypt*

15.1   INTRODUCTION

According to the International Electrotechnical Commission, "insulation coordination comprises the selection of the electric strength of equipment and its application in relation to the voltages which can appear on the system for which the equipment is intended, and taking into account the characteristics of available protective devices, so as to reduce to an economically and operationally acceptable level the probability that the resulting voltage stresses will cause damage to equipment insulation or affect the continuity of service" (IEC, 1976).

Two stages of study must precede a successful insulation coordination, namely, an examination of the different types of overvoltages that the power system encounters, a subject dealt with in Chapter 14, and the general assessment of insulation performance under the stress of those overvoltages. In the light of the outcome of those studies, the insulation coordination effort aims at specifying the following:

1. The electrical strength of the insulation of all system components
2. The phase-to-ground and phase-to-phase clearances
3. The leakage distance of external insulators
4. The ratings, type, number, and location of surge arresters and other possible protective spark gaps.

15.1.1   Types of Insulation

When applying insulation coordination, it is useful to classify the insulation of a power network according to location and dielectric performance. Insulation is classified according to location as external or internal.

1. *External insulation*: clearances and insulator surfaces in the open air. They are, therefore, influenced by such atmospheric conditions as pollution and humidity. External insulation can be located outdoors or indoors. In the latter case it is less affected by the ambient weather conditions.
2. *Internal insulation*: internal solid, liquid, or gaseous components of equipment insulation and therefore not exposed to atmospheric conditions. Insulating oils, compressed-gas insulation, and solid insulation of cables, transformers, and electric machines belong to this category.

From the dielectric performance viewpoint, insulation may either be self-restoring or non-self-restoring.

1. *Self-restoring insulation*: regains dielectric integrity following the occurrence of a breakdown. Some insulating films in capacitors produce no permanent leakage path after their breakdown and are therefore called "self-healing" insulation.
2. *Non-self-restoring insulation*: loses its insulating property after a disruptive discharge and must therefore be replaced. Solid insulation as in cables, transformers, and machines belong to this type of insulation.

The coordination of self-restoring insulation, unlike non-self-restoring insulation, lends itself to probabilistic treatment where the evaluation of the insulation's risk of failure is economically justified.

15.2 INSULATION PERFORMANCE UNDER VOLTAGE STRESSES

Power system insulation is subjected to four classes of dielectric stresses: power-frequency normal voltage, temporary overvoltages, switching overvoltages, and lightning overvoltages. The relative durations of these voltages are shown in Figure 15.1. The last three classes of overvoltage are discussed in Chapter 14. Here the response of system insulation to these stresses and the likelihood of failure are outlined.

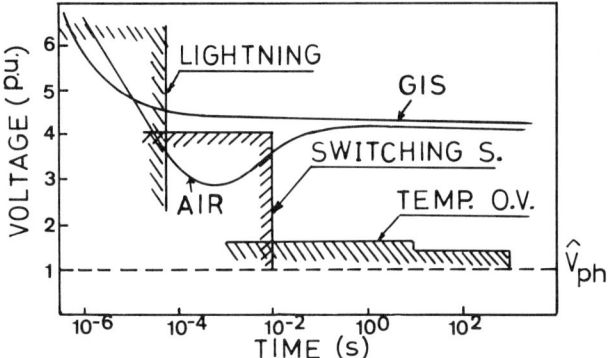

Fig. 15.1 Superposition of V–t characteristics on overvoltage distribution pattern.

15.2.1 Failure Under Power-Frequency Normal Voltage

Following are some of the causes of insulation failure under power-frequency voltages:

1. Contamination on the insulation surface can initiate a mechanism that may lead to total breakdown.
2. Thermal breakdown may develop under the continuous stress and the consequent dielectric losses.
3. Voids and flaws in insulation may develop corona discharges which would lead to premature aging of insulators and their eventual failure.
4. Normal aging of the insulation may be manifested by the gradual reduction in its withstand capability over a long period of time, and a sudden breakdown may then occur under operating voltage.

Atmospheric pollution is the primary cause of breakdown of external insulation under normal voltages (Khalifa et al., 1988). As discussed in Section 9.9, leakage currents on polluted insulators, if heavy enough, may cause eventual flashover at service voltage. The flashover voltage of a polluted insulator decreases considerably as the conductivity of the pollution layer increases.

15.2.2 Failure Under Impulse Voltages

Power-line and station insulators are subject to a variety of transient voltages. These voltages—from an insulation flashover point of view—are characterized primarily by their magnitudes and wave shape. The shape of an idealized impulse is made up of its time to crest (or front) and the rate of its decay beyond the crest, which is customarily identified by the impulse tail. Lightning impulses are known to have fronts on the order of a few microseconds, while switching surges have fronts on the order of a few hundred microseconds (Chapter 14). In this section the various factors that influence the flashover voltage of external insulation, and are thus relevant to insulation dimensioning, are discussed.

Front Duration

Air gaps and insulators shorter than about 1 m have a flashover voltage that is independent of the impulse front. For longer air gaps and insulators the impulse front has a definite effect on the flashover voltage, as discussed in Section 4.8.3. The impulse front corresponding to least dielectric strength is commonly known as the critical impulse front, and depending on the strike distance and configuration, it takes values between 100 and 500 µs. The IEC standards choose 250 µs to represent the front of a switching surge where the air insulation displays the least strength (IEC, 1976).

Gap Geometry and Polarity

The physical process governing breakdown is greatly influenced by the field distribution in the gap (Chapter 4). The breakdown voltage of a uniform field gap can be as high as six times that of a rod-plane gap. Therefore, for practical purposes, the term "gap factor" (K_g) has been introduced to express the breakdown voltage of a given gap relative to a positive rod-to-plane gap of the same length, which is known to have the least breakdown voltage. The gap factor is therefore always larger than, or equal to, unity. For example, rod-rod gaps have a gap factor of about 1.7; a conductor-to-tower clearance normally has $K_g = 1.3$.

Figure 15.2 shows the pronounced influence of ground proximity and that of voltage polarity on the breakdown voltage of an air gap (Electrical Power Research Institute, 1979). Positive breakdown voltages (i.e., those when the more highly curved conductor is positive) are lower than negative voltages, as explained in Chapter 4 in detail. The more nonuniform the field in the gap becomes, the larger will be the difference between positive and negative breakdown voltages.

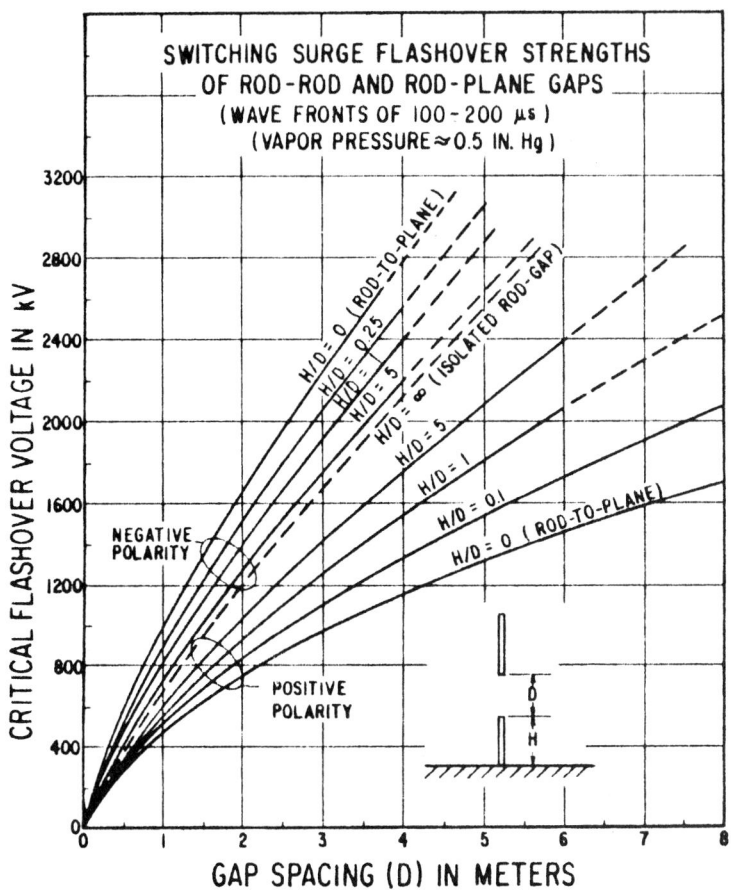

Fig. 15.2   Effect of polarity and ground proximity on air gap flashover.

Breakdown Probability Distribution

If an impulse voltage of a certain shape is applied repeatedly to self-restoring insulation, the outcome of all applications may not be the same. Some impulse application may produce flashover, while others, still with the same amplitude, may be withstood by the insulation. This phenomenon is particularly true for sufficiently large gaps (larger than about a 50-cm rod–plane gap in air at atmospheric pressure). Also, it occurs over a finite range of impulse voltage amplitudes, above which all shots produce breakdowns and below which all shots would cause no breakdown. This probablistic property is very essential to the process of insulation dimensioning. In fact, at very high transmission voltages this property is used to set an economical compromise between the insulation cost and that of the risk of insulation failure.

The probability of breakdown (i.e., the number of breakdowns relative to the total number of applied shots) may be plotted against the applied voltage as shown in Figure 15.3a. Experimental investigations have revealed that for most external insulation the breakdown probability function may be expressed by the well-known cumulative Gaussian distribution function

$$P(V) = \frac{1}{\sigma\sqrt{2\pi}} \int_0^V \exp\left[-\frac{1}{2}\left(\frac{V-V_{50}}{\sigma}\right)^2\right] \qquad (15.1)$$

This function is characterized by two parameters:

V_{50} = 50% breakdown voltage; also called the critical flashover voltage CFO

σ = standard deviation of the representative Gaussian distribution; σ describes the degree of scatter in breakdown voltages about their mean value V_{50}

If the postulate of equation (15.1) — that the breakdown probability fits a Gaussian cumulative function — is accepted, P(V) can alternatively be plotted on a normal probability ordinate scale as seen in Figure 15.3b, in which case P(V) appears linear. The linear units on the ordinate in this case are those of the quantity $(V - V_{50})/\sigma$ rather than P(V).

Time to Breakdown

Upon the application of an impulse voltage, a breakdown, if any, would take place after a certain time lag following the initiation of the impulse. This delay, referred to as the time to breakdown, depends on the gap configuration, the amplitude, and the shape of the voltage impulse. Furthermore, the time to breakdown is a random quantity that may vary markedly from one impulse application to another (Chapter 4). The time to breakdown of insulation is essential for insulation coordination when two or more insulation components are subjected simultaneously to the same impulse stress, particularly when an overvoltage protection device is installed.

Insulation Coordination

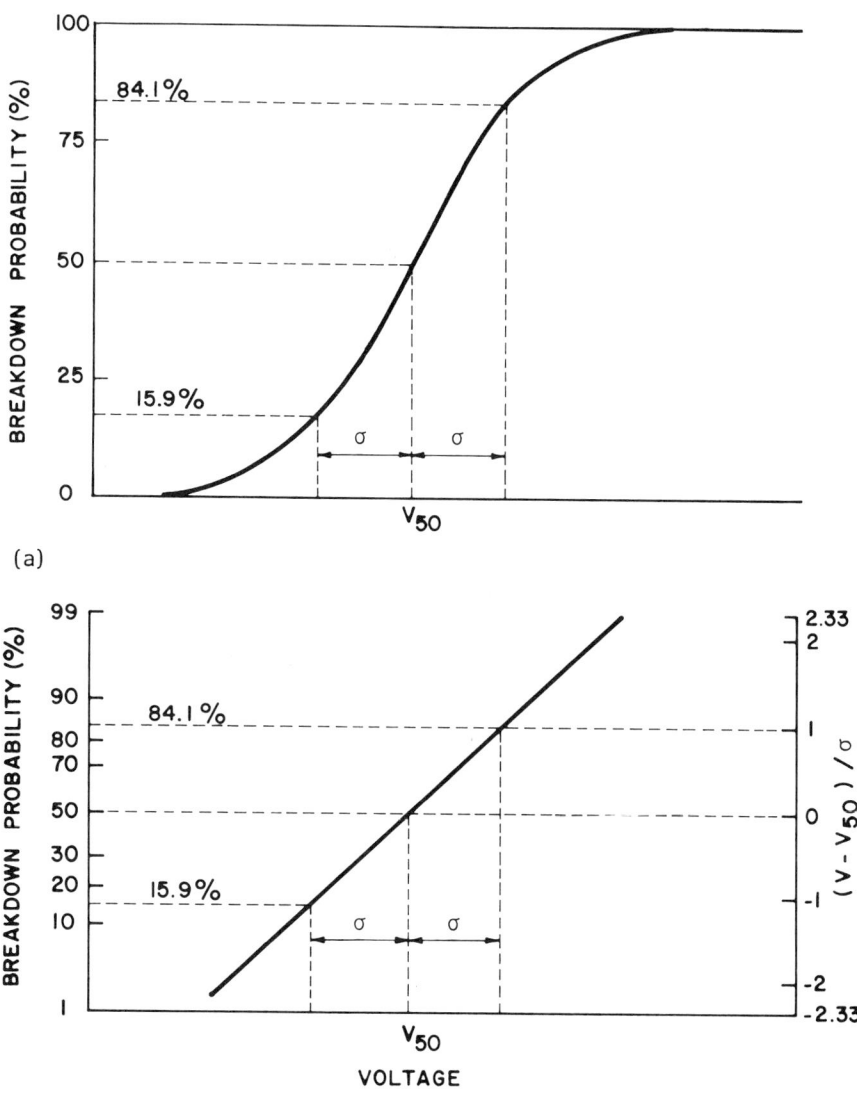

Fig. 15.3 Breakdown probability of external insulation: (a) on linear coordinates, (b) on a Gaussian scale (σ, standard deviation).

15.3 FLASHOVER OF EXTERNAL GAPS IN PARALLEL

A transmission system comprises a large number of insulation components in "parallel" (i.e., subjected to the same normal and abnormal voltages). In a substation, the number of line-to-ground insulation structures may be as high as 100. The number of bus insulators per phase could vary from 100 to 1000, depending on the station voltage and capacity. On transmission lines, the number of insulation strings connecting the same phase to ground can be as large as a few thousands.

In the study of parallel insulation, two physically oriented assumptions are adopted: first, that upon the incidence of an overvoltage, only one gap at a time is liable to break down; the second, that whenever more than one gap tends to break down simultaneously, only the one with the shortest "time to breakdown" would actually do so.

15.3.1 The Case of Two Parallel Gaps

When the probabilistic properties of the two gaps are given separately, the behavior of a combination of the two gaps may be predicted under a known stress. If the individual breakdown probabilities of the two gaps before the combination were P_1 and P_2, the new breakdown probability of the first gap during the parallel combination would be

$$P'_1 = P_1(1 - P_2) + P_1 P_2 P_{1,2} \qquad (15.2)$$

The first term of equation (15.2) is the probability that gap 1 flashes over whereas gap 2 does not. This is complemented by the probability that the two gaps break down simultaneously ($P_1 P_2$), multiplied by the probability $P_{1,2}$ that in this case gap 1 is faster than gap 2 to develop a breakdown. The breakdown probability of the second gap is, similarly

$$P'_2 = P_2(1 - P_1) + P_1 P_2 P_{2,1} \qquad (15.3)$$

Since the time to breakdown of one gap can only be either shorter or longer than that of the other gap, then

$$P_{1,2} + P_{2,1} = 1 \qquad (15.4)$$

Insulation Coordination

Obviously, the quantity $P_{1,2}$, hence $P_{2,1}$, is a function of the probability density distributions $f_1(t)$ and $f_2(t)$ of the times to breakdown of the two gaps. Such a function can be derived to be

$$P_{1,2} = \int_0^\infty \int_t^\infty f_1(t_1) f_2(t_2) \, dt_1 \, dt_2 \qquad (15.5)$$

In the special case when the two density distributions are normal (Gaussian) with mean values T_1 and T_2 and standard deviations S_1 and S_2, respectively, then

$$P_{1,2} = \frac{1}{2} + \frac{1}{\sqrt{2}} \int_0^{(T_2 - T_1)/\sqrt{\sigma_1^2 + \sigma_2^2}} \exp\left(\frac{-t^2}{2}\right) dt \qquad (15.6)$$

The overall breakdown probability of the system of two gaps is given by

$$P_c = P_1' + P_2' = P_1 + P_2 - P_1 P_2 \qquad (15.7)$$

which is characteristically independent of the times to breakdown.

15.3.2 Flashover of Identical Gaps in Parallel

If the probability of withstand at a given voltage level for a single air gap is W, the probability of withstand for n identical gaps in parallel is then

$$W_n = W^n \qquad (15.8)$$

In terms of breakdown probabilities, equation (15.8) can be rewritten as

$$1 - P_n = (1 - P)^n$$

or

$$P_n = 1 - (1 - P)^n \qquad (15.9)$$

where P is the breakdown probability of one gap and P_n is that of n gaps. Experimental investigations showed that the relationships shown previously hold accurately. Statistically, if the distribution of the breakdown probability of a single gap P is accepted to be cumulative Gaussian, the distribution of P_n can no longer be Gaussian.

It is concluded from equations (15.8) and (15.9) that the probability of withstand of a transmission system decreases as the number of insulators in parallel increases. For example, with a probability of withstand of a single insulator of 99%, a system of 20 parallel insulators will have a withstand probability of only 82%. If the string insulators of successive towers are roughly considered to be in parallel, the attenuation of the traveling overvoltage wave should first be considered in evaluating the line's flashover probability.

15.3.3 Minimum Time to Breakdown

The time to breakdown (TBD) of a system of parallel gaps (or insulators) is also a random quantity (Chapter 4). The knowledge of the distribution of this time is vital to insulation coordination and the selection of a protective device. Testing the system of parallel gaps for the determination of its TBD distribution can be impractical in view of the varying number of components of the system. The TBD of a system of parallel gaps can, alternatively, be related to the characteristics of the individual components which make up that system.

The probability density distribution of the collective TBD is composed of the TBD probability densities of the individual components, each being weighed by the corresponding probability of occurrence of a breakdown. To illustrate this relation, the case of two generally unidentical gaps is discussed.

Given two parallel gaps with breakdown probabilities P_1 and P_2 and the TBD distributions $f_1(t)$ and $f_2(t)$ under a given impulse stress, the TBD distribution of the combination is

$$f_c(t) = \frac{\phi[P_1, P_2, f_1(t), f_2(t)]}{\int_0^\infty \phi[P_1, P_2, f_1(t), f_2(t)]\, dt} \tag{15.10}$$

where, in view of equation (15.2),

$$\phi = P_1(1 - P_2)f_1(t) + P_2(1 - P_1)f_2(t) + P_1 P_2 f_{12}(t)$$

Insulation Coordination

and the denominator is introduced to fulfill the constraint

$$\int_0^\infty f_c(t)\, dt = 1$$

Applying the same constraint — that a density distribution integrated over all times equals unity — to the constituents of the denominator gives

$$f_c(t) = \frac{P_1(1 - P_2)f_1(t) + P_2(1 - P_1)f_2(t) + P_1 P_2 f_{12}(t)}{P_c} \quad (15.11)$$

where $P_c = P_1 + P_2 - P_1 P_2$ is the breakdown probability of the combination, as was shown in equation (15.7). The first two terms of $f_c(t)$ are the contributions of the individual TBD of the two gaps whenever one gap breaks down alone. In the third term, which represents the contribution of a simultaneous breakdown in the two gaps, $f_{12}(t)$ is the density distribution of the minimum TBD of the two gaps since only the gap with the shorter TBD will break down at a time. It can be shown that, $f_{12}(t)$ is related to the individual TBD values by

$$f_{12}(t) = [1 - F_1(t)][1 - F_2(t)]\left[\frac{f_1(t)}{1 - F_1(t)} + \frac{f_2(t)}{1 - F_2(t)}\right] \quad (15.12)$$

where F_1 and F_2 are the cumulative distributions of f_1 and f_2, respectively.

To extend the analysis above to a large number of equal gaps n, different combinations of gaps will have to be considered (e.g., singles, twos, threes, etc.). The distribution of the minimum TBD of only r gaps (out of a total of n gaps) can be shown to be given by

$$f_{(r)}(t) = [1 - F(t)]^{r-1} r f(t) \quad (15.13)$$

The overall TBD of the n-gap system can now be deduced in a manner similar to that of two gaps [equation (15.11)]. Considering the binomial combinations of breakdowns in the n gaps, the overall distribution of the TBD of a system of n equal gaps will be

$$f_c(t) = \frac{1}{P_c} \sum_{r=1}^{r=n} \frac{n!}{(n-r)!} P^r (1-P)^{n-r} f_{(r)}(t) \qquad (15.14)$$

It is worth mentioning that whenever the individual breakdown probabilities are very small, all combinational contributions to $f_c(t)$ could be neglected and $f_c(t)$ becomes equal to $f(t)$ of any one gap. On the other hand, as the number of gaps in parallel is increased and/or the individual breakdown probability increases, an ultimate distribution of the system's TBD is approached, which can be shown to be

$$f_{ult}(t) = 2[1 - F(t)] f(t) \qquad (15.15)$$

15.4 PRINCIPLES OF INSULATION COORDINATION

The procedure leading to well-coordinated system insulation begins by evaluating the stresses to which insulation is subjected. The insulation dimensions are then decided in view of existing protective devices such that a predetermined level of safety against insulation failure is ensured. Depending on whether the coordinated insulation is self-restoring or non-self-restoring, a statistical or conventional approach is adopted. In the statistical approach a calculated risk of insulation failure is taken, whereas in the conventional approach the coordinator is set to eliminate the possibility of system failure (Diesendorf, 1974).

15.4.1 Statistical Insulation Coordination

With the advent of extra-high-voltage transmission and in anticipation of ultra high voltage, two phenomena came into existence and justified the use of a statistical approach to insulation coordination. First, overvoltages, particularly those of internal origin, exhibited a distinct random nature where they varied markedly in magnitude and form. Second, long insulators and air clearances displayed widely scattered probabilities of breakdown and associated times to breakdown, as explained earlier.

The overvoltages that stress the system insulation may be expressed by a joint statistical distribution of the magnitude V_m and time to crest t [i.e., $f(V_m, t)$]. At the same time, the breakdown probability of insulation may also be expressed as a function of the same two quantities, to be $P(V_m, t)$. Therefore, overvoltages whose magnitudes fall within an interval dV_m, and with times to

Insulation Coordination

crest falling within dt, should produce an insulation risk of failure given by

$$dR = P(V_m, t) \, [f(V_m, t) \, dV_m \, dt] \tag{15.16}$$

The overall risk of insulation failure can thus be written as (Anis et al., 1978)

$$R = \int_0^\infty \int_0^\infty P(V_m, t) f(V_m, t) \, dV_m \, dt \tag{15.17}$$

Due to the complexity of equation (15.17) and the difficulty in acquiring the necessary bivariate data, it is customary to consider only the overvoltage magnitude when calculating the risk of failure, that is,

$$R' = \int_0^\infty P(V_m) f(V_m) dV_m \tag{15.18}$$

in which $P(V_m)$ is the breakdown probability distribution, as discussed earlier in this chapter. Figure 15.4 gives a graphical interpretation of equation (15.18).

Insulation can theoretically be made sizable enough to eliminate any risk of insulation failure. This attitude, however, is economically unwise since the excessive cost of insulation may soon exceed the cost of system interruption due to insulation failure. It is therefore necessary to define a reasonable safety factor and design the insulation accordingly. This factor sets a relation between the insulation's withstand threshold as obtained from $P(V_m)$ and a representative overvoltage given by $f(V_m)$. The following quantities are defined for this purpose (IEC, 1976).

Statistical withstand voltage (SWV): peak value of a switching (or lightning) impulse test voltage at which the insulation exhibits under specified conditions a probability of withstand equal to 90%.

Statistical overvoltage (SOV): switching (lightning) overvoltage applied to the equipment as a result of an event of one specific type on the system, the peak value of which has a probability of being exceeded of 2%.

Statistical safety factor (SSF): ratio, for a given type of event, of the appropriate statistical switching (lightning) impulse withstand voltage and the statistical overvoltage, established on the basis of a given risk of failure, taking into account the statistical distributions of withstand voltages and overvoltages.

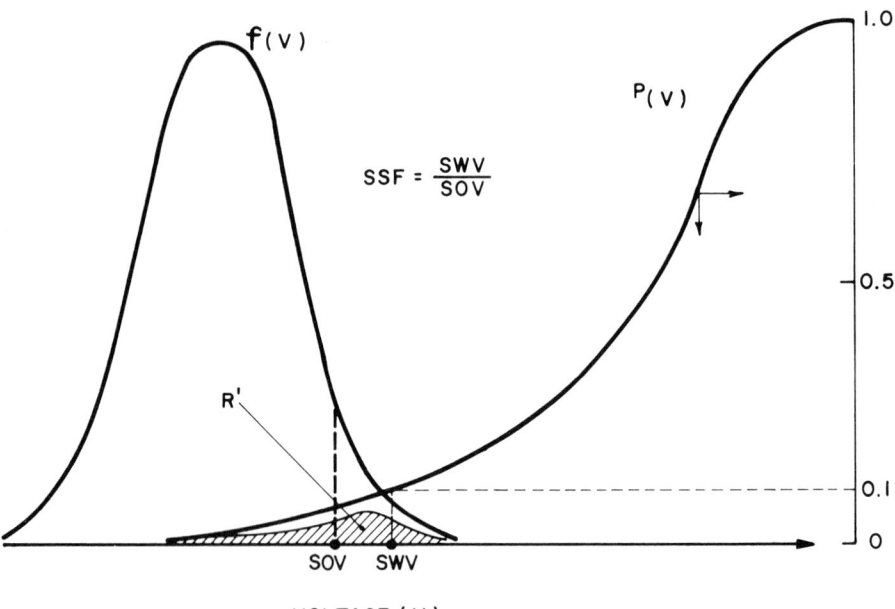

Fig. 15.4 Computation of the risk of insulation failure.

$$\text{SSF} = \frac{\text{SWV}}{\text{SOV}} \qquad (15.19)$$

In Figure 15.5 the effect of the statistical safety factor on the consequent risk of insulation failure is demonstrated (IEC, 1976). It appears that the relationship above is almost invariant for all EHV transmission networks — a fact that permits the use of SSF as a reliable design factor.

Example

The following procedural example for a disconnecting switch illustrates the application of statistical insulation coordination:

Highest voltage for equipment (line voltage) = 765 kV (rms)

Phase-to-earth voltage V_{ph} = 442 kV (rms)
= 625 kV (peak)
= 1.0 p.u.

SOV (from system studies) = 1255 kV (peak)
= 2.0 p.u.

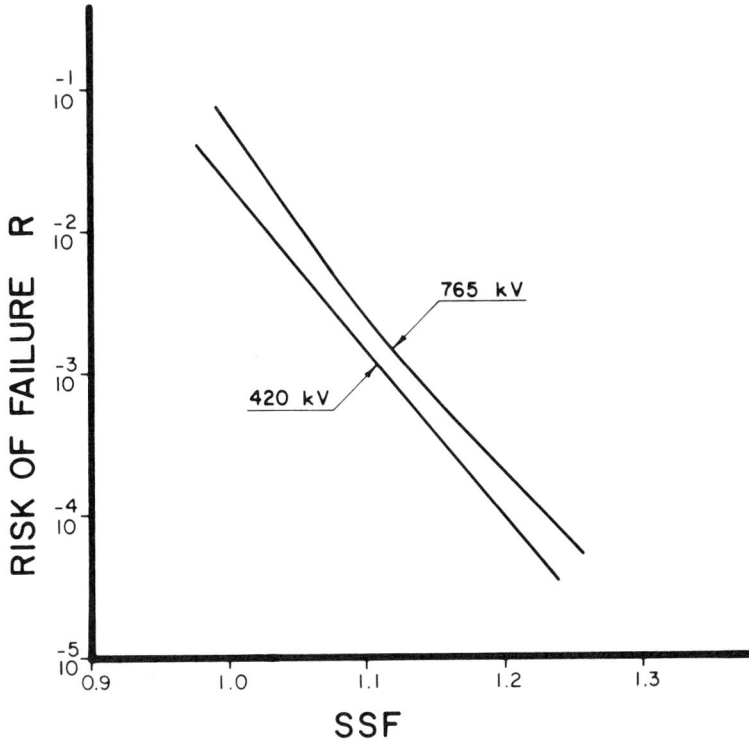

Fig. 15.5 Effect of the statistical safety factor on the risk of insulation failure.

Maximum acceptable risk of insulation failure to ground $R = 10^{-4}$
Choice of insulation's SWV:

SWV			
kV	p.u.	SSF	R
1300	2.08	1.04	$1 \times 10^{-2} > 10^{-4}$
1425	2.28	1.14	$7 \times 10^{-4} > 10^{-4}$
1550	2.48	1.24	$7 \times 10^{-5} < 10^{-4}$

Conclusion. A rated switching withstand voltage for equipment insulation of 1550 kV is selected. The associated rated lightning withstand voltage is 2400 kV.

15.4.2 Conventional Insulation Coordination

With non-self-restoring insulation the risk of insulation failure should at all times be avoided. This is normally achieved by placing an overvoltage protective device, usually a surge arrester, in the vicinity of the equipment to be protected. The conventional insulation coordination approach seeks the impulse voltage level at which the equipment insulation will not show any disruptive discharge. This procedure is done in view of the characteristics of existing surge arresters.

Figure 15.6 demonstrates the steps leading to the determination of the insulation's basic impulse insulation level (BIL) and basic switching impulse insulation level (BSL). The former expresses the insulation withstand under a standard impulse of 1.2/50 µs,

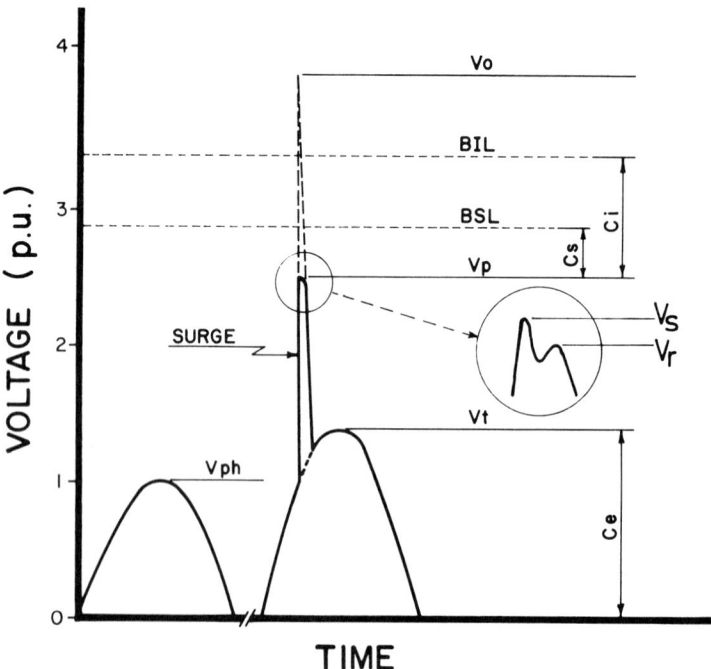

Fig. 15.6 Determination of BIL and BSL. The symbols are explained in the text.

Insulation Coordination

while the latter refers to a 250/2500-µs impulse. If the peak value of the conductor-to-earth voltage in a three-phase system is defined as 1 p.u., the most common temporary overvoltage (i.e., the voltage of a healthy phase under single line-to-ground fault conditions) is equal to C_e. This "earth fault factor" is approximately 1.4 for an earthed system. The surge withstand voltage of the system, which is also referred to as the basic impulse insulation level (BIL), is approximately 3.4 p.u. for a transmission system with a rated voltage U_m greater than 200 kV, while the switching impulse insulation level (BSL) is about 2.8 p.u. The arrester will limit an overvoltage wave to a certain protection level V_p under normal operating and temporary overvoltages conditions.

The protection level of a protective device is defined as the highest permissible surge voltage that may appear at the terminals of the equipment to be protected; for switching impulses it is taken directly as the maximum sparkover voltage, V_s, in Figure 15.6. For lightning impulses, however, the protection level is taken as the maximum of the following three values:

1. Maximum sparkover voltage
2. Maximum residual voltage at specified discharge current
3. Maximum sparkover voltage in the front divided by 1.15

The residual voltage V_r is built up in the arrester after sparkover by subsequent flow of current; it can be larger or smaller than V_s, depending on that current. Safety margins (C_i and C_s) exist between the insulation levels BIL and BSL on the one hand and the arrester protection level V_p on the other; in accordance with IEC recommendations, this must be at least equal to

$$C_s = 1.15 \qquad C_i = 1.25$$

Example

The following procedural example (for a transformer) demonstrates the application of the conventional approach to insulation coordination.

Highest voltage for equipment (line voltage) = 765 kV (rms)

Phase-to-earth voltage V_{ph} = 442 kV (rms)
= 625 kV (peak)
= 1.0 p.u.

Computed (or measured) temporary overvoltage to ground V_t = 605 kV (rms)
= 855 kV (peak)
C_e = 1.37 p.u.

If temporary overvoltages are not known, take
$C_e = 1.4$ p.u.)
Minimum $C_s = 1.15$
Minimum $C_i = 1.25$

Characteristics of surge arrester

Rated voltage (available rating immediately above the temporary overvoltage level of 605 kV)	= 612 kV (rms)
Maximum switching impulse sparkover voltage	= 1230 kV (peak)
Maximum lightning impulse sparkover voltage	= 1400 kV (peak)
Maximum residual voltage at specified discharge current	= 1400 kV (peak)
Maximum sparkover voltage in the front	= 1660 kV (peak)
Protection level V_p:	
to switching impulse	= 1230 kV
(i.e., protection factor)	= 1.97 p.u.
to lightning impulse	= 1443 kV
(i.e., protection factor)	= 2.3 p.u.

Recommended insulation

Minimum switching impulse voltage ($C_s V_p$)	= 1415 kV
Rated switching impulse voltage BSL (nearest higher standard)	= 1425 kV
	= 2.28 p.u.
Minimum lightning impulse voltage ($C_i V_p$)	= 1800 kV
Rated lightning impulse voltage BIL (nearest higher standard)	= 1800 kV
	= 2.88 p.u.

15.5 OVERVOLTAGE PROTECTIVE DEVICES

The duty of overvoltage protection is to ensure that transients arriving at an installation from the power lines or originating in the installation itself are reduced to a level bearing a definite relationship to the rated voltage of the network. One of the oldest known means of providing protection against overvoltages, and one that is still used to some extent, is the spark gap. As a rule it consists of two metal electrodes pointing toward one another, one being connected to the high voltage, the other to ground. Frequently, the spark gap is referred to by a descriptive name (e.g., arcing horns). Evidently, the breakdown voltage depends on the gap width and weather conditions (IEC, 1973).

Insulation Coordination

By allowing the gap to flash over, the propagation of undesired voltages in the network can be prevented. However, the spark gap is not really a very efficient means of protection against overvoltages. The main drawbacks are the time lag that occurs before the gap sparks over, the variation of the sparkover voltage with the polarity and surrounding conditions, and the fact that once an arc has started to burn, it continues even after the overvoltage has disappeared, causing a line-to-ground short circuit on the network.

15.5.1 Magnetically Blown Surge Arresters

The surge arrester does not exhibit the disadvantages described previously. Over the wide range of overvoltages met in practice, the surge arrester possesses an almost unchanging sparkover voltage characteristic, thereby affording high operational safety, consistency, and reliability. Its follow current, resulting from the service voltage, is interrupted automatically.

A major requirement of surge arresters is that they must be able to discharge high energy without changes in their protective levels or damage to themselves or adjacent equipment. This property is decided by the arrester's thermal capacity. Unlike lightning strokes, which have short durations (hundreds of microseconds), switching surges generally have long durations (thousands of microseconds). The arrester is then required to discharge a considerable amount of energy, a situation that simple arresters with "unassisted" spark gaps cannot handle. Therefore, only heavy-duty surge arresters with magnetically blown spark gaps are recommended to operate on switching surges.

Figure 15.7 shows schematically the arrangement of a modern, magnetically blown arrester unit. The main components (i.e., nonlinear resistance elements R_a and the spark gaps E) are mounted in an airtight porcelain housing. The electrodes of the tongue-shaped spark gaps are embedded in the disk-shaped chamber K. The movement of the point of origin of the arc, which is influenced by the shape of the spark gaps, prevents local overheating or serious pitting of the electrodes. In normal operating conditions the control current i_0 flowing through the grading resistors R_s ensures that the voltage is distributed almost evenly among the arrester elements. When the discharge dies down with a rapid change in the current, a follow current flows, initially determined by the system voltage and the value of the discharge resistors.

The blow-out coil presents no great impedance to the follow current, which changes only slowly, so the current commutates from the resistor R_b to the blowout coil. The current, now flowing

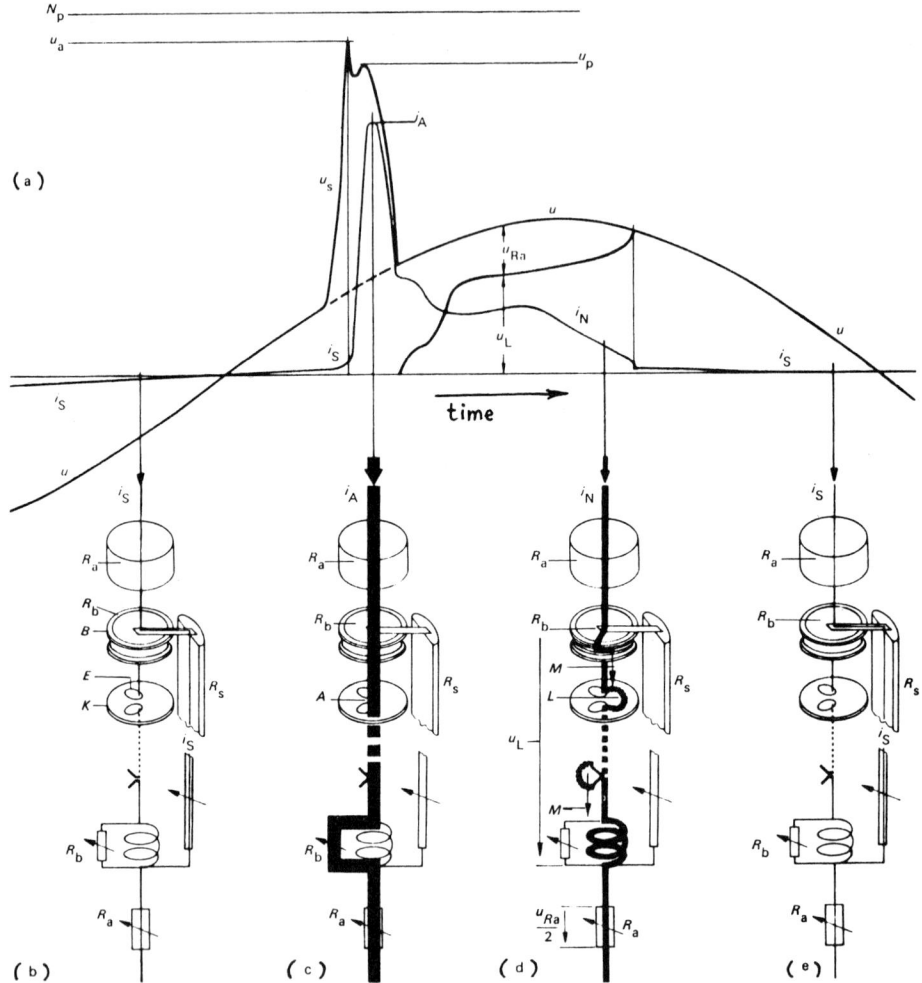

Fig. 15.7 Operation of a magnetically blown arrester: (a) arrester voltage and current; (b) and (e) normal operation; (c) passage of surge current; (d) passage of follow current. N_p = guaranteed protection level, U_a = sparkover voltage, U_p = residual voltage during diversion, U_s = surge voltage, U = service voltage at arrester assembly, U_{Ra} = voltage drop across R_a resistors during quenching, U_L = arc voltage during quenching, i_A = surge current, i_N = follow current, i_S = control current, R_b = bypass resistor, and R_S = grading resistor. [From Ragaller (1980).]

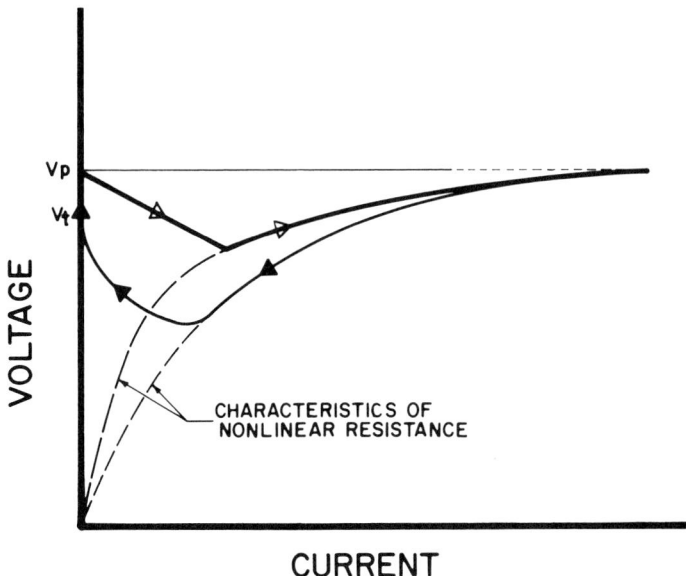

Fig. 15.8 Arrester's current-voltage characteristics.

through the coil, creates a strong magnetic field which extends throughout the stack of spark gaps. This tends to draw the arc L and extend it. A high arc voltage U_L thus builds up. This arc voltage helps to extinguish the follow current even before its natural zero. In Figure 15.8 the current/voltage characteristics of the arrester are shown.

15.5.2 Metal-Oxide Surge Arresters

A recently developed ceramic material, based on zinc, bismuth, and cobalt oxides, can be used to make resistors with a much higher degree of nonlinearity ($I = cV^\alpha$; where $\alpha > 20$) over a large current range. With such resistors one can design arresters having voltage/current characteristics close to ideal. The electrical properties of the material make it possible to dispense with the series-connected spark gaps and thereby to produce solid-state arresters suitable for system protection up to the highest voltages. To enhance the energy absorption capacity it is also possible to connect disks in parallel. The most important property of these

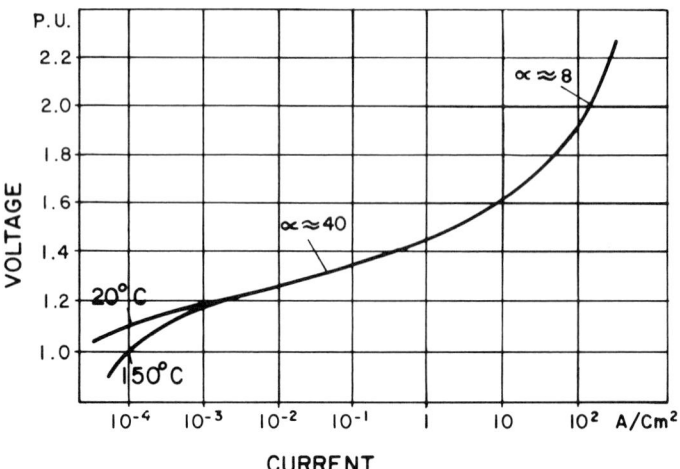

Fig. 15.9   Characteristics of metal-oxide arrester disks.

resistors is their current-voltage characteristics, which are illustrated in Figure 15.9 for a disk of 80 mm diameter and 32 mm thickness. The voltage arising for a standard lightning current surge of 10 kA (8/20 µs wave) is just over twice the reference dc voltage (Knecht and Menth, 1979).

The arrester, which is a series-connected stack of such disks, operates in a very simple fashion. It is dimensioned so that the peak value of the phase-to-earth voltage in normal operation never exceeds the sum of the reference voltages of the series-connected disks. The resistive losses in the arrester in normal operation are therefore very small. When an overvoltage occurs the current will rise with the wavefront according to the characteristics of Figure 15.9 without delay. No breakdown occurs but rather, a continuous transition to the conducting state. At the end of the voltage transient the current is reduced closely following the I-V curve (i.e., in contrast to the conventional arrester, there is no follow-current).

A considerable advantage of the zinc oxide arrester is its very simple construction. Meanwhile, the absence of spark gaps results in a continuous flow of current through the device so that theoretically the danger of thermal runaway is present. However, very stable resistors have been developed, so these apprehensions are almost eliminated. Furthermore, the absence of spark gaps makes the voltage grading system unnecessary. The zinc oxide arrester is inherently self-regulating in that the current flow of 0.5 to 1 mA at normal supply voltage leads to reliable operation even in polluted conditions. An extremely stable I−V characteristic would guarantee constant losses for the entire life of the arrester as well as a constant protection level (Ieda et al., 1987).

15.6 INSULATION COORDINATION OF GAS-INSULATED SUBSTATIONS

The temporal development of breakdown in compressed SF_6 in a GIS is significantly different from that in open-air insulation. The former case is characterized by a weak field nonuniformity, short clearances, and high gas pressure, as well as the strong electronegativity of the gas. In contrast, open-air insulation may exhibit strong field nonuniformity, long gaps, and it is always at atmospheric pressure. These differences produce sharply distinct voltage-time characteristics as seen in Figure 15.1. While the withstand of air gaps drops to a minimum with increasing impulse front time before gaining strength under longer-fronted impulses, GIS insulation shows a steadily decreasing strength with increasing impulse time.

In Figure 15.1 the V-t characteristics of insulation are superimposed on the frequency-of-occurrence diagram for the various types of overvoltages. It is concluded that lightning overvoltages are more significant to GIS insulation dimensioning, while conductor spacings and other open-air insulation for EHV and UHV systems are basically determined by switching overvoltages and to a lesser extent by lightning overvoltages. Due to the rather flat insulation withstand dependence of the GIS on the front time of overvoltage, the withstand of the GIS against temporary overvoltages is satisfactory, as the insulation must be designed for the much higher lightning impulse.

For a conventional substation it is normally deemed sufficient if the transformers, being the most valuable equipment and difficult to replace, are well protected against excessive overvoltage stresses. In contrast, in the case of GIS, even if a part of the GIS can be replaced comparatively quickly and on site, a failure would lead to a great disturbance in the operation of the system. Good overvoltage protection is thus needed for all gas-insulated parts of the substation. This causes the need to protect also the distance from the line entrance to an open circuit breaker, which in turn demands installation of lightning arresters at the line entrance where overhead lines are connected.

15.6.1 Incident Lightning Surges

Design values of stroke current amplitude and steepness are about 70 kA and 70 kA/μs, respectively. The structures comprising towers and overhead lines have, at least during the first microsecond after a stroke has hit the line, a wave impedance of about 100 Ω. It is therefore understood that the steepness of generated surges can be up to 7000 kV/μs at the stroke origin. When the wave travels along the line, its peak and front steepness will be reduced due to corona discharges (see Chapter 5). The wave impedance of an overhead line is considerably greater than that of GIS (typically 75 Ω), and thus reflection will occur at the entrance to the GIS. At an open gap (disconnector or circuit breaker) the surge will be totally reflected, meaning the doubling of amplitude.

As explained in Chapter 14, after entering one part, a, is transmitted into the substation and one part, $(1 - a)$, is reflected back along the line. The entered wave is reflected at the open end and when reaching the GIS entrance will be reflected again. The growth of the overvoltage will follow the equation (Eriksson and Holmborn, 1978)

$$\frac{U}{U_0} = 2 - (2 - a)(1 - a)^n \tag{15.20}$$

where n is the number of reflections. As n increases, U/U_0 approaches 2.

With finite surge steepness, S, however, the voltage at the entrance to the GIS is given as a function of time by

$$V = Sat \tag{15.21}$$

before the first reflection. After n reflections, it will take the form

$$V = Sa \left\{ t + \sum_{m=2}^{m=n} (1 - a)^{m-2}(2 - a)[t - (m - 1)T] \right\} \tag{15.22}$$

where

$$2(n - 1) < \frac{t}{T} < 2n$$

The voltage at the open end at time t will be

Insulation Coordination

$$U = S \sum_{m=1}^{m=n} 2(1 - \chi)^{m-1}[t - (2m - 1)T]$$

where

$$2(n - 1) < \frac{t}{T} < (2n - 1)$$

and where S is the steepness of wave on the overhead line, T the wave traveling time in a GIS, L the length of GIS from entrance to the open circuit, and n the number of reflections. $T = L/300$ µs. The voltage increase is dependent on three parameters: steepness (S), transmission coefficient (Chapter 14) (a), and length of substation L.

15.6.2 Effect of Arrester Presence

With the presence of a surge arrester, the sequence of events differs. When the voltage has reached the sparkover voltage of the arrester, it will take a further travel time T between arrester and the open bus end. Meanwhile, a negative wave originates at the arrester and travels to the open bus. Thus the total voltage is reduced. Therefore, in case of surges with steeper fronts and with longer GIS buses, the voltage buildup on the system is greater. On the other hand, for shorter bus lengths, the voltage increase by repeated reflections becomes rapid, resulting in a higher sparkover voltage of the arrester, whereas the voltage buildup reaches a relatively lower level (Fig. 15.10).

The arrester sparkover voltage for steep surges is of great interest to the insulation coordination of GIS since the increase of the GIS withstand level is limited for steep surges. Due to the flat SF_6 V-t characteristics, it is necessary to have a low surge arrester sparkover voltage also for front times less than 1 µs.

In Figure 15.11 the average rate of rise of the voltage across an arrester at the line entrance has been plotted versus the distance from entrance to an open gap in the GIS. With a longer bus length the surge steepness across the arrester will decrease.

15.6.3 Insulation Coordination with Arresters

The first important decision to make is selection of the surge arrester rated voltage V_r (reseal voltage). A choice of a reseal

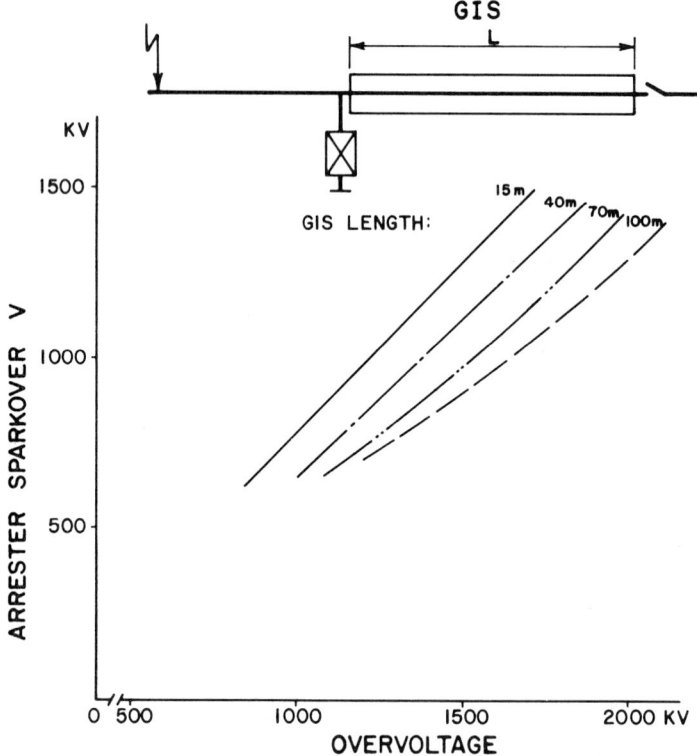

Fig. 15.10 Overvoltage versus sparkover voltage of GIS. Enclosed arrester at line entrance; 2000 kV/μs incoming surge.

voltage equal to the temporary overvoltage results in reseal voltages 1.3 to 1.6 times normal phase-to-ground voltage (IEC, 1976). Based on the rated voltage (V_r) the following arrester characteristics can be obtained:

Front of wave sparkover voltage ($\sim 2.3\ V_r$ if unknown)
1.2/50-μs sparkover voltage ($\sim 2.1\ V_r$ if unknown)
Residual voltage at 10 kA ($\sim 1.9\ V_r$ if unknown)

the largest of which determines the protection level of the arrester.

Insulation Coordination

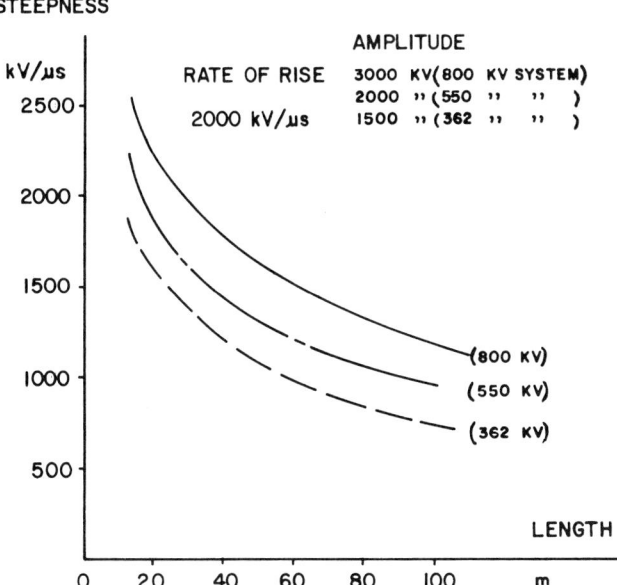

Fig. 15.11 Rate of rise of voltage across arrester versus GIS length.

Example

The following procedural example demonstrates the determination of the BIL of GIS equipment.

Rated system voltage (line voltage) = 420 kV (rms)
Rated phase voltage V_{ph} = 243 kV (rms)
          = 344 kV (peak)
Temporary overvoltage factor = 1.4 p.u.

Arrester specifications

Arrester rated voltage $C_e \times V_{ph}$ = 340 kV (rms)
Protection level V_p (determined as stated above) = 2.2 p.u.
          = 755 kV (peak)

System characteristics

Maximum lightning overvoltage in the substation (computed at open end of GIS as explained) $V_{ov} = 2.7$ p.u.
$= 929$ kV

Safety margin $C_i = BIL/V_{0v} = 1.4$ (The factor C_i is usually based on experience.)

BIL of GIS $= 1300$ kV

REFERENCES

Anis, H., Radwan, R., and El-Morshedy, A. (1978). *Proc. IEEE Winter Power Meeting*, January/February, Paper A-78-154-7, New York.

Diesendorf, W. (1974). *Insulation Coordination in High Voltage Electric Power Systems*, Butterworth & Company (Publishers) Ltd., London.

Electrical Power Research Institute (1979). *Transmission Line Reference Book 345 kV and Above*, Project UHV, Electrical Power Research Institute, Palo Alto, Calif.

Eriksson, R. and Holmborn, H. (1978). *Proc. 1st Int. Symposium on Gaseous Dielectrics*, Knoxville, Tenn., pp. 314–337.

Ieda, M., Mizutani, T., Suzuki, Y., and Ohki, A. (1987). *Proc. CIGRE Symposium Report 300.01*, Vienna, May.

IEC (1973). *High Voltage Test Techniques*, Publication 60, International Electrotechnical Commission, Geneva.

IEC (1976). *Insulation Coordination*, Publication 71, International Electrotechnical Commission, Geneva.

Khalifa, M., El-Morshedy, A., Gouda, O., and Habib, S. (1988). *Proc. IEE*, 135C:24–30.

Knecht, B. and Menth, A. (1979). *Brown Boveri Review*, 66(11): 739–742.

Maller, V. and Naidu, M. (1981). *Advances in High Voltage Insulation and Arc Interruption in SF_6 and Vacuum*, Pergamon Press Ltd., Oxford.

Ragaller, K., ed. (1980). *Surges in High-Voltage Networks*, Plenum Press, New York, pp. 251–281.

16
High-Voltage Generation

M. ABDEL-SALAM *Assiut University, Assiut, Egypt*

16.1 INTRODUCTION

In the field of electrical engineering and applied physics, high voltages (dc, ac, and impulse) are required for several applications. To name just a few examples:

1. Electron microscopes and x-ray units require high dc voltages on the order of 100 kV or more.
2. Electrostatic precipitators, particle accelerators, and so on, require high dc voltages of several kilovolts and megavolts.
3. Testing the insulation of power apparatus requires high ac voltages of up to millions of volts, depending on their normal operating voltages, as discussed in Chapter 18.
4. Simulation of overvoltages that occur in power systems requires high impulse voltages of very short and longer durations.

In fact, one main concern of high-voltage engineers is for insulation testing of various power-system components under power-frequency ac, dc, switching, and lightning impulse voltages. Normally, in high-voltage testing, the currents are limited to a small value up to about 1 A under ac or dc voltages and a few amperes under impulse voltages. However, when testing surge arresters or short-circuit testing of switchgear, currents several orders of magnitude higher are required. Methods of generating high voltages and high impulse currents are discussed in this chapter.

16.2 GENERATION OF HIGH ALTERNATING VOLTAGES

For generating ac test voltages of less than a few hundred kV, a single transformer can be used. For higher voltages, a single transformer construction would entail undue insulation problems. Also, expenses, transportation, and erection problems connected with large testing transformers become prohibitive. These drawbacks are avoided by cascading several transformers of relatively small size with their high-voltage windings effectively connected in series.

16.2.1 Single-Unit Testing Transformers

Single-unit testing transformers do not differ from single-phase power transformers with regard to the design of core and windings in relation to the kVA output. However, particular attention is given to heavy insulation of the high-voltage winding and to low magnetizing currents for minimum distortion of the output voltage waveform. The equivalent impedances of testing transformers are usually within about 5%.

The iron core as well as one terminal of the low-voltage and high-voltage windings are usually maintained at earth potential. The other terminal of the high-voltage winding would be insulated to the full output voltage. Considerable economy is achieved, however, if the center point rather than one terminal of the high-voltage winding is earthed. Thus each terminal of the high-voltage winding would be insulated to only half the output voltage. A compensating winding is installed close to the core to reduce the high leakage reactance between the high- and low-voltage windings. The high-voltage winding is arranged in layers that are carefully insulated and the potential distribution over it is carefully controlled.

16.2.2 Cascaded Transformers

Cascaded transformers were first used by Petersen, Dessauer, and Welter (Fig. 16.1) (Kind, 1978). The low-voltage winding (L) of each upper stage is fed from an excitation winding (E) in the stage immediately below it. The (L) and (E) windings are therefore rated at currents higher than those of the H windings; the excitation windings at the lower stages evidently carry higher loading than those of the upper stages. If the power carried by the third stage of Figure 16.1 is P, the power carried by the second and first stages are, respectively, 2P and 3P. The total short-circuit impedance of a cascade transformer could easily be related to the impedance of the individual stages (Kind, 1978).

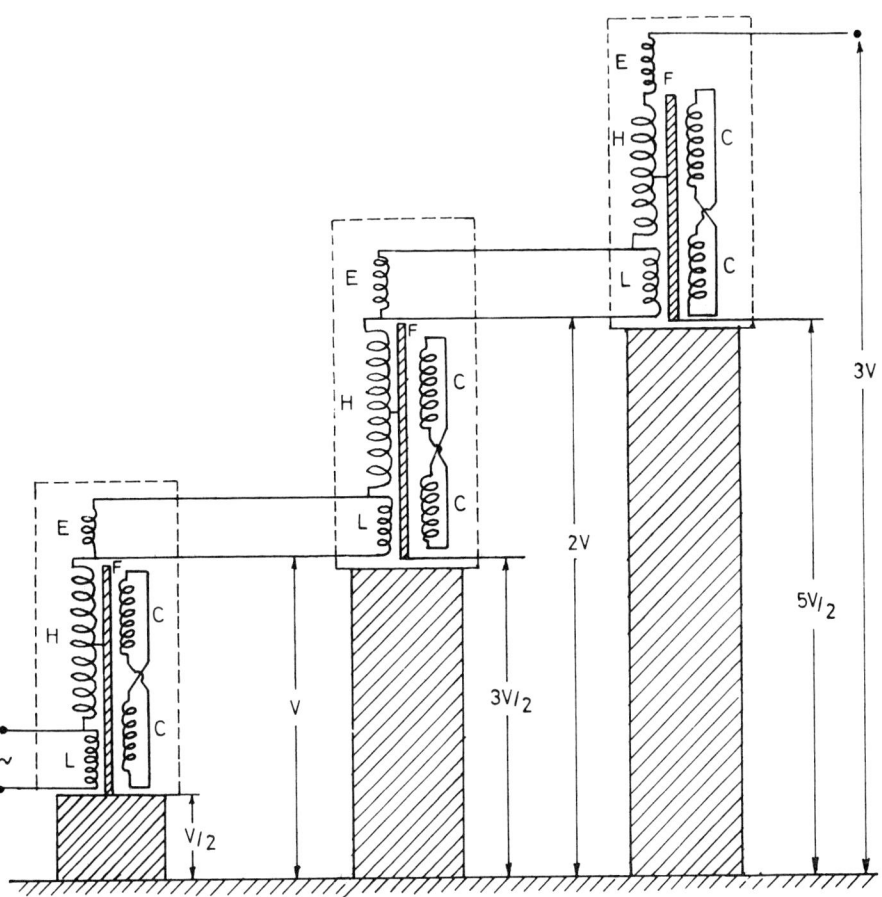

Fig. 16.1 Cascaded transformers with compensating windings and extra insulation. H, high-voltage winding; L, low-voltage winding; E, excitation winding; C, compensating winding; F, iron core.

If the output voltage of the first stage is V, the full output voltage is its multiple according to the number of stages. The iron core and the container (if metallic) of each stage other than the first are accordingly insulated from earth. This scheme is more expensive and requires more space. To reduce the size and cost of the insulation, transformers with center-tapped high-voltage winding are cascaded as shown in Figure 16.1. The tank of the first stage is kept at V/2, while those of the second and third stages are at 3V/2 and 5V/2 (Fig. 16.1).

Cascade transformers of ratings up to 10 MVA and voltages up to 2.25 MV are available for both indoor and outdoor testing (Fig. 16.2). These transformers being driven by a full-rated voltage regulator are referred to as "straight circuits," as distinguished from resonant circuits.

16.2.3 Resonant Transformers

The test circuits discussed in Section 16.2.2 consist of a test transformer (single-unit or cascade) and a voltage regulator, both rated to take full power from the supply. A disadvantage of this arrangement is that harmonics can appear in the test voltage either by accentuation of harmonics present in the incoming supply or could otherwise be generated in the transformer itself, due to its nonlinear magnetic performance. Without filtering out these harmonics, such simple straight test circuits would not be entirely suitable for certain tests, such as those for partial discharges. An alternative method that is more economical and sometimes technically superior is offered by resonant circuits.

Parallel Resonance

The addition of parallel reactors either in the primary low-voltage circuit or the secondary high-voltage circuit may partially or completely neutralize the capacitive load current, thus improving the power factor. If a motor-alternator is used as the supply source, the risk of self-excitation of the alternator would thus be eliminated. Input power reductions of 10:1 are feasible, thus reducing drastically the cost of the regulator, reactors, and filters.

Series Resonance

An alternative system is the series resonance circuit. By resonating the circuit through a series reactor L (Fig. 16.3a) at the test frequency (50 or 60 Hz), harmonics are heavily attenuated. The shunt capacitance C usually represents the high-voltage bushing and the test object. Figure 16.3b shows the equivalent circuit of the test transformer. Since $R_e \ll \omega L_e$ and the voltage V_2 is almost in phase with V_1' (Fig. 16.3c), thus

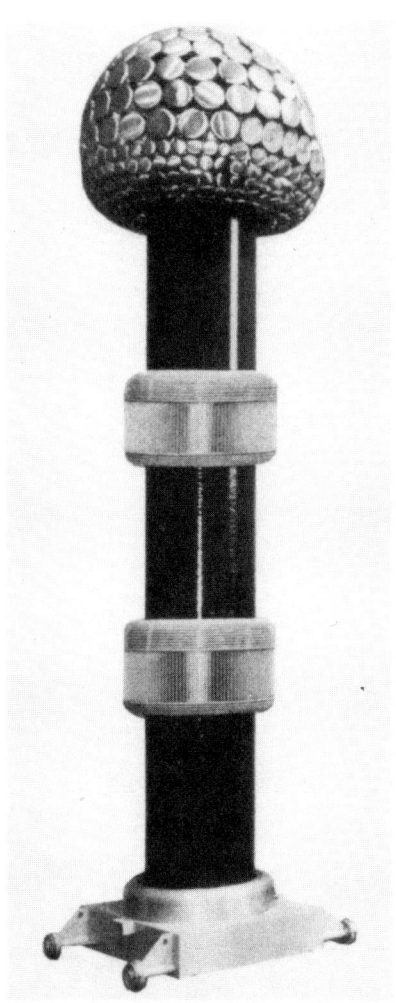

Fig. 16.2   1200-kV Cascaded transformer with insulating containers. Total height about 12 m. (Courtesy of Messwandler-Bau AG.)

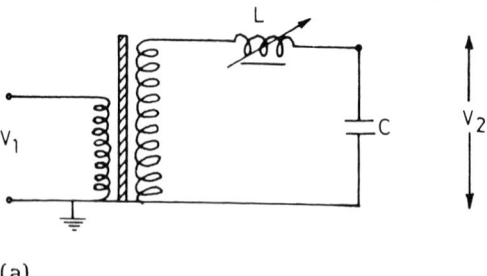

(a)

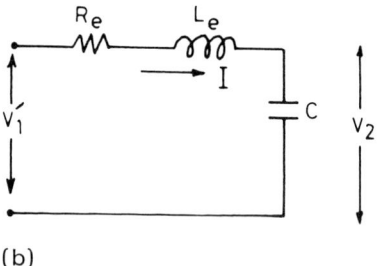

(b)

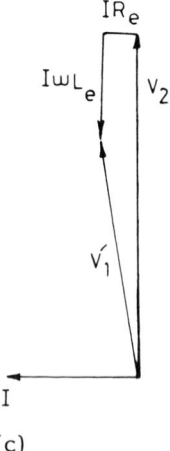

(c)

Fig. 16.3 (a) Simplified circuit of a resonant transformer, (b) equivalent circuit, (c) phasor diagram.

$$V_2 = V'_1 \frac{1}{1 - \omega^2 L_e C} \qquad (16.1)$$

As resonance is approached, $V_2 \gg V'_1$. Thus there is no longer a fixed ratio of primary to secondary voltage. Therefore, the secondary voltage itself should be measured accurately in the tests.

The main advantages of series resonant transformers are:

1. Pure sinusoidal output waveforms
2. Less power requirements from the mains (5 to 10% of straight circuit requirements)

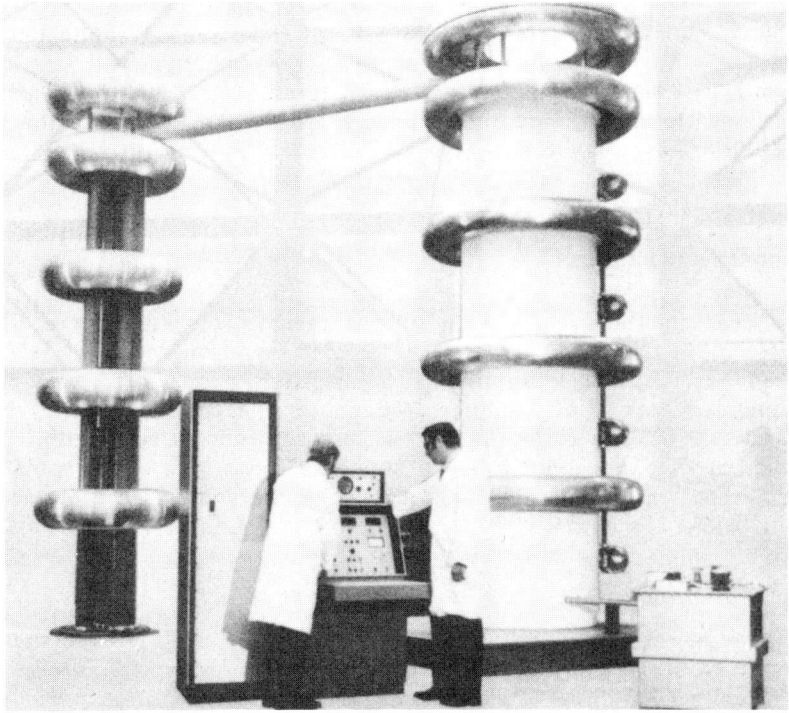

Fig. 16.4   800 kV/8000 kVA series-resonant test system. (Courtesy Hypotronics.)

3. No high-power arcing and heavy current surges occur if the test object fails, as the resonance is heavily disturbed by the resulting short circuit
4. Cascading is also possible for producing very high voltages
5. Simple and compact test setup

The main disadvantage is that the additional variable reactors should withstand the full-test voltage and full-current rating. Series resonance sets are usually brought into resonance by mechanical adjustment of an air gap in the iron core of the reactor. These sets are designed and commonly used for partial discharge testing (Fig. 16.4) and for testing installed gas-insulated systems in the field.

16.3 GENERATION OF HIGH DC VOLTAGES

Besides the Van de Graaff generators, with their special applications (Chapter 5), high-voltage ac and rectifiers are usually used for generating HVDC. The output dc has a trace of ac (ripple), which increases with the current drawn by the load. The "ripple factor," by definition, is considered equal to

$$RF = \frac{V(av)_0 - V(av)}{V(av)_0} \quad (16.2)$$

where $V(av)$ is the average value of the rectified output voltage, and $V(av)_0$ is the value of $V(av)$ at no load. Silicon rectifiers are commonly used. The ac supply to the rectifiers may be of power frequency or may be of a higher frequency. The high frequency is recommended when a ripple of very small magnitude is required without use of expensive filters to suppress them. Different transformer–rectifier circuits for generating high dc voltages are discussed below.

16.3.1 Half- and Full-Wave Rectifier Circuits

In half-wave and full-wave rectifier circuits (e.g., Fig. 16.5), the capacitor C is gradually charged to $V_s(max)$, the maximum ac secondary voltage of the HV transformer during the conduction periods. During the other periods, the capacitor is discharged into the load with a time constant $\tau = CR$. For reasonably small ripple voltage ΔV, τ should be at least 10 times the period of the ac supply.

High-Voltage Generation

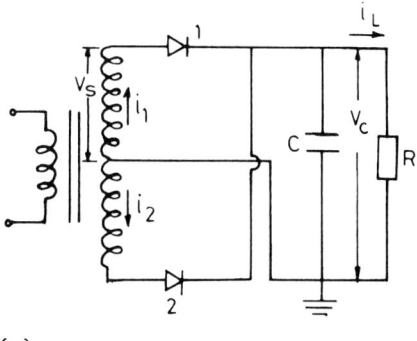

(a)

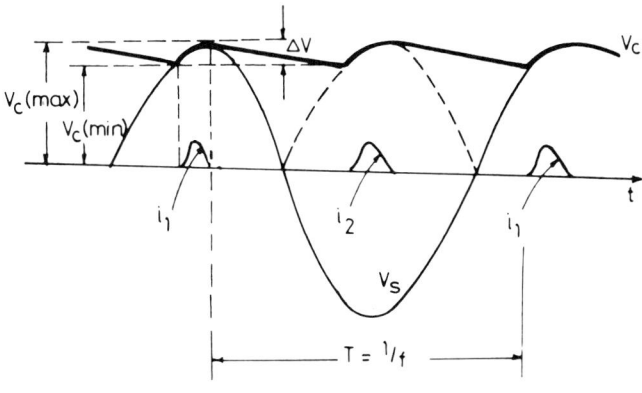

(b)

Fig. 16.5 Single-phase full-wave rectifier with smoothing capacitor: (a) circuit diagram, (b) steady-state voltages and currents with load R.

The rectifier element should have a peak inverse voltage of at least $2V_s(\max)$. To limit the charging current, an additional resistance is connected in series with the secondary of the transformer.

With half-wave rectification and during one period T $(= 1/f)$ of the ac voltage, a charge q is supplied from the transformer to the capacitor within the short conduction period t_1 of the rectifier element and is transferred to the load R during the long nonconduction period t_2, so

$$q = \int_{t_1} i(t) \, dt = \int_{t_2} i_L(t) \, dt$$

$$= \int_{t_2} \frac{V_c(t)}{R} \, dt = I_L T = \frac{I_L}{f} \qquad (16.3)$$

where $t_1 \ll t_2 \sim T = 1/f$ and I_L is the average load current.

Neglecting the voltage drops within the transformer and the rectifiers during conduction, ΔV is easily found from the charge q transferred to the load:

$$\Delta V = \frac{q}{C} = \frac{I_L}{fC} \qquad (16.4)$$

This equation correlates the ripple voltage to the load current and the circuit parameters f and C. The product fC is an important design factor. According to equation (16.2), the voltage ripple factor for half-wave rectification is

$$RF = \frac{(\Delta V/2)}{V_s(\max)} \qquad (16.5)$$

In full-wave rectifier circuits (Fig. 16.5), rectifiers 1 and 2 conduct during alternate half-cycles and charge the capacitor C. The HV transformer requires a center-tapped secondary with a voltage rating of $2V_s$. The ripple voltage according to equation (16.4) is halved.

16.3.2 Voltage-Doubler Circuits

In Figure 16.6a two half-wave rectifier circuits are connected in opposition, thus producing an unsmoothed unidirectional voltage. The source is effectively C_1, C_2 and the transformer in series. The output voltage has a peak value of $3V_s(\max)$ at no load. With the load R connected, the capacitors discharge through it with a corresponding voltage drop. This voltage-doubler circuit may be earthed at any point, provided that the transformer has an adequate insulation.

In the circuit of Figure 16.6b, each of the two capacitors is charged to only $V_s(\max)$ in alternate half-cycles and the total output voltage is $2V_s(\max)$ at no-load, with a small ripple under load.

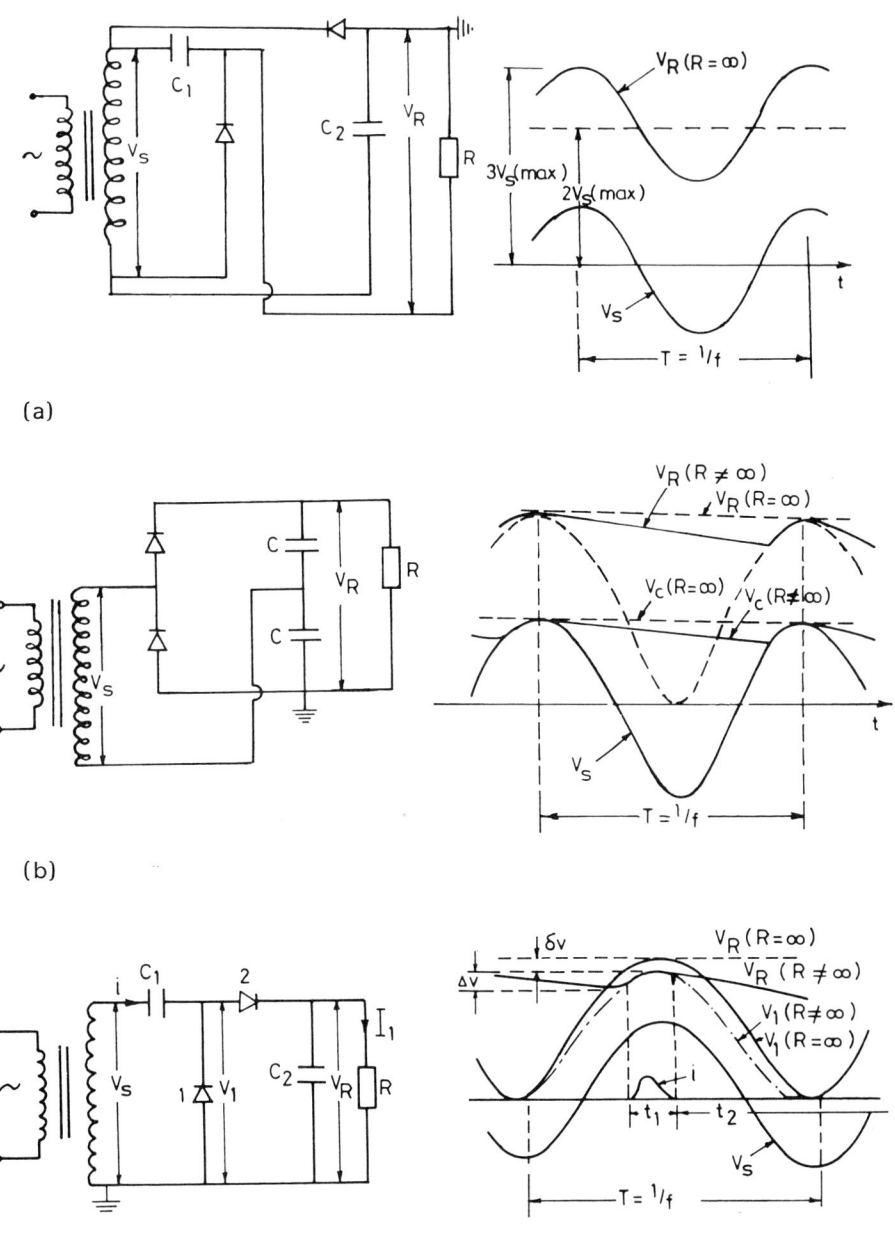

Fig. 16.6 Connection diagrams and steady-state voltage waveforms. (a) output voltage contains a considerable ac component; (b and c) output voltage contains only a small ripple.

As the doubler of Figure 16.6c represents the basic unit of the commonly used Cockcroft-Walton voltage-multiplier circuits, it is discussed in more detail. Let a charge q be transferred from C_2 to the load per cycle when the average load current is I_1. Then $I_1 \sim q/t_2$, where t_2 is the time during which rectifier (2) does not conduct, t_2 is much longer than t_1, the conduction time for rectifier 2. Also, let ΔV be the ripple voltage. Since $q = C_2 \Delta V$ and $t_1 \ll t_2 \sim T = 1/f$, then

$$I_1 = C_2 \Delta V f = qf$$

or

$$\Delta V = \frac{I_1}{C_2 f} \qquad (16.6)$$

At the same time, a charge q is transferred from C_1 to C_2 per cycle, so

$$\delta V = \frac{I_1}{C_1 f}$$

Therefore, the main voltage drop below $2V_s(\max)$ is $(\delta V + \Delta V/2)$ and the average output voltage

$$V_{av} = 2V_s(\max) - \delta V - \frac{\Delta V}{2}$$

$$= 2V_s(\max) - \frac{I_1}{f}\left(\frac{1}{C_1} + \frac{1}{2C_2}\right) \qquad (16.7)$$

According to equation (16.2), the voltage ripple factor is

$$RF = \frac{\delta V + \Delta V/2}{2V_s(\max)} \qquad (16.8)$$

16.3.3 Voltage-Multiplier Circuits

The voltage-multiplier circuit using the Cockcroft-Walton principle is shown in Figure 16.7a. The first stage, composed of rectifiers 1 and 2, the capacitors C_1 and C_2, and the HV transformer, is identical with the voltage doubler shown in Figure 16.6c. For higher output voltages of 4, 6, . . . , 2n of the input voltage $V_s(\max)$, the circuit is repeated with a cascade connection. Thus

High-Voltage Generation

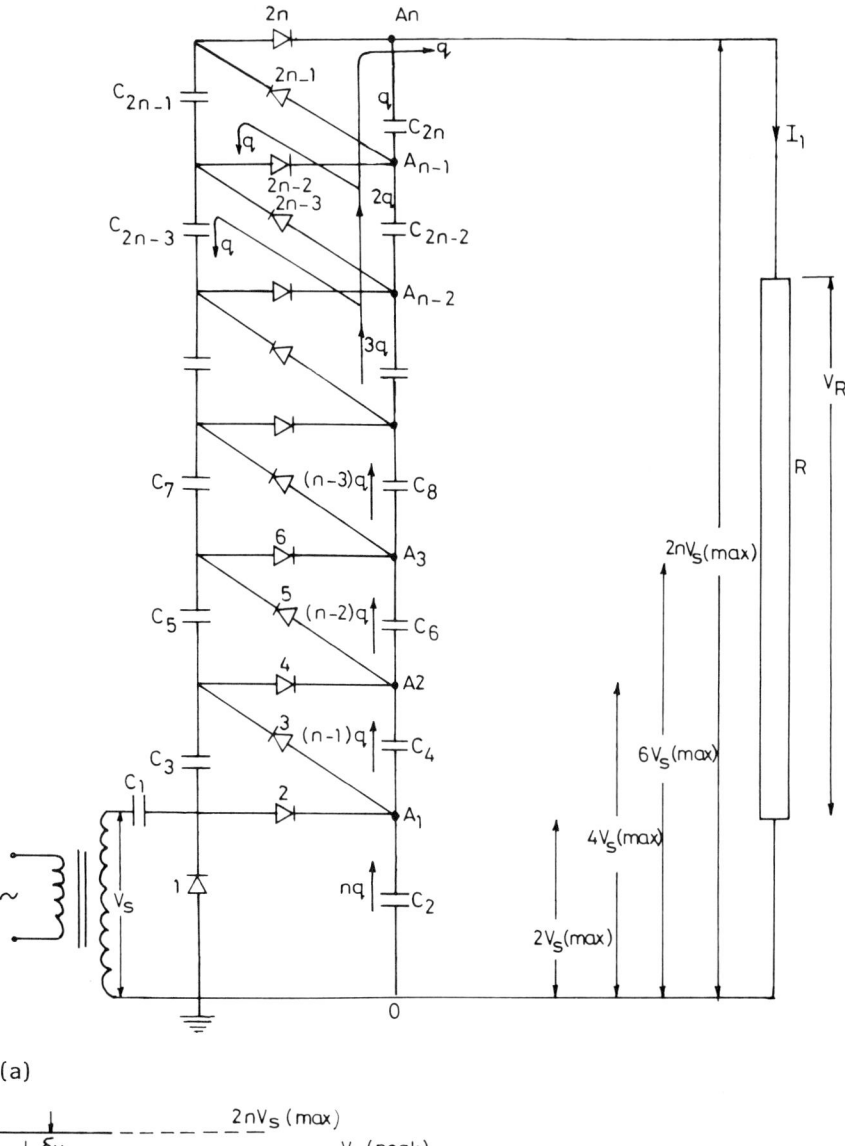

(a)

(b)

Fig. 16.7 Voltage-multiplier circuit according to Cockcroft-Walton: (a) circuit diagram, (b) tracing of load voltage — drop δV and ripple ΔV.

the total output voltage reaches $2nV_s(max)$ above the earth potential (Fig. 16.7a). But the voltage across any individual capacitor or rectifier is $2V_s(max)$, except for C_1, which is charged to $V_s(max)$ only.

Ripple in Multiplier Circuits

Referring to Figure 16.7a, let q be the charge transferred from C_{2n} to the load per cycle with a ripple voltage at capacitor C_{2n}:

$$\Delta V_{2n} = \frac{q}{C_{2n}} \tag{16.9}$$

Simultaneously, C_{2n-2} transfers a charge q to the load and q to C_{2n-1}, so the ripple voltage at capacitor C_{2n-2} is

$$\Delta V_{2n-2} = \frac{2q}{C_{2n-2}} \tag{16.10}$$

Similarly, C_{2n-4} transfers a charge q to the load, q to C_{2n-3}, and q to C_{2n-2} with a ripple voltage at C_{2n-4}.

$$\Delta V_{2n-4} = \frac{3q}{C_{2n-4}} \tag{16.11}$$

Proceeding in the same way, the ripple voltage at C_2 is

$$\Delta V_2 = \frac{nq}{C_2} \tag{16.12}$$

Hence the total ripple voltage will be

$$\Delta V_{total} = q \left(\frac{1}{C_{2n}} + \frac{2}{C_{2n-2}} + \frac{3}{C_{2n-4}} + \cdots + \frac{n}{C_2} \right) \tag{16.13}$$

It is quite evident that the capacitors near the ground terminal (e.g., C_2) are the major contributors to the ripple voltage. If their capacitances are chosen suitably large, the total ripple voltage could be reduced significantly. However, this is not practical, as a transient overvoltage would overstress the smaller capacitors. Therefore, equal capacitances (C) are usually provided, and the corresponding total ripple voltage is

High-Voltage Generation

$$\Delta V_{total} = \frac{q}{C}\frac{n(n+1)}{2} = \frac{I_1}{fC}\frac{n(n+1)}{2} \qquad (16.14)$$

Voltage Drop on Load

In addition to the ripple voltage ΔV, there is a voltage δV, which is the difference between the theoretical no-load and the on-load voltages. With reference to Figure 16.7a, the capacitor C_2 loses nq during each cycle, so the capacitor C_1 has to replenish it. Therefore, C_1 and C_2 charge up to $[V_s(max) - nq/C]$ and $[2V_s(max) - nq/C]$ instead of $[V_s(max)]$ and $[2V_s(max)]$, respectively. Similarly, C_3 can only be charged up to a maximum voltage of $[2V_s(max) - 2nq/C]$ instead of $2V_s(max)$.

Thus one can easily derive the general rule for the total voltage drop at the smoothing column $C_2, C_4, C_6, \ldots, C_{2n}$ (Fig. 16.7a):

$$\delta V_2 = \frac{q}{C} n$$

$$\delta V_4 = \frac{q}{C}[2n + (n-1)]$$

$$\delta V_6 = \frac{q}{C}[2n + 2(n-1) + (n-2)]$$

.
.
.

$$\delta V_{2n} = \frac{q}{C}[2n + 2(n-1) + 2(n-2) + \cdots + 2 \times 3 + 2 \times 2 + 1]$$

Adding all the n voltage drops gives the total voltage drop on load:

$$\delta V_{total} = \frac{q}{C}\left(\frac{2}{3}n^3 + \frac{1}{2}n^2 - \frac{n}{6}\right) = \frac{I_1}{fC}\left(\frac{2}{3}n^3 + \frac{n^2}{2} - \frac{n}{6}\right) \qquad (16.15)$$

In case the generator comprises a large number of stages (i.e., $n > 5$), the first term in equation (16.15) will dominate the others and the load voltage would be approximately (Fig. 16.7b):

$$V_R(peak) = 2nV_s(max) - \frac{I_1}{fC} \times \frac{2}{3}n^3 \qquad (16.16)$$

The output voltage V_R(peak) can be increased by increasing either C or the frequency f. An upper limit of 500 to 1000 Hz is set by the high-voltage appearing across the circuit inductances and by the high capacitive currents to be drawn. An increase of supply frequency is in general more economical than an increase in the capacitance values. Small values of C provide a dc supply with limited stored energy which might be an essential design factor for special breakdown investigations of dielectrics. For fast response to load changes and supply variations, high supply frequency and small stored energy are prerequisites.

As seen in Figure 16.7 for a given load, V_R(peak) could be increased with the number of stages n. This reaches an optimum value beyond which it drops if n is chosen too large. For given values of I_1, V_s(max), f, and C, the highest value of V_R(peak) is achieved with the "optimum" number of stages, that is, when

$$\frac{dV_R(peak)}{dn} = 0 \quad \text{or} \quad n_{op} = \sqrt{\frac{V_s(max)fC}{I_1}} \quad (16.17)$$

Thus, for a multiplier circuit with f = 50 Hz, C = 0.1 μF, V_s(max) = 200 kV, and I_1 = 8 mA, the optimum number of stages $n_{op} \simeq 12$. Using equations (16.14) and (16.15) (Fig. 16.7b), the mean output voltage

$$V_R(av) = 2nV_s(max) - \delta V_{total} - \frac{\Delta V_{total}}{2}$$

$$= 2nV_s(max) - \frac{I_1}{fC}\left(\frac{2}{3}n^3 + \frac{n^2}{2} - \frac{n}{6}\right) - \frac{I_1}{fC}\frac{n(n+1)}{4}$$

(16.18)

Therefore, the ripple factor is expressed as

$$RF = \frac{2\delta V_{total} + \Delta V_{total}}{4nV_s(max)} \quad (16.19)$$

16.3.4 Cascade Rectifier Circuits with Cascaded Transformers

The multiple charge transfer with the voltage multiplier of Cockcroft-Walton type demonstrated the limitation in the dc output concerning the ripple voltage and voltage drop. This disadvantage can be waived by cascading voltage doublers without a need for

High-Voltage Generation

additional isolating transformers for every new stage. This can be achieved if the different stages are energized by specially designed cascaded transformers. The transformer in each stage comprises a tertiary low-voltage winding which excites the primary winding of the next upper stage. In this way, low ripple and voltage drops can be achieved when load currents are high. However, there are limitations with regard to the number of stages, as the lower transformers have to supply the energy for the upper ones.

16.3.5 Deltatron Circuits

A very sophisticated cascade transformer HV dc voltage multiplier circuit is shown in Figure 16.8. The very small ripple factor, high stability, fast response, and small stored energies are the main advantages of this circuit. However, the output of the circuit is limited to about 1 MV and a few milliamperes and needs excessive insulation for the higher stages.

The circuit consists primarily of a cascade connection of transformers having no iron cores and fed from a high-frequency source (Fig. 16.8). These transformers are coupled by series capacitors C_s to compensate for their leakage inductances. Also, shunt capacitors C_p compensate for the transformers magnetizing inductances. The string of cascaded transformers is loaded by a terminating resistor, R_t (Fig. 16.8). The voltage remains nearly constant over the cascaded transformers but with a small phase shift with respect to the input voltage.

The usual Cockcroft-Walton cascade circuit is connected to each stage (Fig. 16.8). As the frequency is high, the capacitors C of the cascades can be made very small, with a correspondingly low energy stored and fast circuit response to load changes or supply variations. The small ripple voltage characterizing this circuit is caused not only by the small storage capacitors, but also by the phase shift between the input voltages of the different stages provided by their transformers.

16.3.6 Electrostatic HVDC Generators

Van de Graaff Generators

Van de Graaff generators have been developed to produce very high voltages of 5 to 6 MV with output currents of microamperes (Section 5.6). They are useful for energizing particle accelerators. The main advantages of these generators are their ripple-free high dc voltages, which can easily be reached with precision and flexibility.

A modification of Van de Graaff's generator was made by Felici (1953) to eliminate its mechanical shortcomings as regards the belt

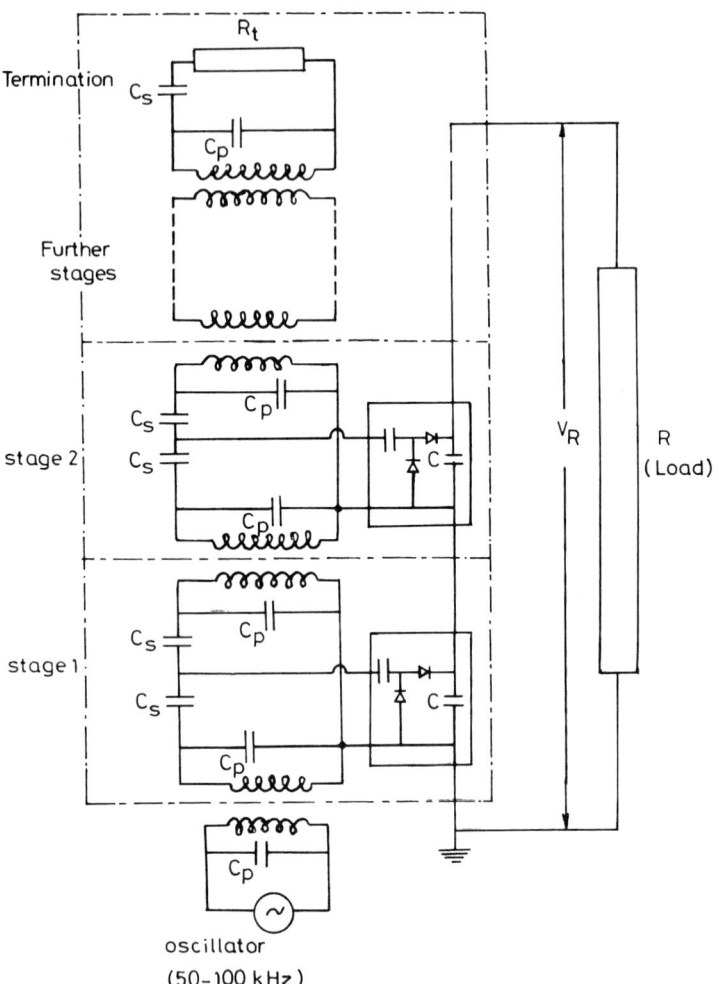

Fig. 16.8 Deltatron-circuit operating principle.

and its vibrations. In Felici's generator, the belt is replaced by an insulating cylinder surrounding a gas discharge chamber and rotating at a high speed. The chamber is subjected to a high dc field and ions of both polarities are drawn from it by a pair of electrodes connected to the load. Such a source could produce an output of a few hundred kilovolts at a few milliamperes.

Variable-Capacitance Generators

These generators are designed to convert mechanical energy into electrical energy using the variable-capacitance principle. They consist of a stator with interleaved rotor vanes forming a variable capacitor and operates in vacuum.

The current through a variable capacitor is given by

$$I = C \frac{dV_1}{dt} + V_1 \frac{dC}{dt} \tag{16.20}$$

where C is the capacitance being charged to a voltage V_1. The power input into the circuit at any instant

$$P = V_1 I = CV_1 \frac{dv_1}{dt} + V_1^2 \frac{dC}{dt} \tag{16.21}$$

If dC/dt is negative, mechanical energy is converted into electrical energy, and vice versa. With the capacitor charged from a dc voltage, $dV_1/dt = 0$ and the power input will be

$$P = V_1^2 \frac{dC}{dt} \tag{16.22}$$

When the rotor-to-stator capacitance is maximum (C_m), the charge between the rotor and the stator is therefore $Q_m = C_m V_1$. As the rotor rotates, the capacitance C decreases and the voltage across C ($= C_m V_1/C$) increases, and further rotation of the rotor causes eventually the current to flow from the generator to the load.

A generator of this type with an output voltage of 1 MV and a field gradient of 1 MV/cm in a hard vacuum and having 16 rotor poles, 50 rotor plates of 1.2 m maximum and 0.6 m minimum diameter, and a speed of 4000 rpm could develop a maximum power of 7 MW. This vacuum-insulated generator was first discussed by Trump (1947) and recently investigated by Philp (1977).

16.3.7 Regulation and Stabilization of HVDC Generators

The output of a dc generator changes with the load current as well as with the supply voltage. To maintain a constant voltage at the load terminals, it is essential to have a regulator circuit. It is necessary to keep the change in voltage between ±0.1% and ±0.001%, depending on the applications.

A dc voltage regulator consists of a detector that senses the voltage deviations from the desired value, and a controller actuated by the detector in such a manner as to correct the deviations (Naidu and Kamaraju, 1982). The regulators are generally of a series or a shunt type (Dorf, 1986).

If ΔV is the change in the dc output voltage V as a result of a change ΔV_1 in the ac supply voltage V_1, the stabilization ratio ψ_s is defined as

$$\psi_s = \frac{\Delta V_1/V_1}{\Delta V/V}$$

$$= \frac{\Delta V_1}{\Delta V} \frac{V}{V_1} \quad (16.23)$$

ψ_s is greater than unity and it indicates how many times better (V) is in stability compared with (V_1).

The "regulation" ψ_r of a dc voltage regulator is defined as the fractional change in (V) caused by a fractional change in the output current (I), that is,

$$\psi_r = \frac{\Delta V/V}{\Delta I/I}$$

$$= \frac{\Delta V/\Delta I}{V/I} = \frac{R_0}{R} \quad (16.24)$$

where R_0 is the effective internal resistance of the regulator as seen from its output terminals, and R is the resistance of the load.

16.4 GENERATION OF IMPULSE VOLTAGES

The standardized lightning impulse waves are represented by the general equation

$$e(t) = A[\exp(-\alpha_1 t) - \exp(-\alpha_2 t)] \quad (16.25)$$

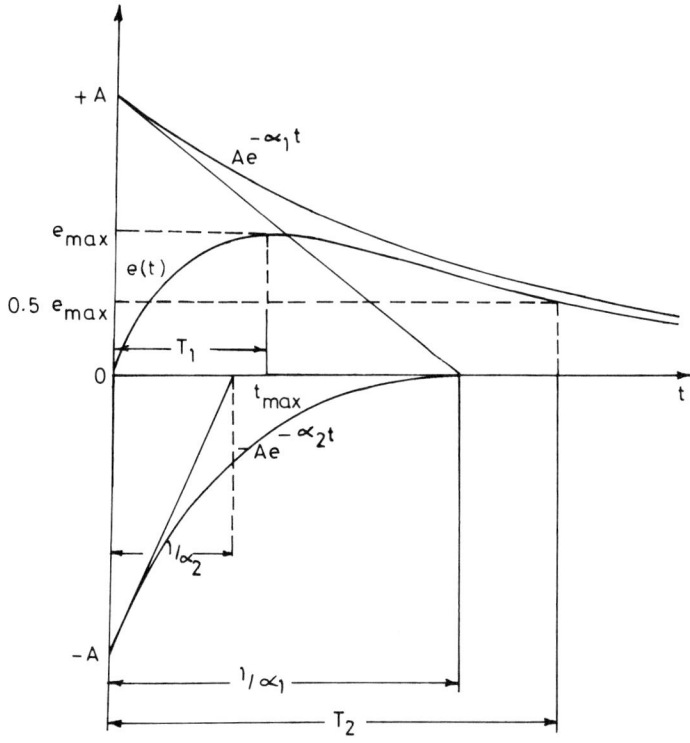

Fig. 16.9 Impulse wave and its components as a double exponential.

The graph of this expression is shown in Figure 16.9. The tolerances allowed in the front and tail durations are ±30% and ±20%, respectively. However, the tolerance allowed in the peak value is ±3%. The standard lightning impulse wave has a front duration of 1.2 μs and a wavetail duration of 50 μs, and is described as a 1.2/50-μs wave. The standard wave front and tail are defined in Figure 16.17 (IEC, 1973).

16.4.1 Circuit for Producing Lightning Impulses

The double-exponential waveform expressed by equation (16.25) may be obtained in the laboratory by any connection of the $R-C$ circuit as exemplified in Figure 16.10a to d) or by a series $R-L-C$ circuit under overdamped conditions (Fig. 16.10e).

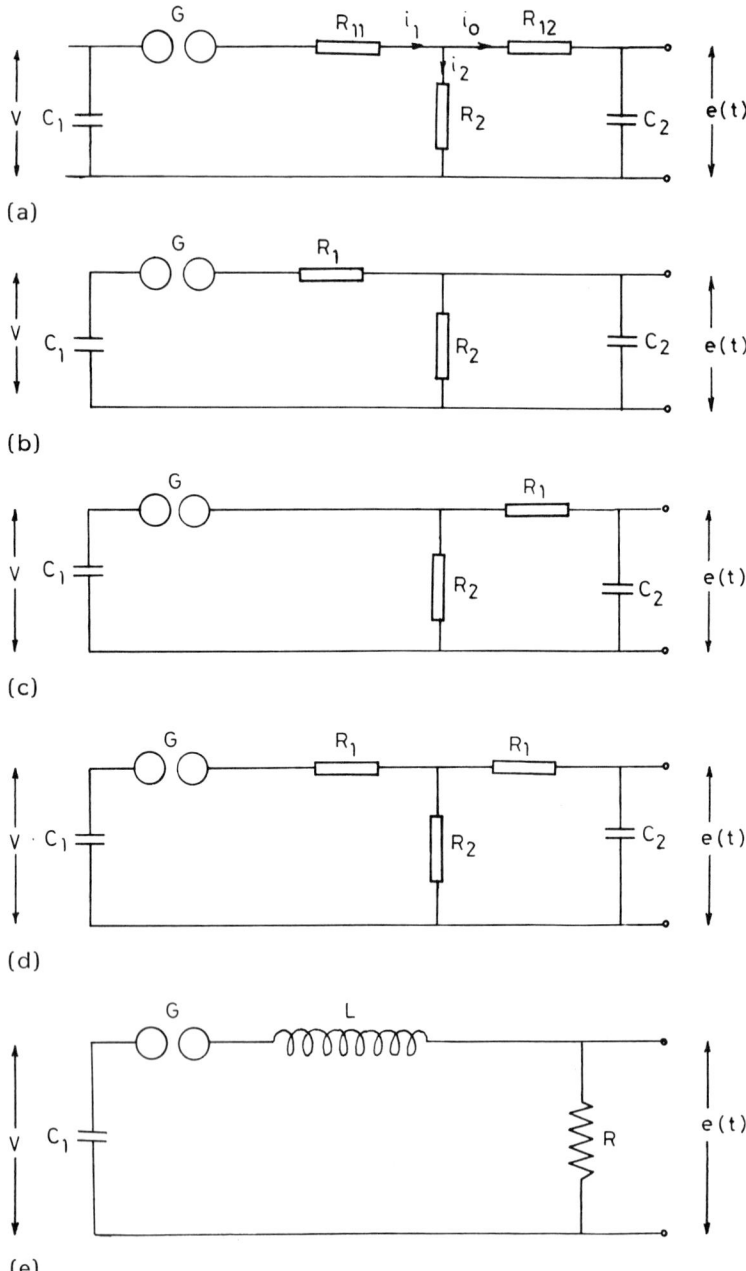

Fig. 16.10 Circuits for producing impulse voltage waves.

High-Voltage Generation

The circuits commonly used for impulse generators are those of Figure 16.10b and c. The advantages of these circuits are that the wavefront and wavetail are independently controlled by changing R_1 and R_2 separately, different from the circuit of Figure 16.10e. Second, being capacitive, the test objects admittedly form a part of C_2.

The magnitudes of the different circuit elements together determine the required rapid charging of C_2 to the peak value e_{max}. However, a long wavetail calls for a slow discharge. This is achieved if R_2 is much larger than R_1 (Fig. 16.10b to d). The smaller the time constant of the circuit, the faster is the rate by which the output voltage e(t) approaches its peak value. The peak value e_{max} cannot exceed the value determined by the distribution of the initially available charge VC_1 among C_1 and C_2, so the voltage efficiency

$$\eta = \frac{e_{max}}{V} < \frac{C_1}{C_1 + C_2} \tag{16.26}$$

For high efficiencies, C_1 should be chosen much larger than C_2. The exponential decay of the impulse voltage on the tail occurs with a time constant of about $C_1(R_{11} + R_2)$, $C_1(R_1 + R_2)$, $C_1 R_2$, and $C_1(R_1 + R_2)$ in circuits a, b, c, and d of Figure 16.10, respectively. The impulse energy consumed is

$$W = \frac{1}{2} C_1 V^2$$

which is an important characteristic parameter of the impulse generator.

It is sufficient to analyze the general circuit a of Figure 16.10, being a general one. The other circuits (b to d) are simply special forms of circuit a. Analysis of the inductive circuit e is that of a simple R-L-C circuit.

Circuit Analysis

With reference to circuit a of Figure 16.10, on breakdown of the gap G, the capacitor C_1 discharges through the circuit while its voltage decreases from its initial magnitude V_0.

$$i_1 = i_2 + i_0 \tag{16.27}$$

$$V = \frac{-1}{C_1} \int i_1 \, dt + V_0 \tag{16.28}$$

$$e = \frac{1}{C_2} \int i_0 \, dt \qquad (16.29)$$

Hence

$$i_1 = \left(\frac{R_{12}}{R_2} i_0 + \frac{e}{R_2} \right) + i_0 \qquad (16.30)$$

Using simple mathematical manipulation, the following relation for e could be reached:

$$A_1 \frac{d^2 e}{dt^2} + A_2 \frac{de}{dt} + A_3 e = 0 \qquad (16.31)$$

with the initial conditions

$$e \Big|_{t=0} = 0$$

and

$$\frac{de}{dt} \Big|_{t=0} = \frac{C_1 V_0}{A_1}$$

Hence

$$e(t) = \frac{C_1 V_0}{A_1 (\alpha_2 - \alpha_1)} [\exp(-\alpha_1 t) - \exp(-\alpha_2 t)] \qquad (16.32)$$

The peak value e_{max} and the corresponding period t_{max} can be evaluated by differentiating equation (16.32).

In a normalized form, the impulse wave represented by equation (16.25) could be written as (Auget and Lanovici, 1982)

$$e(t) = \eta \frac{\alpha V_0}{\sqrt{\alpha^2 - 1}} [\exp(-\alpha_1 t) - \exp(-\alpha_2 t)] \qquad (16.33)$$

where

$$\alpha_1 = \frac{\alpha - \sqrt{\alpha^2 - 1}}{\theta} \qquad (16.34)$$

High-Voltage Generation

$$\alpha_2 = \frac{\alpha + \sqrt{\alpha^2 - 1}}{\theta} \qquad (16.35)$$

and

$$\theta = \sqrt{C_1 C_2 (R_{11} R_{12} + R_{11} R_2 + R_{12} R_2)} \qquad (16.36)$$

$$\eta = \frac{C_1}{C_1(1 + R_{11}/R_2) + C_2(1 + R_{12}/R_2)} \qquad (16.37)$$

$$\alpha = \frac{1}{2} \frac{R_2 C_1}{\eta \theta} \qquad (16.38)$$

Table 16.1 gives the values of the parameters θ, η, and α and the corresponding elements of the circuits (Kind, 1978; Kuffel and Zaengl, 1984). Evidently, in circuit b, $R_{12} = 0$ and $R_{11} = R_1$; in circuit c, $R_{11} = 0$ and $R_{12} = R_1$; and in circuit d, $R_{12} = R_{11} = R_1$.

Dimensioning of Circuit Elements

For a given wave shape, the choice of the resistances to control the wavefront and wavetail durations is not entirely independent but depends on the ratio C_1/C_2. However, there is a limitation on the ratio C_1/C_2. For circuit b of Figure 16.10 there is an upper limit, while for circuit c there is a lower limit. Table 16.2 gives the relationship between the time constants $1/\alpha_1$ and $1/\alpha_2$ and their ratios to the durations T_1 and T_2 for typical impulse waves. Table 16.3 gives the corresponding limiting values of C_1/C_2.

Effect of Circuit Inductance

Any residual inductance in the circuit elements would produce some oscillations in the wavefront and wavetail. It is noted that with increasing series inductance, the wavefront sensitively increases, but the magnitude of the peak value is affected only slightly. The lengthening of the wavefront by increasing the circuit inductance provides a convenient method for generating long-front (switching) impulses. On the other hand, an inductance across the load would drastically change the waveform, as depicted in Figure 16.11.

16.4.2 Multistage Impulse Generators

This is the best way for generating voltage impulses of very high amplitudes (millions of volts) using a dc source of a moderate

Table 16.1 Parameters θ, η, and α and the Wave-Shaping Elements for Different Circuits Producing Impulse Waves

	θ	η
Circuit (a)	$\sqrt{C_1 C_2 (R_{11} R_2 + R_2 R_{12} + R_{11} R_{22})}$	$\dfrac{1}{(1 + R_{11}/R_2 + C_2/C_1 (1 + R_{12}/R_2))}$
Circuit (b)	$\sqrt{C_1 C_2 R_1 R_2}$	$\dfrac{1}{1 + C_2/C_1 + R_1/R_2}$
Circuit (c)	$\sqrt{C_1 C_2 R_1 R_2}$	$\dfrac{1}{1 + C_2/C_1 (1 + R_1/R_2)}$
Circuit (d)	$\sqrt{C_1 C_2 R_1 (R_1 + 2R_2)}$	$\dfrac{1}{(1 + R_1/R_2)(1 + C_2/C_1)}$
Circuit (e)	$\sqrt{LC_1}$	

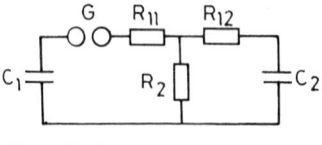

Circuit (a)

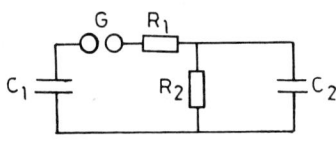

Circuit (b)

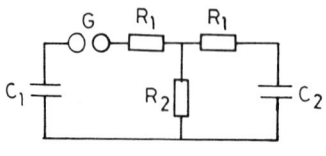

Circuit (d)

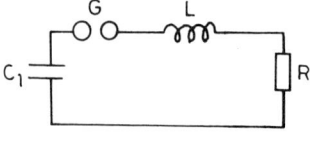

Circuit (e)

α	X	R_1 (Ω)	R_2 (Ω)
$\dfrac{R_2 C_1}{2\theta\eta}$			
$\dfrac{R_2 C_1}{2\theta\eta}$	$\dfrac{1}{\alpha^2}\left(1+\dfrac{C_1}{C_2}\right)$	$\dfrac{\alpha\theta}{C_1}(1-\sqrt{1-X})$	$\dfrac{\alpha\theta}{C_1+C_2}(1+\sqrt{1-X})$
$\dfrac{R_2 C_1}{2\theta\eta}$	$\dfrac{1}{\alpha^2}\left(1+\dfrac{C_1}{C_2}\right)$	$\dfrac{\alpha\theta}{C_1}(1-\sqrt{1-X})$	$\dfrac{\alpha\theta}{C_1+C_2}(1+\sqrt{1-X})$
$\dfrac{R_2 C_1}{2\theta\eta}$	$\dfrac{1}{4\alpha^2}\left(1+\dfrac{C_1}{C_2}\right)$	$\dfrac{2\alpha\theta}{C_1+C_2}(1-\sqrt{1-X})$	$\dfrac{2\alpha\theta}{C_1+C_2}\sqrt{1-X}$
$\dfrac{R}{2}\sqrt{\dfrac{C_1}{L}}$		$L(H) = \dfrac{\theta^2}{C_1}$	$R(\Omega) = \dfrac{2\alpha\theta}{C_1}$

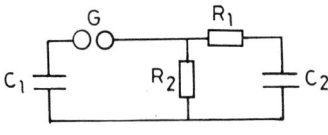

Circuit (c)

Table 16.2 $1/\alpha_1$, $1/\alpha_2$, and αT for Different Standard Waves

T_1/T_2 (μs)	$1/\alpha_1$ (μs)	$1/\alpha_2$ (μs)	α_2/T_1 (μs)	$\alpha_1 T_2$ (μs)
1.2/5	3.48	0.8	1.49	1.44
1.2/50	68.2	0.405	2.06	0.73
1.2/200	284.0	0.381	3.15	0.7
250/2500	2877.0	104.0	2.4	0.87

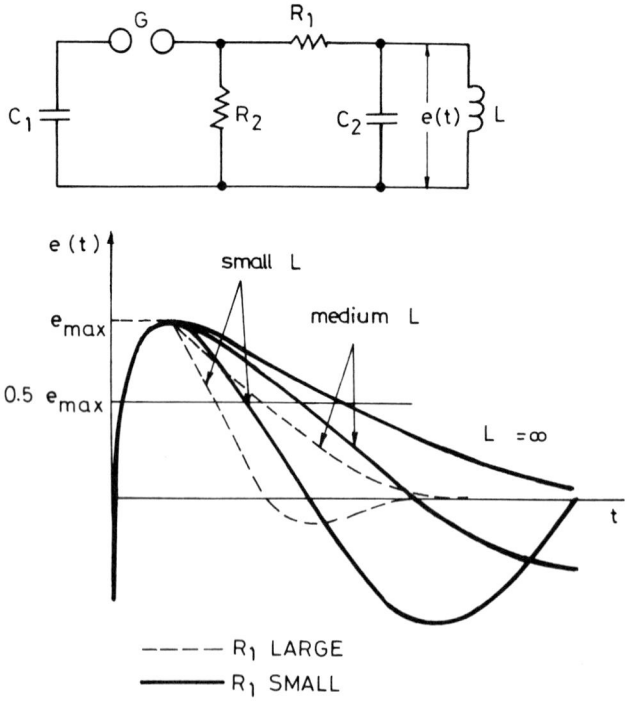

Fig. 16.11 Effect of varying the inductance across the output terminals on the generated impulse voltage wave shape, e(t).

Table 16.3 Limiting Values of C_1/C_2 for Different Standard Waves

Circuit	Value measured	T_1/T_2 (µs)			
		1.2/5	1.2/50	1.2/200	250/2500
Figure 16.10c	Min (C_1/C_2)	1.57	0.025	0.0054	0.157
Figure 16.10b	Max (C_1/C_2)	—	40.0	185.19	6.37

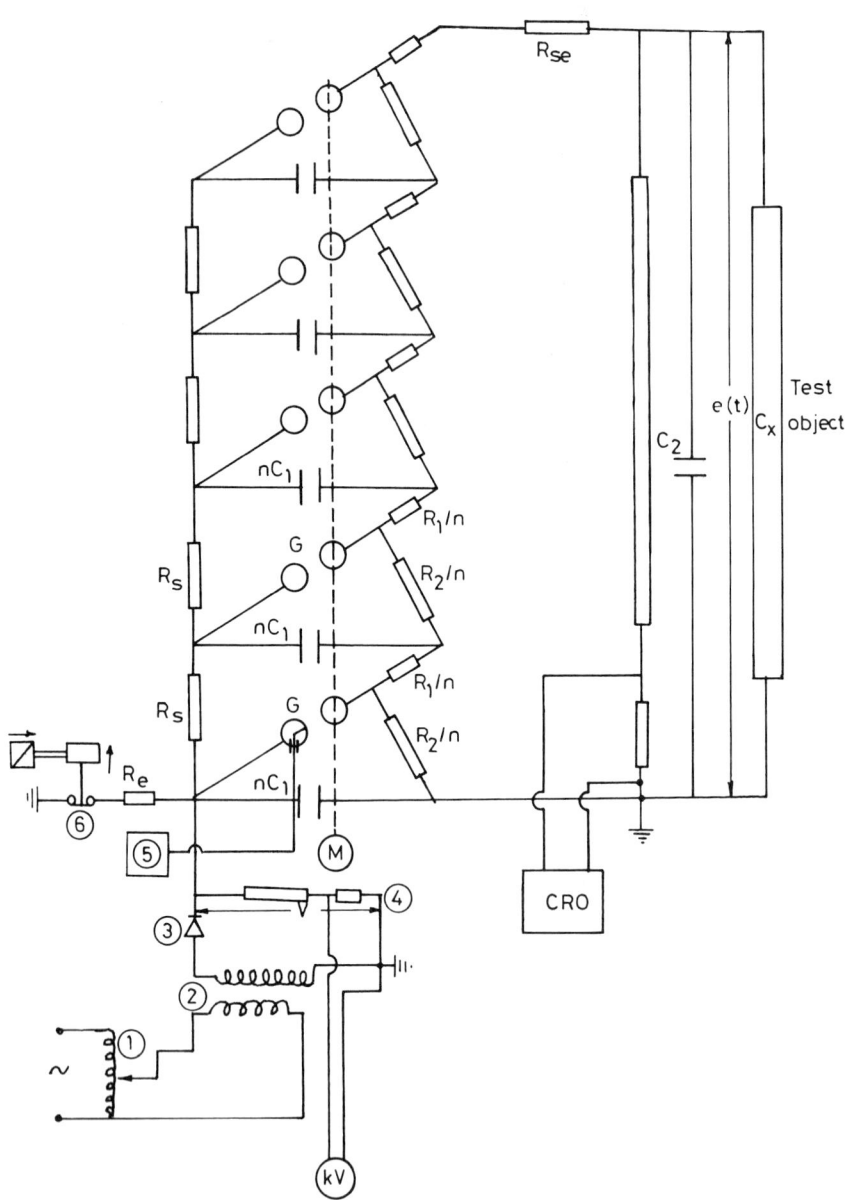

Fig. 16.12 Multistage impulse generator incorporating the waveshaping resistors within the generator proper. 1, Regulating transformer; 2, HV transformer; 3, rectifier; 4, measuring resistive divider for the charging voltage; 5, triggering unit; 6, earthing device; M, motor for control of sphere gaps; R_1/n, front resistors; R_2/n, tail resistors; R_e, discharging resistor; R, external part of front resistor.

High-Voltage Generation

output. A bank of capacitors are charged in parallel and then discharged in series, as originally proposed in 1923 by Marx. Since then, various modifications have been implemented.

A typical modification of a Marx circuit is shown in Figure 16.12. The resistances R_1 and R_2 are incorporated in the individual stages. R_1/n is connected in series with the gap G of each stage. R_2 is also divided into n parts connected across the stage capacitors. These modifications save the space and cost of the generator and result in smaller resistors and higher voltage efficiency.

The gap spacing is chosen such that the breakdown voltage of the gaps G is slightly higher than the charging voltage V. Thus all the capacitors are charged to the voltage V. When the impulse generator is to be discharged, the gaps G are made to spark over simultaneously by some external means. Thus all the capacitors get connected in series and discharge through the wave-shaping resistors R_1 and R_2 into the test object. The discharge time constant C_1R_1 is very small (microseconds) compared to the charging time constant nC_1R_s; thus no significant discharge takes place through the charging resistors R_s. Multistage impulse generators are usually specified by their total output voltage, the number of stages (n), and the stored gross energy (Hyltén-Cavallius, 1986).

Control of Multistage Impulse Generators

In multistage impulse generators, the spark gaps are arranged such that sparking of the lowest gap induces sparking of the other gaps, the voltage across them being only slightly lower than their breakdown voltage. Thus the impulse is generated at the instant a triggering signal is received by the lowest gap.

Up to the 1950s, the lowest gap was a three-electrode gap, as proposed by White in 1947, but since then, trigatron gaps have been in common use. The trigatron gap comprises a high-voltage spherical electrode and an earthed main hemispherical electrode with the main gap between. A trigger electrode with the shape of a metal rod is located inside and insulated from the main electrode by an annular clearance (Fig. 16.13). The trigatron is included in a pulse circuit such as the one shown in Figure 16.13. Closing the switch S produces between the trigger and the main electrode a voltage pulse sufficient to break down the annular clearance. The ionization of the discharge and the disturbance of the field in the main gap causes its complete breakdown. This breakdown initiates the breakdown of gaps G of the other stages of the multistage impulse generator (Fig. 16.12). Thus the instant of generating the impulse can be controlled. The switch S could be replaced by a synchronizing circuit so as to generate the impulse at the required instant with respect to the power-frequency ac, thus enabling ac and impulse high voltages to be superimposed on a test specimen

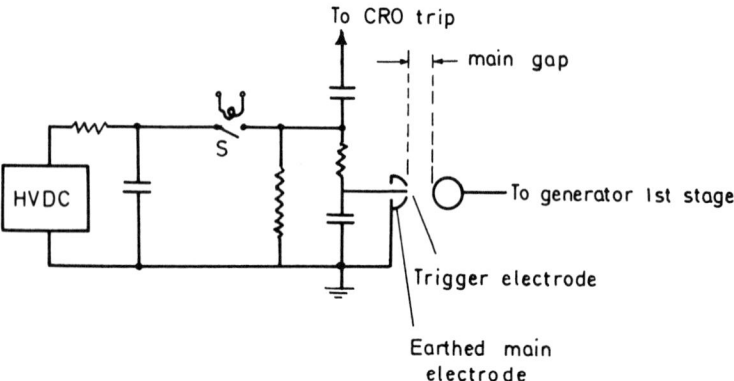

Fig. 16.13 Impulse generator tripping circuit using a Trigatron spark gap.

as required. A trigatron gap connected across the load could be similarly triggered at the appropriate subsequent moment to produce chopped impulse waves.

A modification of the trigatron gap was suggested in which the triggering spark is produced by breaking the current in a low-voltage inductive circuit at a pair of contacts embedded in the earthed main electrode. When the contacts are remotely separated, the energy stored in the circuit inductance is released in the spark, which initiates the breakdown of the main gap as mentioned before.

16.5 GENERATION OF SWITCHING IMPULSES

The switching waveforms experienced in power systems are not unique. They have front rise times up to several milliseconds and considerably longer tails. They may contain oscillatory components with a frequency ranging from a few hertz to a few kilohertz (Chapter 14). Switching surges are accompanied by energies much larger than lightning impulses.

Several circuits have been proposed in the literature for generating switching impulses. They are grouped as follows:

1. Impulse generator circuits are modified to give longer-duration wave shapes by a proper choice of the front and tail resistors. In order to produce unidirectional damped oscillations, an inductance L is connected in series with the conventional impulse generator circuit (Fig. 16.14). These oscillations may have a

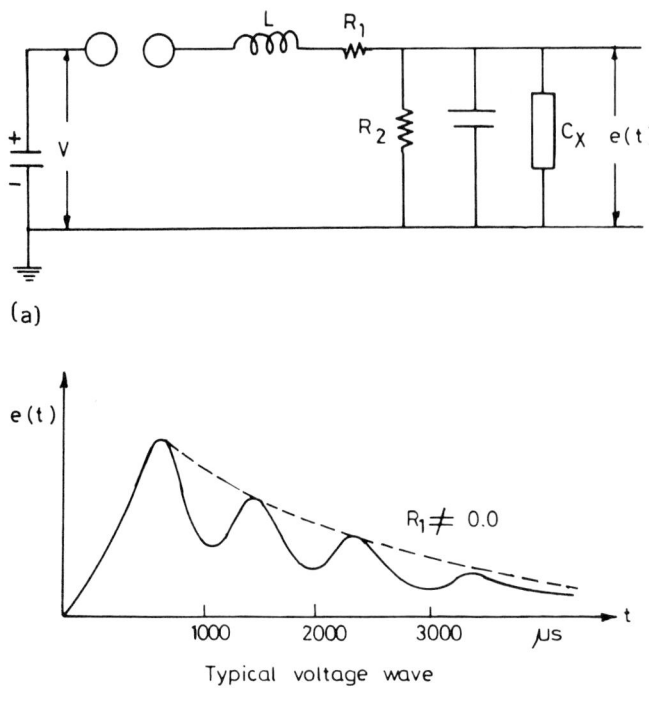

Fig. 16.14 Circuit for producing switching impulse voltage and its output waveform. C_x, test object.

frequency of 1 to 10 kHz, depending on the circuit parameters. A sphere gap may be included in parallel with the test object for producing chopped waves, the same as with lightning impulses. The main drawback of these circuits is that the efficiency of the generator gets reduced to 50% or even less.

2. A testing transformer is excited by a voltage impulse to give long-duration or oscillatory waves in its secondary. The capacitor C_1 of Figure 16.15 is charged to a moderate dc voltage (20 to 25 kV) and then discharged into the low-voltage winding of a power transformer. The high-voltage winding is connected in parallel to a load capacitance C_2 and the test object. Through autotransformer action, switching impulses of proper waveform can be generated across the test object. The disadvantages of this technique are associated with the considerable amount of high-frequency distortion in the output waveform. Furthermore, the

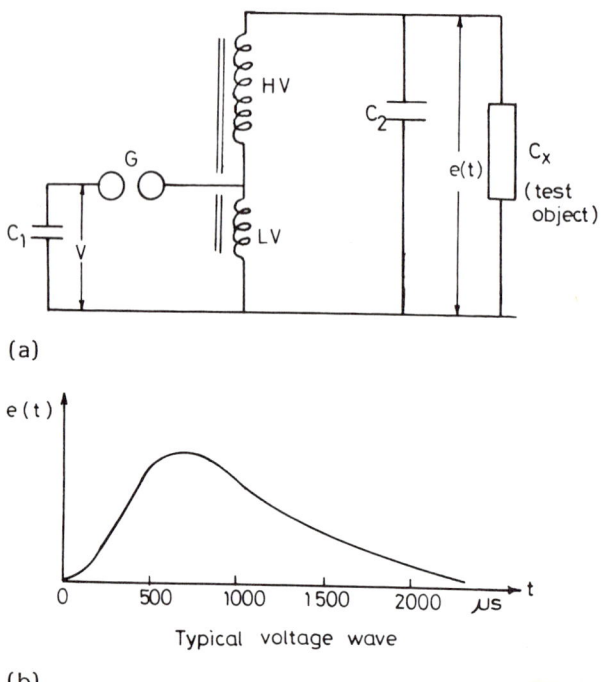

Fig. 16.15 Circuit for producing switching impulse voltage using a test transformer energized from a charged capacitor.

size of the capacitor C_1 for producing a reasonable output voltage may be large. Variations of the front of the generated impulse was achieved by adding inductances of different magnitudes in series with C_1. Such a method proved tedious and in most cases uneconomical.

A preferable method is to energize the transformer from an ac supply for less than half a cycle. The energization of the transformer is initiated via a thyristor or an ignitron to be triggered at the required phase angle. The transformer remains energized until the supply current passes through a zero value. Then the energy stored in the transformer is discharged through the test object. For generating a much higher output voltage, several transformer units were arranged in cascade (Anis et al., 1975). The advantages inherent in this method are mainly its small high-frequency distortion in the output voltage and the elimination of the energy storage capacitors C_1 (Fig. 16.15).

High-Voltage Generation

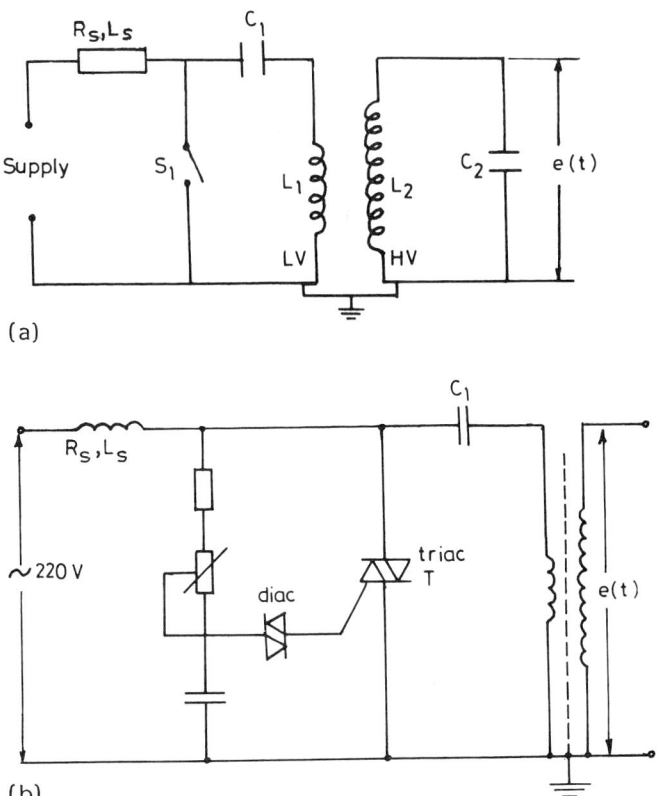

Fig. 16.16 (a) Tesla coil equivalent circuit (S_1, switch); (b) switch S_1 is replaced by a triac.

3. For producing damped oscillations, the source of high voltage is a Tesla coil, which consists of an air-cored transformer. Its LV side is connected to the dc or ac supply circuit through a capacitor C_1 and a series element (R_s, L_s) (Fig. 16.16a). C_2 is the equivalent capacitance of the high-voltage winding and the test object. On closing the switch S_1, an oscillatory current will flow in the LV circuit, inducing high oscillatory voltage in the HV circuit of the Tesla coil. The frequency of the induced oscillations can be made the same by tuning the two circuits (i.e., when $L_1 C_1 = L_2 C_2$). The generator above has been modified by Sandhaus (1976). The switch S_1 was replaced by a triac for better control of the firing instant (Fig. 16.16b).

16.6 GENERATION OF IMPULSE CURRENTS

Lightning strokes involve both high-voltage and high-current impulses on transmission systems. Therefore, generation of impulse currents of the order of several hundreds of kiloamperes finds applications in testing lightning arresters. They are also used in electric-arc and electric-plasma studies. The waveshapes in common use are the double exponential and the rectangular waves (Fig. 16.17).

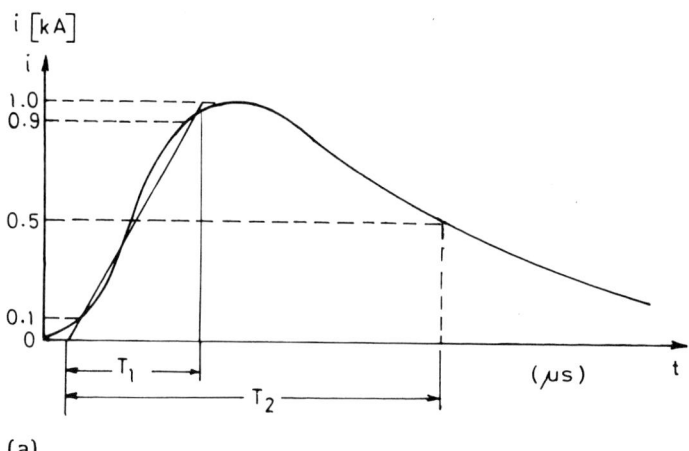

(a)

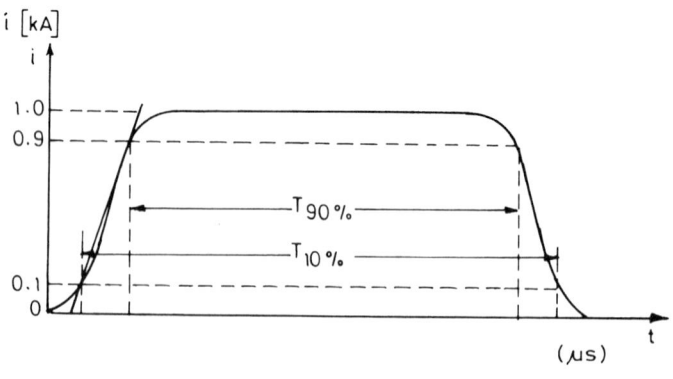

(b)

Fig. 16.17 Impulse current double-exponential (a) and rectangular (b) waveforms.

16.6.1 Double-Exponential Current Impulses

The impulse wave shape is defined according to the IEC recommendations. T_1 is the front duration and T_2 is the time to half peak. According to the IEC, standard waves are 4/10 and 8/20 µs (IEC, 1973).

To produce one of the standard current waveforms, a circuit such as the one shown in Figure 16.18 is normally used. The basic difference between this circuit and those generating impulse voltages is that here the capacitances are much larger and the resistors are much smaller in magnitude (Figs. 16.10 and 16.12).

If the capacitor C is charged to a voltage V and discharged when the gap G is triggered (Fig. 16.18), the current i can be shown to vary with time according to the relation

$$i(t) = \frac{V}{\omega L} \exp(-\gamma t) \sin \omega t \qquad (16.39)$$

where $\gamma = R/2L$ and $\omega = \sqrt{1/LC - R^2/4L^2}$. The equivalent resistance of the test object should be chosen $R < 2\sqrt{L/C}$ for underdamped oscillation in the current wave. The time taken for the current i to rise from zero to the first peak is

$$T_1 = \frac{1}{\omega} \sin^{-1} \frac{\omega}{\sqrt{LC}} \qquad (16.40)$$

If the test object is an ideal surge arrester, the following approximate expressions are used for determining the waveforms of the impulse current (Haefely, 1982).

$$\text{Peak current} = \frac{V - V_r}{\sqrt{L/C}} \qquad (16.41)$$

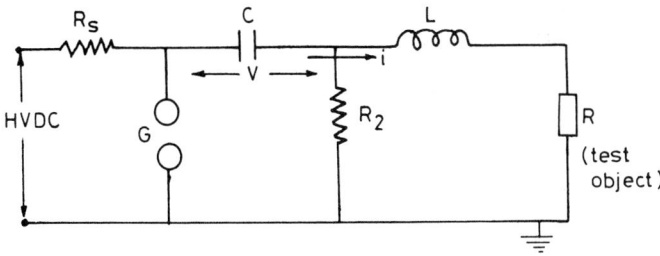

Fig. 16.18 Circuit for producing double-exponential current impulses.

where V_r is the residual voltage of the surge arrester.

Front time $T_1 \simeq 1.25 \sqrt{LC}$ (16.42)

Time to half-peak value $T_2 \simeq 2.5 \sqrt{LC}$ (16.43)

16.6.2 Rectangular Current Impulses

The rectangular impulse wave as defined by the IEC recommendations is shown in Figure 16.17. The duration $T_{10\%} < 1.5 T_{90\%}$ (IEC, 1970). The rectangular current impulses are generated by discharging an artificial transmission line with lumped L and C elements into the test object through a sphere gap. If the line is charged to a dc voltage V and discharged through the test object of resistance R, the current pulse is given by $I = V/(R + Z_0)$, where Z_0 is the surge impedance of the line. A pulse voltage of $RV/(R + Z_0)$ is developed across the test object. The duration $T_{90\%}$ of the current wave is estimated as

$$T_{90\%} \simeq 2 \frac{n-1}{n} \sqrt{L/C}$$ (16.44)

where n is the number of L-C stages (Modrusan, 1977).

REFERENCES

Anis, H., Giao Trinh, N., and Train, D. (1975). *IEEE Trans.*, Pas-94:187.
Auget, M. and Lanovici, M. (1982). *Haute tension*, Vol. XXII, Presses Polyechniques Romandes, Lausanne, Switzerland.
Dorf, R. C. (1986). *Modern Control Systems*, Addison-Wesley Publishing Company, Inc., Reading, Mass.
Felici, N. J. (1953). *Direct Current*, 1:122.
Haefely (1982). *Haefely Manual*, Publication BE-701, E. Haefely, Basel, Switzerland.
Hyltén-Cavallius, N. (1986). *High Voltage Laboratory Planning*, E. Haefely, Basel, Switzerland.
IEC (1970). *Nonlinear Resistor Type Arrester for AC Systems*, Publication 99-1, International Electrotechnical Commission, Geneva.
IEC (1973). *High Voltage Test Techniques. II — Test Procedures*, Publication 60-2, International Electrotechnical Commission, Geneva.

Kind, D. (1978). *An Introduction to High Voltage Experimental Techniques*, Friedr Vieweg & Sohn Verlagsgesellschaft mbH, West Germany.

Kuffel, E. and Zaengl, W. S. (1984). *High Voltage Engineering Fundamentals*, Pergamon Press Ltd., Oxford.

Modrusan, M. (1977). *Bulletin SEV/VSE*, 68:1304.

Naidu, M. S. and Kamaraju, V. (1982). *High Voltage Engineering*, Tata McGraw-Hall Publishing Co. Ltd., New Dehli.

Philp, S. F. (1977). *IEEE Trans.*, EI-12:130.

Sandhaus, S. (1976). *Nouveaux aspects de l'ergonomie, Revue mensuelle suisse d'odontostomatologie*, 86:1312.

17
High-Voltage Measurements

H. ANIS *Cairo University, Giza, Egypt*

17.1 INTRODUCTION

This chapter describes various methods for measuring high voltages and high transient currents. Also described are means of measuring electric fields normally associated with high voltages. Table 17.1 lists the most common methods of measuring high voltage, together with their modes of usage. Cathode-ray oscilloscopes are not included in the table, as their versatility with their probes and accessories make them suitable for almost all kinds of measurements.

17.2 SPHERE GAPS

When sphere gaps are used in high-voltage measurement, one of the spheres is earthed and the other is connected to high voltage. The sphere gap arrangement is shown in Figure 17.1. A horizontal sphere gap arrangement is sometimes preferred at lower measured voltages. The procedure is to establish for a particular test circuit a relation between the peak voltage, determined by sparkover between the spheres, and the reading of a voltmeter on the primary or input side of the high-voltage source. This relation should be within 3% (IEC, 1973).

Standard values of sphere diameter are 6.25, 12.5, 25, 50, 75, 100, 150, and 200 cm. Table 17.2 gives the clearance limits in sphere gap arrangements, stipulated by international standards (IEC, 1960).

The effect of humidity is to increase the breakdown voltage of sphere gaps by up to 3% (Kuffel and Zaengl, 1984). Temperature and pressure, however, have a significant influence on breakdown

Table 17.1 Methods of Measuring High Voltage

Method of measurement	DC Mean	DC Peak	AC rms	AC Peak	AC Waveform	Impulse Peak	Impulse Waveform
Sphere gaps		x		x		x	
Peak voltmeter				x			
Electrostatic voltmeter	x (rms)		x				
Voltage transformer			x	x	x		
Resistor in series with milliammeter	x		x				
Resistive divider	x	x	x	x	x	x	x
Capacitive divider			x	x	x	x	x

High-Voltage Measurements 451

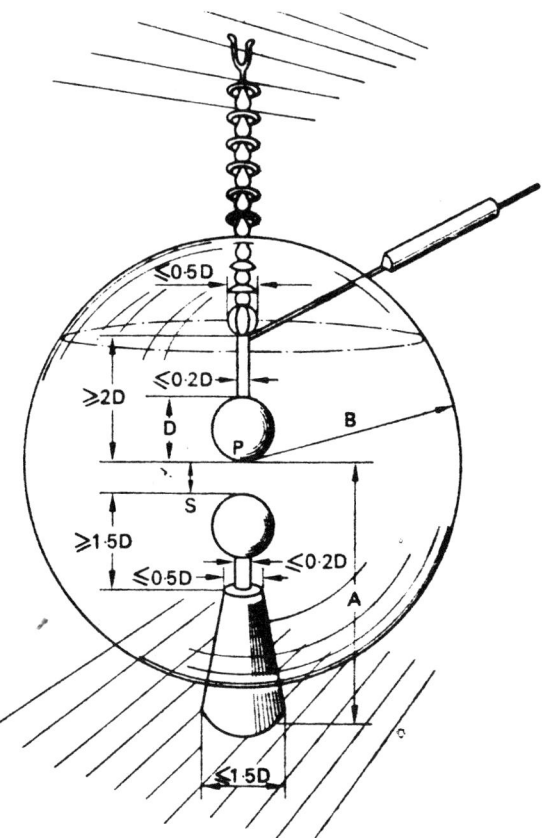

Fig. 17.1 Vertical sphere gap.

voltages. The breakdown voltage V_s is related to its value under normal atmospheric conditions by

$$V_s = kV_n \qquad (17.1)$$

where k is a factor related to the relative air density (RAD) δ described in Chapter 5. Table 17.3 gives the relation between the RAD and the correction factor k.

Under impulse voltages, the voltage at which there is a 50% breakdown probability is recognized as the breakdown level.

Table 17.2 Clearances Around the Sphere Gap (Fig. 17.1)

Sphere diameter, d (cm)	Minimum value of A	Maximum value of A	Minimum value of B
Up to 6.25	7D	9D	14S
10–15	6D	8D	12S
25	5D	7D	10S
50	4D	6D	8S
75	4D	6D	8S
100	3.5D	5D	7S
150	3D	4D	6S
200	3D	4D	6S

Table 17.3 Correction Factor for Atmospheric Pressure and Temperature

δ	0.70	0.75	0.80	0.85	0.90	0.95	1.0	1.05	1.10
k	0.72	0.76	0.81	0.86	0.90	0.95	1.0	1.05	1.09

Table 17.4 Peak Values of Breakdown Voltages[a] (kV) of Sphere Gaps With One Sphere Grounded at Atmospheric Reference Conditions[b]

Sphere gap (cm)	Sphere 6.25		12.5		25		50	
	a	b	a	b	a	b	a	b
0.5	16.9		16.5					
1	31.4		31.2					
1.5	44.7	45.1	44.7					
2	57.5	58	58.0					
2.5	68.5	70	71.5		71.5			
3	78.0	80.5	84.0		84.5			
3.5	(86.0)[c]	(90.0)	95.5	96.5				
4	(93.5)	(99)	106	108	110			
4.5	(99.0)	(106)	117	120				
5	(105)	(113)	127	132	135	136	136	
5.5	(110)	(120)	136	143				
6	(114)	(125)	144	152	158	160		
6.25	(115)	(125)	148	157				
7			(159)	(170)	181	184		
7.5							199	
8			(171)	(186)	203	207		
9			(182)	(200)	222	229		
10			(192)	(210)	240	250	259	
11			(200)	(225)	257	268		
12			(208)	(230)	271	286		
12.5			(210)	(235)	277	294	315	317
15					(309)	(331)	367	374
17.5					(336)	(362)	413	425
20					(362)	(389)	452	472
22.5					(379)	(409)		
25					(393)	(426)	520	545
30							(575)	(610)
35							(620)	(660)
40							(660)	(705)
45							(690)	(730)

diameter[b] (cm)							
75		100		150		200	
a	b	a	b	a	b	a	b
	136						
200	199						
260	261	262		262		262	
321	322						
380	381	383	384				
435	440						
483	497	500		500		500	
575	595	605	610				
655	685	700	715	730	735	735	740
725	755	785	800				
(785)	(820)	862	885	940	950	960	965
(835)	(875)	925	965				

Table 17.4 (Continued)

Sphere gap (cm)	6.25		12.5		25		Sphere 50	
	a	b	a	b	a	b	a	b
50							(720)	(760)
60								
70								
75								
80								
90								
100								
110								
120								
130								
140								
150								
160								
170								
180								
190								
200								

[a] (a) Values for ac, dc, and negative impulses. (b) Values for positive lightning and switching impulses [if different from (a)].

[b] 25°C and 101.3 kPa.

[c] Values in parentheses are accurate only ±5% as their corresponding gaps exceed 1/2 sphere diameter.

diameter[b] (cm)							
75		100		150		200	
a	b	a	b	a	b	a	b
(880)	(925)	1000	1020	1110	1130	1160	1170
(955)	(1005)	(1090)	(1130)	1260	1290	1320	1360
(1010)	(1050)	(1180)	(1220)	1370	1410	1460	1520
(1025)	(1070)	(1210)	(1260)	1420	1460	1510	1590
		(1240)	(1290)	1460	1500	1570	1670
		(1300)	(1350)	(1550)	(1600)	1690	1790
		(1340)	(1390)	(1630)	(1690)	1810	1900
				(1700)	(1760)	(1910)	(2000)
				(1770)	(1830)	(1990)	(2030)
				(1840)	(1900)	(2070)	(2160)
				(1890)	(1950)	(2140)	(2240)
				(1930)	(1990)	(2210)	(2310)
						(2280)	(2370)
						(2330)	(2430)
						(2370)	(2470)
						(2420)	(2510)
						(2450)	(2540)

Table 17.4 lists the standardized disruptive discharge voltages from the results of a large number of international experiments (IEC, 1960). The accuracy of the values of Table 17.4 are generally within 3%, except for the bracketed figures, where the accuracy is about ±5%.

17.3 PEAK VOLTMETERS

The measuring circuit of a peak voltmeter is shown in Figure 17.2 and is made up of two diodes, a properly chosen capacitor, and a milliammeter that can be recalibrated for peak voltage. The arithmetic mean value of the rectified current as indicated by the instrument is given by

$$I = \frac{1}{T} \int_0^{T/2} i_1(t) \, dt = \frac{C}{T} \int_{-v}^{+v} dv$$

$$= \frac{C}{T} \left[v\left(\frac{T}{2}\right) - v(0) \right] \tag{17.2}$$

If the voltage is symmetric about zero and its peak value = V, then

$$\bar{I} = 2fCV \tag{17.3}$$

Therefore,

$$V = \frac{\bar{I}}{2fC} \tag{17.4}$$

The current waveform should ideally be monitored on an oscilloscope to determine if there will be more than one zero crossing in a half-cycle. A peak voltmeter may be of either the analog or the digital type (Malewski and Dechamplain, 1980).

Referring to equation (17.4), it appears that in addition to the instrument error, the error in the measured voltage dV is a function of errors in the current di, frequency df, and capacitance dC. It can be proven that

$$dV = di + df + dC \tag{17.5}$$

Impulse peak voltmeters are different in that they contain active

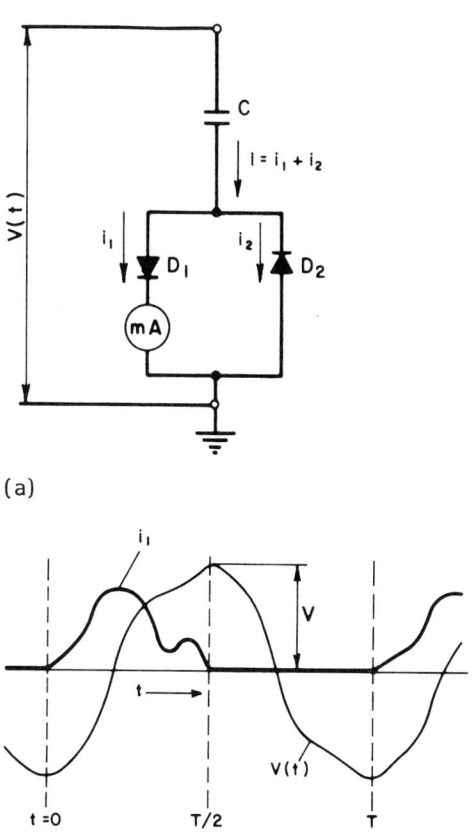

Fig. 17.2 Peak voltmeter circuit.

circuits and a memory device and possibly a storage oscilloscope (Hyltén-Cavallius, 1988).

17.4 ELECTROSTATIC VOLTMETERS

An electrostatic voltmeter utilizes the force existing between two opposite plates. The force is created by the process in which a

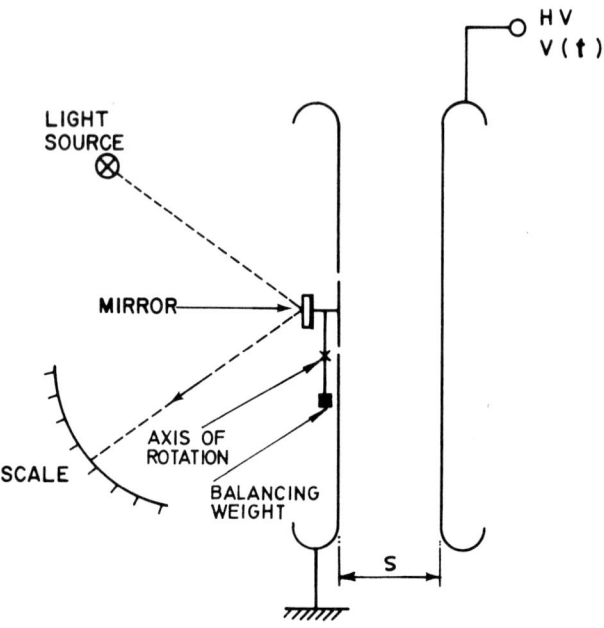

Fig. 17.3 Electrostatic voltmeter.

change in stored electrostatic energy is converted into mechanical work. Referring to Figure 17.3, the electrostatic voltmeter is seen to be made up of two parallel plates. One is fixed and the other has a very small movable part that is restrained by a spring. The force of attraction F(t) created by the applied voltage causes the movable part — to which a mirror is attached — to assume a position at which a balance of forces takes place. An incident light beam will therefore be reflected toward a scale calibrated to read the applied voltage magnitude. Assuming the capacitance between plates to be C, the stored electrostatic energy W in the system will be $W(t) = (1/2)\, CV^2(t)$. A change $dW(t)$ in stored energy will be faithfully converted into mechanical work. Therefore.

$$dW(t) = -F(t)\, dS \tag{17.6}$$

where dS is the change in S, the separation between plates (Fig. 17.3). The attraction force as a function of time is thus

$$|F(t)| = \frac{dW(t)}{dS} = \frac{1}{2} V^2(t) \frac{dC}{dS} \qquad (17.7)$$

The actual force is the arithmetic mean of expression (17.7), given by

$$\bar{F} = \frac{1}{2}\left(\frac{dC}{dS}\right)\left[\frac{1}{T}\int_0^T V^2(t)\,dt\right] \qquad (17.8)$$

where T is the period of variation, if any. Equation (17.8) relates the force to the rms value of the applied voltage V:

$$\bar{F} = \frac{1}{2}\left(\frac{dC}{dS}\right)V^2 \qquad (17.9)$$

In the most common attracted disk type of electrostatic voltmeter, the factor dC/dS can be simply evaluated by recalling that

$$C = \frac{A\varepsilon_0}{S} \qquad (17.10)$$

where A is the area of the plate and ε_0 is the permittivity of free space. Therefore, the force will be given by

$$|\bar{F}(t)| = \frac{A\varepsilon_0}{2S^2} V^2 \qquad (17.11)$$

If V is in volts and A and S^2 are in m^2 the force is in newtons. According to equation (17.11), the factor $A\varepsilon_0/S^2$ is used to control the range of measurement.

17.5 VOLTAGE TRANSFORMERS

Voltage transformers are used to measure high ac voltage accurately. The voltage on the secondary side is closely proportional in amplitude to and almost in phase with the voltage on the primary side. Two alternatives are, however, preferred at very high voltages, namely, capacitive and cascaded voltage transformers (IEEE Committee, 1981).

17.5.1 Capacitive Voltage Transformers

A capacitive voltage transformer consists of a capacitive divider used in conjunction with a conventional auxiliary transformer which steps further down the divider output voltage, typically, about 10 kV, to the desired secondary value (Fig. 17.4). By adjustment of the inductance L, which may consist wholly or partly of leakage inductance of the auxiliary transformer, to equal $1/(2\pi f)^2(C_1 + C_2)$, the voltage drop across C_2 due to the current drained from the divider is largely compensated so that the overall ratio is nearly independent of burden and is the product of the divider and transformer ratios.

17.5.2 Cascaded Voltage Transformers

In cascaded voltage transformers the primary winding is distributed over a series of coils wound on separate cores which are mounted in oil-filled porcelain containers, stacked in series. The secondary coil is wound on the lowest core. The power-frequency voltage is thus uniformly distributed over a number of coils, each of which need not be insulated from its core for more than 100 kV.

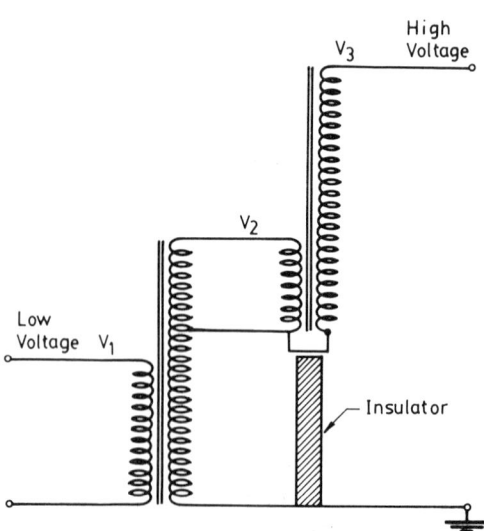

Fig. 17.4 Capacitor voltage transformer circuit.

High-Voltage Measurements 463

17.6 AMMETER IN SERIES WITH A HIGH IMPEDANCE

An impedance can be used in series with a microammeter or a milliammeter for the measurement of high voltages. Neglecting the impedance of the instrument, the current through the instrument will be proportional to the applied voltage. Evidently, if the impedance is a resistor, the current is in phase with and faithfully represents the voltage. The instrument may be replaced by an oscilloscope if the voltage wave form is to be recorded.

Wire-wound or thin-film resistors may be used. Wire-wound resistors are usually preferred for their superior stability in service, and their extremely small temperature coefficient, which reaches ±0.01%/K. Freedom from corona discharge and improved cooling may be achieved by immersing the resistor in insulating oil.

17.7 POTENTIAL DIVIDERS FOR AC AND DC

A potential divider consists of two impedances, Z_1 and Z_2, connected in series, to which the voltage to be measured is applied; the components that constitute the impedances are referred to as the high- and low-voltage arms of the divider. Connection between the low-voltage arm and the measuring instrument must be made through a shielded coaxial cable to avoid the adverse effects of stray capacitance between that connection and the high-voltage arm. High-voltage dividers generally consist of either resistors or capacitors, but sometimes a combination of resistors and capacitors, in either series or parallel, is used depending on the type of voltage to be measured.

17.7.1 Resistive Potential Dividers

A resistive potential divider is usually employed for the measurement of direct voltages. If, however, the ripples on the dc voltage are to be recorded as well, a resistive/capacitive potential divider will then be more suitable (Kind, 1978).

17.7.2 Resistive Dividers for AC Voltage Measurement

The divider resistive arms should—in case of alternating voltage—be looked upon as impedances Z_1 and Z_2, which generally possess resistive and reactive components. Therefore,

$$Z_1 = R_1 + jX_1$$

and

$$Z_2 = R_2 + jX_2$$

Under ac conditions the divider ratio in magnitude is

$$\left|\frac{Z_1 + Z_2}{Z_2}\right| = \left[\frac{(R_1 + R_2)^2 + (X_1 + X_2)^2}{R_2^2 + X_2^2}\right]^{1/2} \qquad (17.12)$$

The output voltage V_2 will be shifted in phase from the measured voltage V_1 by

$$\theta = \tan^{-1}\frac{X_1 + X_2}{R_1 + R_2} - \tan^{-1}\frac{X_2}{R_2} \qquad (17.13)$$

Therefore, for the output voltage V_2 to represent V_1 faithfully, this phase shift should vanish. This, according to (17.13), amounts to ensuring that

$$\frac{X_1}{R_1} = \frac{X_2}{R_2} \qquad (17.14)$$

The high-voltage arm will consist of many resistor elements stacked in series. As the length of the stack increases, so does its capacitance to surrounding conductors. The influence of earth capacitance C_e on the effective impedance can be derived by considering the general case of a stack earthed at one end and having inherent series resistance $Z = R_1$ and admittance to earth $Y = j\omega C_e$, both assumed uniformly distributed along its length (Fig. 17.5a). The currents i_0 and i_1 at the earthed and high-voltage ends of the stack, respectively, due to an applied voltage V are then

$$i_0 = \frac{V}{R_1}\frac{\xi}{\sinh \xi} \qquad (17.15)$$

$$i_1 = \frac{V}{R_1}\frac{\xi}{\tanh \xi} \qquad (17.16)$$

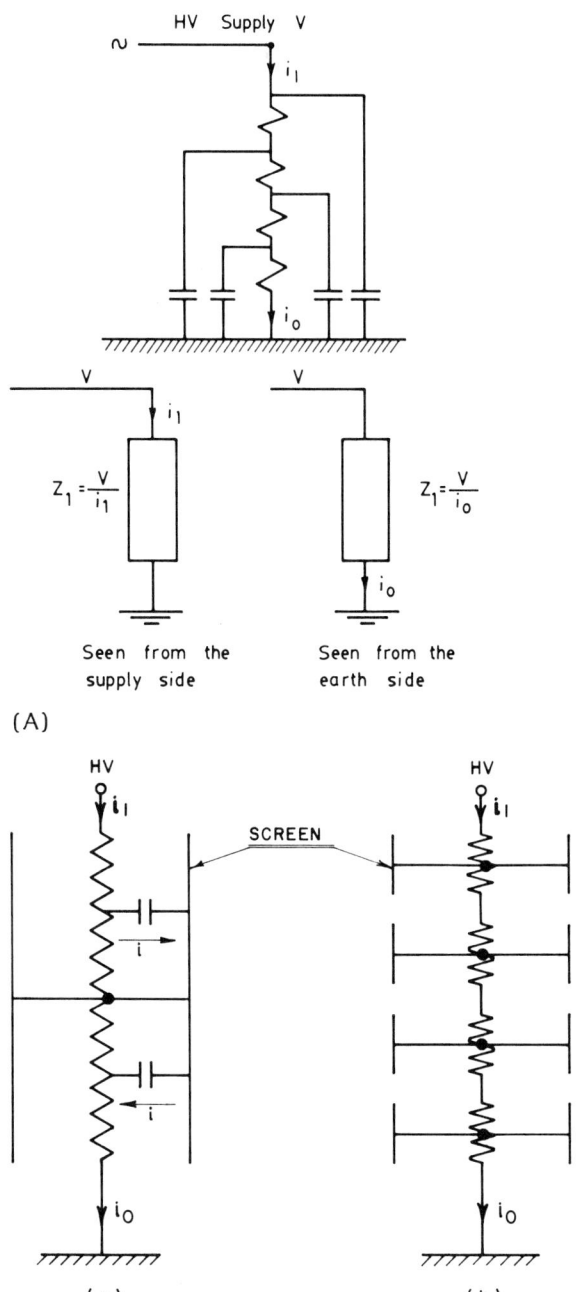

Fig. 17.5 (A) Unscreened high-voltage resistance, (B) screened high-voltage resistance, the screen being one piece (a) or subdivided (b).

where

$$\xi = \sqrt{ZY}$$

$$= \sqrt{R_1 C_e \omega}$$

The effective impeance Z_1 (V/i_0 or V/i_1) of the stack may be derived by expanding the hyperbolic functions in equations (17.15) and (17.16), giving

At the earthed end: $Z_1 \underset{\sim}{\sim} R_1 \left(1 + \dfrac{j\xi^2}{6}\right)$

(17.17)

At the high-voltage end: $Z_1 \underset{\sim}{\sim} R_1 \left(1 - \dfrac{j\xi^2}{3}\right)$

When this resistance is used as the high-voltage arm of a resistive divider whose low-voltage-arm resistance is R_2, the effective divider ratio is thus

$$\frac{i_0 R_2}{V} = \frac{R_2}{R_2 + R_1(1 + j\xi^2/6)} \quad (17.18)$$

It appears from equation (17.18) that the divider ratio is sensitive to frequency. Therefore, resistive dividers are not suitable for high-frequency alternating voltages.

Unscreened Resistors

Most high-voltage resistors — also, high-voltage arms of resistive dividers — may be considered as a vertical column of length L (m) and diameter 2r standing on a grounded plane (Fig. 17.5). For this configuration the equivalent stray capacitance to earth is given approximately by

$$C_e = \frac{111L}{2 \ln(L/r) - 1.1} \quad \text{pF} \quad (17.19)$$

When $\ln(L/r)$ is greater than about 3, the capacitance depends primarily on L and not on r; it is also not much affected by the shape and disposition of the earthed surface since if earth is looked upon as a coaxial cylinder of radius L(m), its capacitance is only slightly different, that is,

High-Voltage Measurements

$$C_e = \frac{111L}{2\ln(L/r)} \quad \text{pF} \qquad (17.20)$$

Screened Resistors

Screening is the process of canceling or neutralizing the effect of stray capacitances. It is basically accomplished by surrounding the high-voltage resistor by a conducting screen maintained at the mean potential of the resistor (Fig. 17.5B). Capacitive currents will flow between the screen and the resistor in one direction within the upper portion of the resistor and in the opposite direction within the lower portion. The two currents i_0 and i_1 viewed from either end of the resistor will see an effective impedance Z_1 of value

$$Z_1 \sim R_1 (1 - jwR_1 C_s/12) \qquad (17.21)$$

which is intermediate between the two values of Z_1 in the case of unscreened resistors given by equation (17.17). It is preferred, however, to divide the resistor into identical smaller units, each contained in a screen that is maintained at the mean potential of the unit (Fig. 17.5B). The action poses less risk of flashover between the resistor and the screen extremities. At 50 or 60 Hz the practicable voltage limit for a high-precision screened resistor is about 100 kV.

17.7.3 Capacitive Potential Dividers

Resistive potential dividers suffer from two main drawbacks, power losses and stray capacitance to earth. These factors limit their use to voltages below 100 kV at 50 Hz and even lower voltages at higher frequencies. Capacitive potential dividers are therefore more suitable to use with ac voltage, particularly at high voltages and high frequencies. They are limited only by their internal inductances or the dielectric losses of their components (Fig. 17.6a).

The low-voltage capacitor C_2 will normally consist of a fixed unit of low-loss angle (air, mica, or polystyrene dielectric). It will, in general, be shunted by a high resistance R_2, which may either be introduced deliberately to avoid the accumulation of a random charge on C_2 or be inherent in the measuring instrument. High-voltage capacitors C_1 can be either screened or unscreened.

Unscreened Capacitors

Unscreened capacitors which may be used as the high-voltage component of a capacitive divider normally take the form of a stack

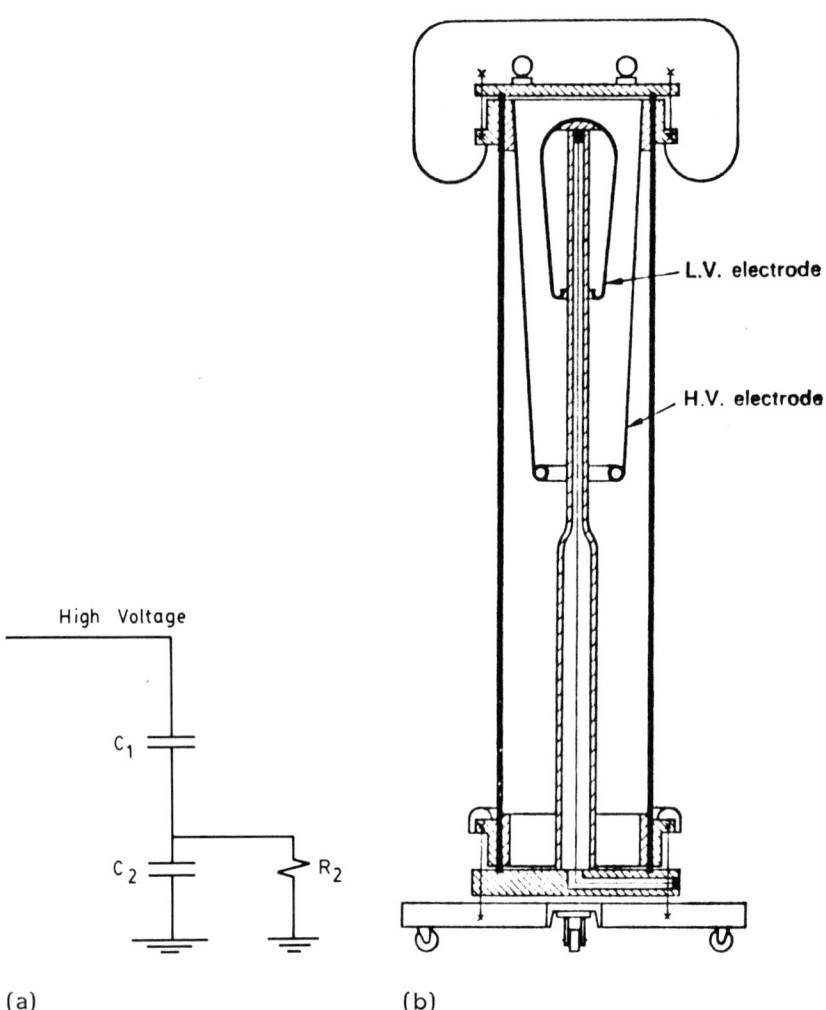

Fig. 17.6 (a) Capacitive voltage divider, (b) typical screened capacitor.

of identical cylindrical units. In this case, if the series capacitance of the stack is C_1 and the stray capacitance to earth is C_e, then, substituting C_e/C_1 for ξ^2 in equations similar to (17.17) and (17.18), with C_e not greater than C_1, the approximate effect of C_e is to decrease the effective capacitance at the earthed end by $C_e/6$ and to increase it at the other end by $C_e/3$. According to equation (17.19), which also applies here, $C_e \sim 20$ pF/m in practice, so that the correction due to it in any particular case can be estimated.

Screened Capacitors

For voltages up to about 30 kV (peak), capacitors with parallel-plate electrodes can be used and screened by being enclosed in a metal case. When filled with carefully dried gas they have almost zero power factor and a highly stable value of capacitance. At higher voltages a capacitor with coaxial cylindrical electrodes as that shown in Figure 17.6b may be used as the high-voltage component. The outer cylinder is flared at the ends to avoid discharges. The capacitance C_1 is calculable from the electrode dimensions:

$$C_1 = \frac{111L}{2 \ln(r_1/r_2)} \quad \text{pF} \qquad (17.22)$$

where L is the effective length of the LV electrode (m), r_1 the radius of outer cylinder (m), and r_2 the radius of inner cylinder (m).

The maximum stress on the dielectric occurs at the surface of the inner electrode and for a given voltage and size of outer cylinder, is minimal when the ratio r_1/r_2 is equal to e (= 2.718). At atmospheric pressure with smooth clean electrode surfaces, the capacitor can function with voltage gradients at the inner electrode surface of about 14 kV/cm peak without partial discharges. Higher voltage gradients can be attained by increasing the gas pressure (compressed-gas capacitors).

17.8 DIVIDERS FOR IMPULSE VOLTAGES

With impulse voltages, the complete waveform is to be recorded. Therefore, a potential divider for impulse voltages should have a good impulse fidelity. Both the divider and the connection leads must be considered when this impulse fidelity is to be assessed.

Instead of exploring the frequency characteristics of a divider, it is more common to examine its response to a step voltage applied at its high-voltage terminals. If a unit step voltage that is defined by

$$v_{-1}(t) = \begin{cases} 0 & t < 0 \\ 1 & t > 0 \end{cases} \quad (17.23)$$

is applied to a divider, its response (output) is u(t). The output with any other input voltage $v_1(t)$ can be found accordingly. If the Laplace transform of the input voltage $v_1(t)$ and of the step response u(t) are, respectively, $V_1(s)$ and $U(s)$, the Laplace transform of the output voltage may be given by

$$V_2(s) = V_1(s)\, U(s)\, s \quad (17.24)$$

In the time domain the output voltage would be

$$v_2(t) = L^{-1}[V_2(s)] \quad (17.25)$$

Equation (17.25) can be solved analytically whenever u(t) takes an analytical form. Otherwise, numerical methods can be invoked to evaluate $v_2(t)$ based on the superposition theorem:

$$v_2(t) = u(0)v_1(t) + \int_0^t v_1(\lambda) u'(t-\lambda)\, d\lambda \quad (17.26)$$

For faithful reproduction of $v_1(t)$, the wave shape of the response u(t) should be as close as possible to the applied step voltage $v_{-1}(t)$. A reasonable measure of the faithfulness of the response is the total area enclosed between the real and the ideal responses (Fig. 17.7). This area will have the dimensions of time and is thus referred to as the response time of the divider. On the basis of this representation, two criteria are used simultaneously to evaluate the response faithfulness.

1. The total response time (net enclosed area) must be as small as possible.
2. The response should settle down to the correct value in a time much shorter than the rise time of the voltage wave to be measured.

17.8.1 Resistive Dividers

With unscreened resistive dividers the considerations in Section 17.7.2 relating to the effect of stray capacitance to earth (C_e) of the high-voltage arm of a resistor divider apply where impulses are concerned. With an applied voltage step of amplitude V,

High-Voltage Measurements 471

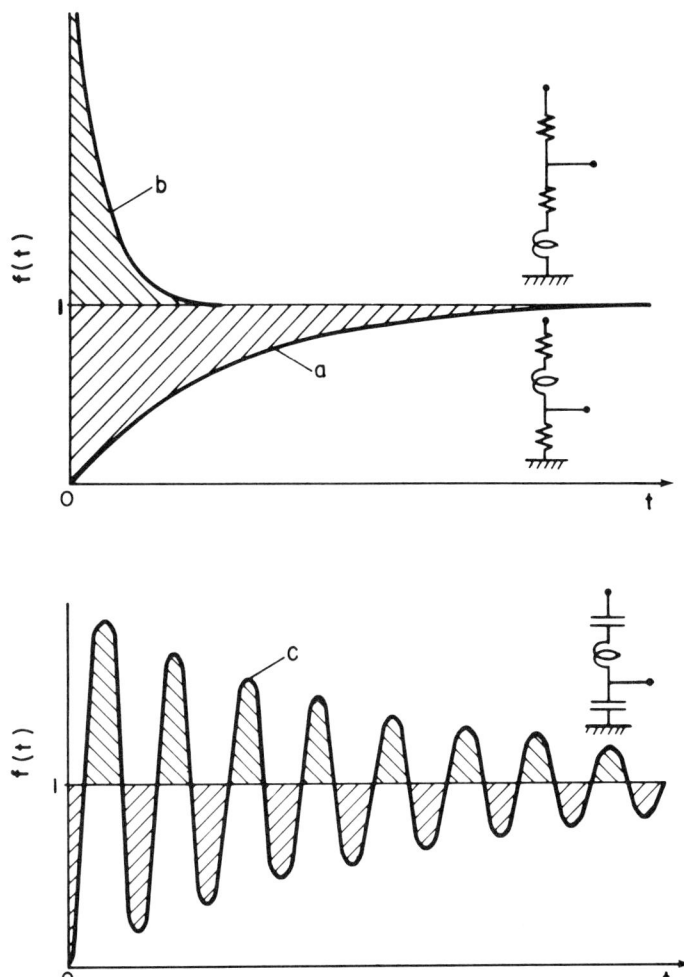

Fig. 17.7 Possible responses of impulse dividers: (a) for resistive divider with inductive high-voltage arm, (b) for resistive divider with inductive low-voltage arm, (c) for capacitive divider with inductive high-voltage arm.

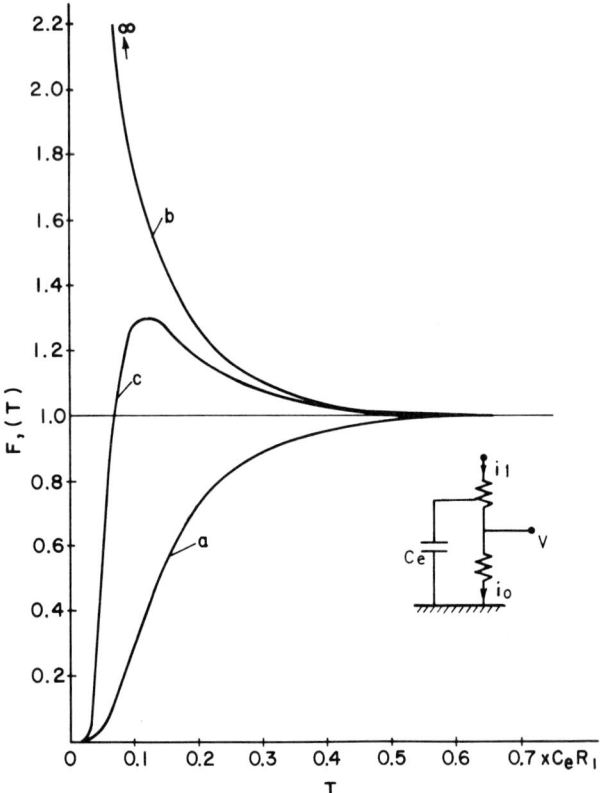

Fig. 17.8 Response of resistive divider to a unit step voltage: (a) current at earthed end i_0, (b) current at high voltage end i_1, and (c) divider output v when $L_2/R_2 = R_1 C_e/6$.

equations (17.15) and (17.16) still hold where $Z = R_1$ and $Y = sC_e$ (s being the operator d/dt). The calculated values of the response times of the resistor are $(C_e R_1/6)$ and $(-C_e R_1/3)$ at the earthed and high-voltage ends, respectively. These are identical with the time constants derived from equation (17.17). A typical divider response is shown in Figure 17.8 together with its voltage constituents. An upper limit on the value of R_1 should be imposed such that the response time of the high-voltage arm is made much shorter than the rise time of the measured impulse voltage. When a much shorter divider response is desired, the response time of the low-voltage arm should be made equal to that of the high-voltage arm. Since the low-voltage arm is mainly inductive, its time constant is L_2/R_2. Curve c of Figure 17.8 shows the case where $L_2/R_2 = C_e R_1/6$, thus reducing the total response time to zero. Only a small reduction in the final settlement time, however, results.

17.8.2 Capacitive Dividers

In a capacitive divider the high- and low-voltage arms can be assumed to have capacitances C_1 and C_2 in series with lead resistances r_1 and r_2 inductances and L_1 and L_2 in series, and also leakage resistances R_1 and R_2 across C_1 and C_2. The response of the divider to a step voltage will be an oscillation, about unity, of frequency

$$f = \frac{1}{2\pi} \sqrt{\frac{C_1 + C_2}{(L_1 + L_2)\, C_1 C_2}} \qquad (17.27)$$

which is damped out at a time constant of $(L_1 + L_2)/(r_1 + r_2)$. Unless the leakage resistances are ensured to be inversely proportional to the capacitances (i.e., $R_1 C_1 = R_2 C_2$), the eventual settlement output would drop, after some time, to a value of $R_2 C_2 / R_1 C_1$.

Additional damping resistances are usually connected in the lead on the high-voltage side of a capacitive potential divider. A total series resistance value is aimed at which critically damps out the oscillation; it should be nearly $2\sqrt{(L_1 + L_2)(C_1 + C_2)/C_1 C_2}$. The resistances should be well distributed along the divider. Such mixed capacitive/resistive divider acts under high-frequency voltages as a resistive divider, and under low-frequency voltages as a capacitive divider.

Mixed capacitive/resistive potential dividers are also used where the introduced resistors, R_1 and R_2, are connected across the capacitors C_1 and C_2, respectively. Their response is determined initially by the capacitances and ultimately by the resisors. If $R_1 C_1$ is made equal to $R_2 C_2$, the overall response will be ideal except for an initial effect of the inherent inductance.

17.8.3 Response Time of the Measuring System

The response time may be defined in view of linearly rising voltages (ramps) as that time beyond which the difference in amplitude between the input and output voltages remains constant. If a linearly rising voltage is applied to the measuring system, equation (17.26) can be used to predict the output. Let the input voltage be

$$v_1(t) = kt$$

where k is a constant slope. The output voltage according to equation (17.26), is

$$v_2(t) = k \int_0^t u(\lambda) \, d\lambda \qquad (17.28)$$

which may be expanded to

$$v_2(t) = k \left[t - \int_0^t [1 - u(\lambda)] \, d\lambda \right]$$

The difference between the ideal and actual normalized responses v_1 and v_2, respectively, is then

$$v_1(t) - v_2(t) = k \int_0^t [1 - u(\lambda)] \, d\lambda$$

For all t above a certain value, the difference above becomes a constant proportional to k. Therefore,

$$k \int_0^\infty [1 - u(\lambda)] \, d\lambda = kT$$

according to which the response time T is found to be (Kuffel and Zaengl, 1984)

$$T = \int_0^\infty [1 - u(\lambda)] \, d\lambda \qquad (17.29)$$

as defined earlier.

17.9 THE MEASURING CABLE

The output of a potential divider is connected to the oscilloscope via a coaxial cable. For a loss-free cable the velocity v of propagation of the electromagnetic wave is equal to $\sqrt{1/LC}$, where L and C are the inductance and capacitance per unit length of the cable.

High-Voltage Measurements

When the electromagnetic waves are required to travel along a cable, this expression gets simplified to the form

$$v \simeq \frac{3 \times 10^8}{\sqrt{\varepsilon_r \mu_r}} \quad \text{m/s}$$

where ε_r is the relative permittivity of the insulation surrounding the conductor and μ_r is the relative permeability of the conductor material. Generally, $\mu_r = 1.0$ for most of the conductors used in cables (IEC, 1962).

17.9.1 Cable Termination

To avoid errors due to reflections at its ends, the cable should be terminated at one end, or preferably both ends, by a resistance equal in value to its characteristic impedance Z_0, which for low-loss cables and very high frequencies is a pure resistance. The cable is thus essentially compatible with a resistor divider, to which it may be connected as shown in Figure 17.9. Matching is achieved by making

$$Z_0 = R_3 + \frac{R_1 R_2}{R_1 + R_2}$$

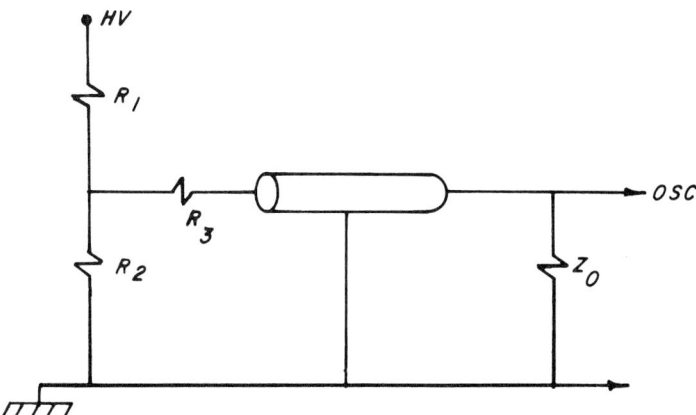

Fig. 17.9 Measuring cable connection to a resistive divider.

Normally, R_2 is much smaller than R_1. Therefore,

$$Z_0 \sim R_3 + R_2 \tag{17.30}$$

The low-voltage-arm resistance R_2 is now shunted by a resistance R_3 in series with the input impedance to the cable Z_0. Therefore, the divider ratio becomes

$$\frac{V_2}{V_1} = \frac{Z}{Z + R_1} \tag{17.31}$$

where

$$Z = \frac{R_2(Z_0 + R_3)}{Z_0 + R_2 + R_3}$$

which according to (17.30) becomes

$$Z = \frac{R_2(2Z_0 - R_2)}{2Z_0} \tag{17.32}$$

With a capacitive divider, the cable input resistance R_3 is chosen with a value $= Z_0$; its far end is not shunted. Thus with a unit function voltage applied to the divider, a voltage $C_1/2(C_1 + C_2)$ is initially injected into the cable, is doubled by reflection at the open end, and is absorbed without significant reflection when it returns to the input end. Thus the voltage at the open end jumps to $C_1/(C_1 + C_2)$.

In cases where the divider capacitance is intentionally made small, the cable capacitance C_c is considered and its far end is loaded by Z_0 and C_3 in series. When $C_1 + C_2 = C_3 + C_c$, the initial and final values of the response at the oscilloscope end are equal and a flat-topped overshoot occurs.

With a divider consisting of resistance and capacitance in parallel, the same cable connection is used. This is justified by the fact that the initial portion of the divider's response is controlled by the capacitive components. If the high- and low-voltage capacitances C_1 and C_2 are shunted by resistances R_1 and R_2 and the time constants R_1C_1 and $R_2(C_2 + C_3 + C_c)$ are adjusted to be equal and also large compared with the cable delay time, the response of the divider will be practically the same as if R_1 and R_2 were absent.

High-Voltage Measurements

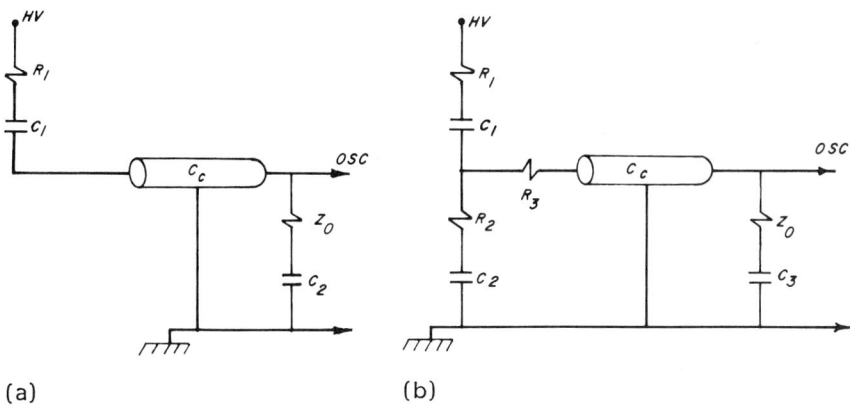

Fig. 17.10 Measuring cable connection to a capacitive/resistive divider.

If the mixed divider has its resistances and capacitances in series, the cable arrangement can be such that the low-voltage arm is essentially placed at the output of the delay cable (Fig. 17.10a). The time constants $R_1 C_1$ and $Z_0 (C_2 + C_c)$ should be made equal to ensure equal initial and final response values. The arrangement is generally suitable whenever the cable capacitance is much smaller than C_2. When this condition is not valid, the circuit of Figure 17.10b could be used in which matching is ensured at the cable's sending end by making

$$Z_0 = R_3 + \frac{R_1 R_2}{R_1 + R_2}$$

that is,

$$Z_0 \approx R_3 + R_2 \qquad (17.33)$$

17.10 MEASUREMENT OF TRANSIENT CURRENTS

Two methods are generally used to measure transient currents: the use of resistive, low-ohmic shunts and the measurement of currents by their inductive effects.

17.10.1 Resistive Shunts

A resistive, low-ohmic shunt is designed as a four-terminal device (Fig. 17.11) into which current I is fed via current terminals x and x' and from which a voltage V is tapped via potential terminals Y and Y'. The impedance V/I under ac conditions can be expressed as $R + j\omega L$, where R and L are the effective values of its resistive and inductive components. The shunt should therefore be designed so that over the frequency band concerned in any measurements in which it is involved, R is substantially equal to the dc resistance and $\omega L/R \ll 1$. Both these conditions require the diameter or thickness of the resistive element, depending on whether it is a wire or a ribbon, to be less than the nominal depth to which alternating current at the upper limit of the frequency band will penetrate a slab of the same material. The skin depth d_s is calculable from the frequency f, resistivity ρ, and relative permeability μ_r of the material by the equation $d_s = \sqrt{\rho/\pi\mu_r\mu_0 f}$. It amounts at 1 MHz to 0.35 and 0.33 mm for constantan and manganin, respectively, which are available in wire or sheet form and are nonmagnetic, for which it can be assumed that $\mu_r = 1$.

Two forms of shunt are in common use: the coiled buffer and the coaxial tube shunts. The former is used in the secondaries of current transformers and other applications where the transient current is relatively limited in magnitude. Coiled bifilar shunts normally have resistances in the order of a few ohms. Coaxial tube shunts may have resistances as small as microohms and are used mainly for measuring large currents (Hebner et al., 1977).

The use of resistive shunts to record transient currents can be represented by the equivalent circuit shown in Figure 17.12a. The impedance Z is the input impedance of the coaxial cable and its matching resistances. To evaluate the adverse effect of the inductive component of the shunt, an incident step current I is assumed. The voltage appearing across the impedance Z is what a

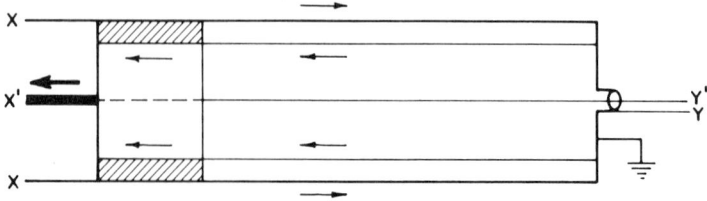

Fig. 17.11 Resistive shunt with the shape of a coaxial tube.

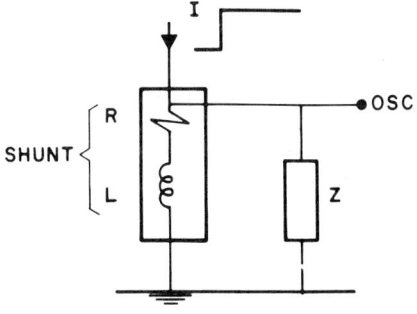

(a)

(b)

Fig. 17.12 (a) Equivalent circuit of a resistive shunt and (b) typical response.

recording oscilloscope will show. The time function of the voltage can be shown to take the form

$$v(t) = IZ\left(\frac{R}{R + Z} + \frac{Z}{R + Z} e^{-t/T}\right) \quad (17.34)$$

where $T = L/(R + Z)$ is the time constant. The response of the shunt is sketched in Figure 17.12b.

17.10.2 Current Measurement by Magnetic Coupling

In this method a toroidal coil known as the Rogowski coil surrounds the path of the current to be measured. The time-varying magnetic field induces a voltage at the terminals of the coil which is proportional to the rate of change of current with time:

$$v = M \frac{di}{dt} \tag{17.35}$$

where M is the mutual inductance between the sensing coil and the measured current path. These current sensors have recently been superseded by Hall-effect transducers (Berkebile et al., 1981; Tokoro et al., 1982).

17.10.3 Hall-Effect Current Transducers

If an electric current flows from left to right along a thin strip of metal or semiconductor and if a magnetic field acts vertically upward, the electrons will be deflected toward its back edge. A potential difference is produced between opposite points on the edges, which is known as the Hall voltage. For a constant magnetic field, the Hall voltage is proportional to the current. This property can be used to measure high dc and ac and even transient currents.

In the Hall-effect current transducer the current-carrying conductor is surrounded by an iron-cored magnetic circuit that produces a magnetic field into the air gap. The Hall voltage is amplified and measured or recorded on an oscilloscope. Current probes based on the Hall effect are manufactured to have a bandwidth extending from dc to 50 MHz and more (Holt, 1988).

17.11 MEASUREMENT OF ELECTRIC FIELD

Field measurements play an important role in the construction and problem solving of equipment using high voltages or subjected to high electric fields (IEEE Working Group, 1983). Most electric field measuring devices are based on the principle that the electric field is proportional to the electric flux density D. When this changes with time due to natural phenomena or by mechanical movement, a displacement current results. The displacement current density is given by

$$\overline{J}_d = \frac{\partial \overline{D}}{\partial t} \tag{17.36}$$

17.11.1 Passive Capacitance Probe

The capacitance probe is essentially a small disk mounted flush with the surface of an electrode or a screen. In either case the device measures the displacement charge induced on the probe surface due to the electric flux density. The electrode version shown

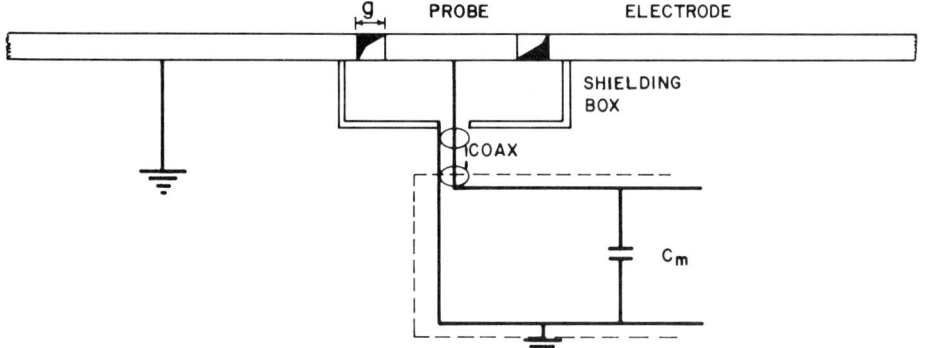

Fig. 17.13 Capacitance probe.

in Figure 17.13 can be used to determine the field at the ground plane of a rod-plane gap. For a probe with an effective head area of A and with an electric flux of Ψ coulombs normal to the probe surface, the electric field is given by

$$E = \frac{D}{\varepsilon_0} = \frac{\Psi}{\varepsilon_0 A}$$

$$= \frac{1}{\varepsilon_0 A} \int i_d \, dt \qquad (17.37)$$

where ε_0 is the permittivity of free space. By measuring the probe current i_d, the field can be obtained if the measured current is integrated with respect to time. The integration process can be performed automatically by inserting a capacitor C_m between the probe and ground as seen in Figure 17.13. The voltage V across the measuring capacitor is proportional to the electric field, which can be calculated from

$$E = \frac{C_m V}{\varepsilon_0 A} \qquad (17.38)$$

The effective area A is a function of the width of the annular ring (g), where the effective radius can be approximated by the probe radius and one-half the ring's width.

17.11.2 Electric Field Mill

An electric field mill consists of a grounded multivane rotor turning in front of a multisegment sensor plate. The grounded rotor alternately exposes the sensor plates to the ambient electrostatic field and shields them (Fig. 17.14). When each sensor plate is shielded, its charge is expelled (Stark, 1979). The induced charge from the displacement current is given by

$$q = \int \overline{D} \, d\overline{A}$$
$$= \varepsilon_0 EA \tag{17.39}$$

assuming a static uniform field. The arithmetic mean value of the current for the sensor plates to be exposed and shielded once is then

$$\frac{1}{T/2} \int_0^{T/2} \frac{dq}{dt} \, dt = f(q_{max} - q_{min})$$

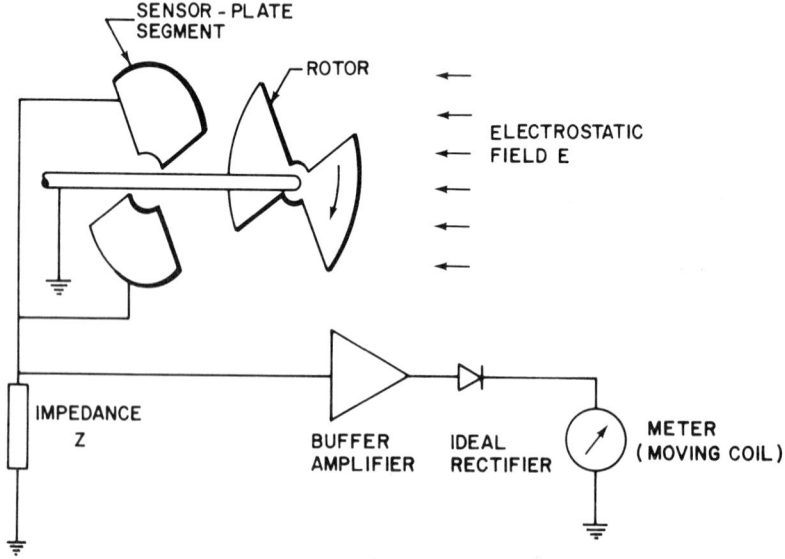

Fig. 17.14 Schematic diagram of a single field mill.

High-Voltage Measurements

where f is the frequency of exposure and shielding. This current is the maximum current and can be expressed in terms of capacitance, where

$$i_{max} = fV(C_{max} - C_{min}) \tag{17.40}$$

The function describing the area exposed is between a triangular and sinusoidal waveform and can be approximated by

$$a = \frac{A(1 + \sin \omega t)}{2} \tag{17.41}$$

ω being the angular velocity. The charge on the sensor is then

$$q = \frac{\varepsilon_0 EA(1 + \sin \omega t)}{2} \tag{17.42}$$

The instantaneous voltage across the impedance Z (seen in Fig. 17.14) is

$$v(t) = Z \frac{dq}{dt}$$

$$= \frac{\varepsilon_0 EA \omega}{2} Z \cos \omega t \tag{17.43}$$

The impedance Z consists normally of a resistance R and a capacitance C in parallel. If $(\omega RC)^2 \ll 1$, the peak voltage is approximated as

$$\hat{V} = \frac{\varepsilon_0 EA \omega R}{2} \tag{17.44}$$

where the field is a function of frequency. If $(\omega RC)^2 \gg 1$, the peak voltage is approximated as

$$\hat{V} = \frac{\varepsilon_0 EA}{2C} \tag{17.45}$$

where the field is not a function of frequency. If the sinusoidal assumption is relaxed and a pure triangular waveform is used, the results obtained are essentially identical to those of equations (17.44) and (17.45). Therefore, if the impedance Z is a capacitance (e.g., 100 nF) and if the output voltage V is fed into a high-impedance amplifier (e.g., $R = 10\ M\Omega$), the restriction for equation (17.45) is satisfied at power frequency.

17.11.3 Free Body Probes

Free body probes, sometimes called electric dipole sensors, are self-contained and normally have no electrical connection to ground. These types of probes generally consist of two electrodes of a specific geometry separated by a small insulating gap. The principle of operation can be shown by examining the probe with two hemispheres. The charge induced on one hemisphere of radius r is given by

$$Q = \int_{S/2} \overline{D}\ d\overline{A}$$

$$= 3\pi r^2 \varepsilon_0 E \qquad (17.46)$$

from which it follows that the short circuit current between the two hemispheres is

$$i(t) = \frac{dQ}{dt} = 3\pi r^2 \varepsilon_0 \frac{dE}{dt}$$

$$3\pi r^2 \varepsilon_0 j\omega E \qquad (17.47)$$

for ac fields. The effective area of the sphere, therefore, becomes

$$A = 3\pi r^2 \qquad (17.48)$$

This means that the effective distance between the poles of the dipole L_e is related to the capacitance of the hemispheres C_a by

$$L_e = \frac{3\pi r^2 \varepsilon_0}{C_a} \qquad (17.49)$$

The open-circuit voltage across the probe is equal to the product of the electric field and the effective distance L_e. The stray capacitance C_s acts with C_2 as a capacitive divider.

High-Voltage Measurements

17.11.4 Electro-Optical Methods

The optical techniques exploit the interaction of an electric field with an electro-optical sensor. There a crystal of a birefringent material suffers a change in its refractive index under the effect of the applied electric field. This Kerr effect could be used for measuring electric fields.

17.12 SOURCES OF ERROR IN HIGH-VOLTAGE MEASUREMENTS

In addition to human and instrument errors, two types of errors are known to exist in high-voltage testing: those caused by electromagnetic (EM) interference and those produced by improper earthing.

17.12.1 Shielding Against EM Interference

Electromagnetic noise may infiltrate into the measuring system and be superimposed on the measured signal. This noise is caused by transient potentials and by strong electromagnetic fields associated with the high rapidly changing voltages and currents. The following measures can be taken to reduce EM noise:

1. Power-line RF filters may be used with oscilloscopes.
2. Signals should be transmitted via shielded and matched coaxial cables.
3. With very steep surge measurements, multiple shielding may be used.
4. If necessary, the oscilloscope may be located away from the source of noise in shielded chambers.

17.12.2 Grounding in High-Voltage Laboratories

High rapidly changing voltages and currents in a high-voltage laboratory produce transient currents in the ground connections. Following are some of the measures that should be taken to avoid these adverse effects:

1. Multiple grounding of the high-voltage test circuit should be avoided.
2. The ground current loop can be interrupted by winding the cable on a ferrite core, thus raising the impedance of that loop.

3. Impulse generators should be connected to very low-resistance ground composed of a grid with rods driven into the ground close to the generator (Chapter 13).
4. The power supplies to the dc source of impulse generators and to the oscilloscope should be made via isolating transformers.

REFERENCES

Berkebile, L., Nilsson, S., and Sun, S. (1981). *IEEE Trans.*, *PAS-100*:1498–1504.
Hebner, R. E., Malewski, R. A., and Cassidy, E. C. (1977). *Proc. IEEE*, 65(11):1524–1548.
Holt, P. (1988). *Electra, 121*, December, 69–75.
Hyltén-Cavallius, M. (1988). *High Voltage Laboratory Planning*, E. Haefely, Basel, Switzerland.
IEEE Committee (1981). *IEEE Trans.*, *PAS-100*:4811–4814.
IEC (1960). *Recommendations for Voltage Measurement by Means of Sphere-Gaps (One Sphere Earthed)*, Publication 52, International Electrotechnical Commission, Geneva.
IEC (1962). *Radio Frequency Cables: General Requirements and Measuring Methods*, Publication 96-1, International Electrotechnical Commission, Geneva.
IEC (1973). *High Voltage Test Techniques*, Publication 62, International Electrotechnical Commission, Geneva.
IEEE Working Group (1983). *IEEE Trans.*, *PAS-102*:3549–3557.
Kind, D. (1978). *An Introduction to High Voltage Experimental Technique*, Friedr Vieweg & Sohn Verlagsgesellschaft mbH, Braunschweig, West Germany.
Kuffel, E. and Zaengl, W. S. (1984). *High Voltage Engineering Fundamentals*, Pergamon Press Ltd., Oxford.
Malewski, R. and Dechamplain, A. (1980). *IEEE Trans.*, *PAS-99*: 636–649.
Roger, A. (1986). *Electronics and Power*, 32(2):140–144.
Stark, W. B. (1979). *Proc. 3rd Int. Symposium on High Voltage Engineering*, Milan, Paper 44-08.
Tokayo, Harumoto, Y., Yamamoto, H., Yoshida, Y., Mukae, H., Shimada, M., and Ida, Y. (1982). *IEEE Trans.*, *PAS-101*, 3967–3976.

18
Testing Techniques

R. RADWAN and A. EL-MORSHEDY *Cairo University, Giza, Egypt*

18.1 INTRODUCTION

This chapter presents some of the techniques and theory involved in the testing of high-voltage equipment either in the laboratory or in the field. The main concern with electric equipment is that its insulation should withstand its operating voltage and the occasional overvoltage transients expected in the system. A further requirement for the satisfactory performance of the insulation is the minimization of internal discharges in voids present within the dielectric, which may cause its deterioration and eventual breakdown. High-voltage dielectric loss and partial discharge testing can, to some extent, help in the selection and design of insulation for electrical equipment.

Equipment insulation may be classified as self-restoring and non-self-restoring. Self-restoring insulation completely recovers its insulating properties after a disruptive discharge. The method of sampling and test techniques differ from one type of insulation to the other. International and national specifications for testing are outlined to meet the users' and manufacturers' requirements. These specifications describe in detail the tests to be carried out on a specific piece of equipment, their procedure, and acceptable limits of test results. Usually, the specifications are not limited to electrical tests but include other tests, such as mechanical and thermal tests.

A new equipment may pass a high-voltage test but may fail in the same test after some time in service. This is due to contamination of the insulators' surfaces, which can lower their dielectric strength while gradual chemical deterioration due to the reaction of the insulation with air is taking place. Corona discharges in voids in the insulation can gradually cause deterioration (Chapter 8). Voltage transients can initiate tracking or carbonization, and can even puncture the insulation and lead to early failure. This shows the importance of periodic maintenance and in-service tests. The following sections are devoted to various types of tests and voltages to be applied to circuit breakers, cables, transformers, high-voltage insulators, and surge arresters.

18.2 CLASSIFICATION OF TESTS

Some of the tests are performed during the early stage of development and production, others after production and installation. Type tests are performed on each type of equipment before their supply on a general commercial scale so as to demonstrate performance characteristics meeting the intended application. These tests are of such a nature that they need not be repeated unless changes are made in the design of the product. Routine tests are made by the manufacturer on every finished piece of product to make sure that it fulfills the specifications. Acceptance and commissioning tests are made by the purchaser and are self-explanatory. Maintenance tests are usually carried out after maintenance or repair of the equipment.

18.3 TEST VOLTAGES

The conventional forms of test voltages in use can be divided into three main groups: (a) direct voltages, (b) power-frequency alternating voltages, and (c), impulse voltages, which are divided into lightning and switching impulses. Also, in the case of testing machine insulation, alternating voltages of low frequency are sometimes used.

Methods of generating these voltages are described in Chapter 16. In this chapter attention is focused on the levels of test voltages and the techniques of voltage application. Also, the test circuits and precautions to be taken in high-voltage laboratories with regard to earthing, clearances, and interference are discussed. Tables 18.1 and 18.2 list the recommended test voltages adopted for testing equipment for rated ac voltages ranging between 1 and 765 kV (IEC, 1976b).

For equipment with rated voltages of 1 to 300 kV, performance under power-frequency operating voltage, temporary overvoltages,

Testing Techniques 489

and switching overvoltages is generally checked by a short-duration power-frequency test. But aging of internal insulation and contamination of external insulation require long-duration power frequency tests. The performance under lightning overvoltage is checked by a lightning-impulse test. For equipment with rated voltages $\geqslant 300$ kV, the performance under switching overvoltages is checked by switching impulse tests. Equipment in systems with effectively earthed neutrals can safely have reduced insulation. Therefore Table 18.2 lists more than one test voltage for each rated ac voltage of the equipment.

18.3.1 Tests with Direct Voltages

Direct voltages are used mainly to test equipment used in high voltage dc transmission systems. It is additionally used in insulation testing of arrangements with high capacitance, such as capacitors and cables. It is also used in fundamental investigations in discharge physics and dielectric behavior.

The value of the test voltage is defined by its arithmetic mean. The test voltage as applied to the test object should not contain ac components corresponding to a ripple factor of more than 5% when normal current is drawn. During the test it is required that the rate of voltage rise above 75% of its estimated final value be about 2% per second (IEC, 1973). The requirements of the test are generally satisfied if no disruptive discharge occurs on the test object when under the test voltage for the specified duration.

18.3.2 Tests with Alternating Voltages

The voltages used in this test generally have a frequency in the range 40 to 60 Hz and a sinusoidal shape. The ratio of its peak to rms values is equal to $\sqrt{2} \pm 5\%$. Partial discharges should not reduce the test voltage. This is usually achieved if the total HV circuit capacitance is within 1000 pF, and the circuit current with the test object short circuited is at least 1 A (IEC, 1973).

For dry tests on small samples of solid insulation or insulating liquids, a short-circuit current on the order of 0.1 A rms may suffice. For tests under artificial pollution, the required short-circuit current depends on the ratio of series resistance R_S to the steady-state reactance X_S of the voltage source, including the generator or supply network at the test frequency. It should be at least 6 A for $R_S/X_S < 0.1$, and at least 1 A for $R_S/X_S = 0.25$ (IEC, 1973).

The value of the test voltage is defined by its peak divided by $\sqrt{2}$. The peak values of voltages can be measured with a sphere

Table 18.1 Recommended Test Voltages for Rated Voltages Less Than 300 kV

Rated voltage (rms) kV	Rated lightning-impulse withstand voltage (peak) kV	Rated power-frequency short-duration withstand voltage (rms) kV
3.6	20[b]	10
	40	
4.4[a]	60[c]	19
	75[d]	
7.2	40[b]	20
	60	
12[a]	60	28
	75	
13.2[a]	95[c]	34
13.97[a]	110[d]	
14.52[a]		
17.5	75[b]	38
	95	
24	95[b]	50
	125	
26.4[a]	150	50
36	145[b]	70
	170	
36.5[a]	200	70
52[a]	250	95
72.5[a]	325	140
123[e]	450	185
123	550	230
145[e]	450	185
145[e]	550	230
145	650	275
170[e]	550	230
170[e]	650	275

Table 18.1 (Continued)

Rated voltage (rms) kV	Rated lightning-impulse withstand voltage (peak) kV	Rated power-frequency short-duration withstand voltage (rms) kV
170	750	325
245[e]	650	275
245[e]	750	325
245[e]	850	360
245[e]	950	395
245	1050	460

[a] Specifications for dielectric tests for the United States and Canada.
[b] For effectively earthed neutral with additional overvoltage protection.
[c] For transformers rating with 500 kVA and below.
[d] For transformers with rating above 500 kVA.
[e] Reduced insulation permissible only for systems with effectively grounded neutral.

gap or a peak voltmeter (Chapter 17). The rated withstand voltage is determined by the same method as that in the direct voltage test.

18.3.3 Tests with Impulse Voltages

A standard lightning impulse voltage has been accepted as an aperiodic impulse that reaches its peak value in 1.2 µs and then decreases slowly in about 50 µs to half its peak value. Switching impulses are characterized by having much longer fronts and total durations (Chapters 14 and 16).

Rated Impulse Withstand Tests

For tests on non-self-restoring insulation, three impulses are applied at the rated withstand voltage level of the specified polarity. The requirements of the tests are satisfied if no failure occurs. For withstand tests on self-restoring insulation, two procedures are in common use:

Table 18.2 Recommended Test Voltages for Rated Voltages Above 300 kV

Rated voltage (rms) kV	Rated switching-impulse withstand voltage (peak) kV	Rated lightning-impulse withstand voltage (peak) kV
300	750	850
		950
	850	950
		1050
362	850	950
		1050
	950	1050
		1175
420	950	1050
		1175
	1050	1175
		1300
		1425
525	1050	1175
		1300
		1425
	1175	1300
		1425
		1550
765	1300	1425
		1550
		1800
	1425	1550
		1800
		2100
	1550	1800
		1950
		2400

1. Fifteen impulses of the rated withstand voltage and the specified shape and polarity are applied. The requirements of the test are satisfied if not more than two disruptive discharges occur (IEC, 1973).
2. The test procedure for determining the 50% disruptive discharge voltage is applied. The test requirements are satisfied if the determined voltage is not less than $1/(1 - 1.3\ \sigma)$ times the rated impulse withstand voltage, where σ is the per-unit standard deviation of the disruptive discharge voltage (IEC, 1973).

The values of the 50% disruptive discharge voltage V_{50} and its standard deviation σ could be calculated using statistical methods. The two test procedures are the multiple-level (or probit) method, and the up-and-down method (IEC, 1973).

Multiple-Level Method

At least 10 impulses are applied at each test voltage level. The voltage interval between levels is approximately 3% of the expected 50% disruptive discharge voltage. The value of 50% disruptive discharge voltage is obtained from a curve of disruptive discharge probability versus prospective test voltage. The accuracy of determination increases with the number of voltage applications at each level.

The values of V_{50} and S could be calculated in terms of the set of measured voltages V_i and their relative deviations $Z_i = (V_i - V_e)/\sigma_e$, where V_e and σ_e are the values initially estimated, and shall be corrected by reiteration. Thus

$$V_{50} = \frac{\overline{V(Z^2)} - \overline{Z} \cdot \overline{ZV}}{\overline{(Z^2)} - (\overline{Z})^2}$$

$$S = \frac{\overline{ZV} - \overline{V} \cdot \overline{Z}}{\overline{(Z^2)} - (\overline{Z})^2} \qquad (18.1)$$

in terms of the average values of V and Z. Some weighting coefficients could be applied to the readings at each voltage level, the confidence limits could be calculated as is well known (ANSI, 1968).

Up-and-Down Method

A voltage V is chosen that is approximately equal to the expected 50% disruptive discharge level. ΔV is the voltage interval and is approximately equal to 3% of V. One impulse is applied at the level V. If this does not cause a disruptive discharge, the next impulse should have the level V + ΔV. If a disruptive discharge occurs at the level V, the next impulse should have the level V − ΔV. This procedure is continued until a sufficient number of observations has been recorded.

To calculate V_{50} and σ, the voltage readings corresponding to either the withstands or discharges could be used. If the total number of either type of events is N with n_i shots (≥ 20) at each voltage level V_i, the lowest level being V_0 and the highest being V_k, then

$$V_{50} = V_0 + \Delta V \left(\frac{A}{N} \pm 0.5 \right) \quad (18.2)$$

$$S = 1.62 \Delta V \left(\frac{NB - A^2}{N^2} + 0.029 \right) \quad (18.3)$$

where $N = \sum_{i=0}^{i=k} n_i$, $A = \sum_{i=0}^{i=k} in_i$, and $B = \sum_{i=0}^{j=k} i^2 n_i$

Evidently, the same statistical manipulation of the test results applies to all similar tests.

Tests with Switching Impulse

The standard switching impulse is an impulse having a time to crest T_1 of 250 µs and a time to half-value T_2 of 2500 µs. It is described as a 250/2500 impulse. The test procedures for switching impulses are, in general, the same as those for lightning impulse testing and similar statistical considerations apply.

With switching impulses, disruptive discharges may occur at random times before the crest. In presenting the results of disruptive discharge tests, the relationship of discharge probability to voltage is generally expressed in terms of the prospective crest value.

18.4 TESTS WITH IMPULSE CURRENTS

Transient currents of large amplitudes are experienced during the discharge of energy-storing devices, lightning strokes, and some short circuits. If these currents have a definite shape, they are referred to as impulse currents. Impulse currents for testing normally take either of two shapes, the double-exponential or the

Testing Techniques

rectangular shape, as defined in Chapter 16. Two standard shapes are in use for impulse currents: the double exponential of 8/20 or 4/10 μs, and the rectangular, with a virtual duration of 500 μs, 1000 μs, or 2000 μs.

18.5 SAFETY PRECAUTIONS IN THE LABORATORY

Extreme caution and safety awareness are essential items of high-voltage test procedures. The following safety features are considered for the safety of the operating personnel, installations, and apparatus:

1. It is essential that the high-voltage equipment be properly designed and manufactured to permit testing without unnecessary danger.
2. The actual danger zone of the high-voltage circuit must be clearly marked and protected from unintentional entry by walls or metallic fences.
3. All doors should be interlocked to remove high voltage automatically when opened.
4. Before touching the high-voltage elements, after test, visible metallic connection with earth must be established.
5. All metallic parts of the setup that do not carry potential during normal service must be grounded reliably.
6. It is preferred that the region of the high-voltage apparatus be matted by a closely meshed copper grid. The earth terminals of the apparatus are connected to it noninductively using wide copper bands.
7. All measuring and control cables and earth connections must be laid avoiding large loops.
8. The measuring signal is transferred to the measuring device (e.g., oscilloscope) via coaxial cables.
9. Shielding of a high-voltage setup by a Faraday cage is necessary for complete elimination of external interference, as sensitive measurements are often required in high-voltage experiments.
10. The clearance between test object and extraneous structure should be at least 1.5S, where S is the flashover distance between the electrodes of the test object. In this case the effect of such structures on the test results will be negligibly small.

A typical test arrangement of a high-voltage laboratory containing an impulse generator, a high-voltage transformer, and a potential divider is shown in Figure 18.1.

Fig. 18.1 Layout of a high-voltage laboratory. (Courtesy of IREQ.)

18.6 NONDESTRUCTIVE TESTING

Nondestructive electrical tests are usually carried out on the equipment insulation to ensure that its electrical characteristics comply with the specifications without destroying it. These tests include partial discharges, radio interference, dielectric loss angle, and insulation resistance measurements (Kind, 1978).

18.6.1 Partial Discharge Measurements

Corona discharges may occur on the surface of an insulator or in voids within its volume. The presence of corona may be detected by several nonelectrical and electrical methods (Nattrass, 1988). The electrical discharge detection methods make use of the current impulses accompanying discharges in the cavity. Before discussing these methods it is appropriate to explain what goes on in a void within an insulating material subjected to an alternating voltage stress.

A specimen of an insulating material containing a gas void can be represented as shown in Figure 18.2a and its equivalent circuit in Figure 18.2b. The capacitance C_v represents the void and C_b the dielectric above and below it. C_a represents the rest of the dielectric. When an alternating voltage V_a in excess of that corresponding to the breakdown threshold of the gas in the void is applied

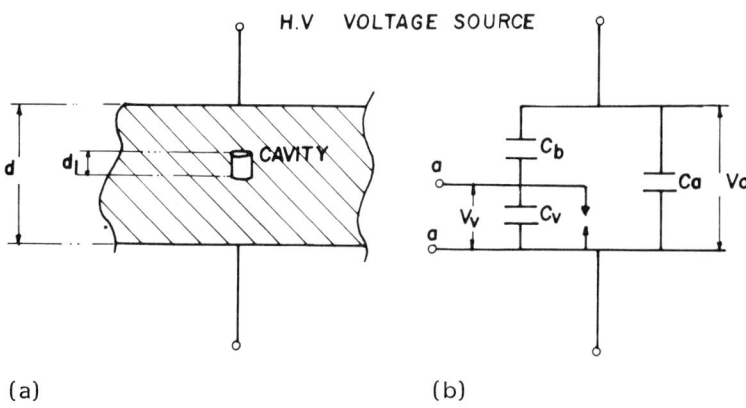

(a)   (b)

Fig. 18.2   (a) Void representation, (b) its equivalent circuit.

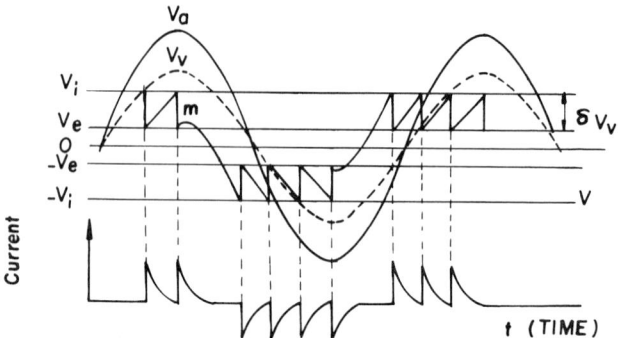

Fig. 18.3 Voltage and current traces of partial discharge in a void.

to the dielectric a partial discharge (PD) will start in it. The process of partial discharge in the void is illustrated in Figure 18.3. The voltage appearing across the void if there were no discharge is V_v and is given by the expression

$$V_v = \frac{V_a}{1 + (d/d_1 - 1)/\varepsilon_r} \qquad (18.4)$$

where d and d_1 are the thicknesses of the insulating specimen and the gas void, respectively, and ε_r is the relative permittivity of the dielectric. Partial discharge in the void will start at a voltage V_i on the positive half-cycle and approximately $-V_i$ on the negative half-cycle. $\pm V_e$ is the voltage at which the discharge stops. The discharge in the void will be accompanied by a sharp current pulse, as indicated in Figure 18.3. This may be repeated several times on the increasing part of the positive half-cycle. At point m the voltage across the void reverses its polarity, since at this instant V_v is decreasing, and discharge will continue with almost regular negative current pulses. The electrical partial discharge detection methods are classified as straight or balanced methods.

Straight Detection Methods

In straight detection methods PD measuring instruments measure the charge released within the discharging sites of the test specimen. The simplest circuit used for such measurement contains a series impedance connected between the test object and ground as illustrated in Figure 18.4.

Testing Techniques

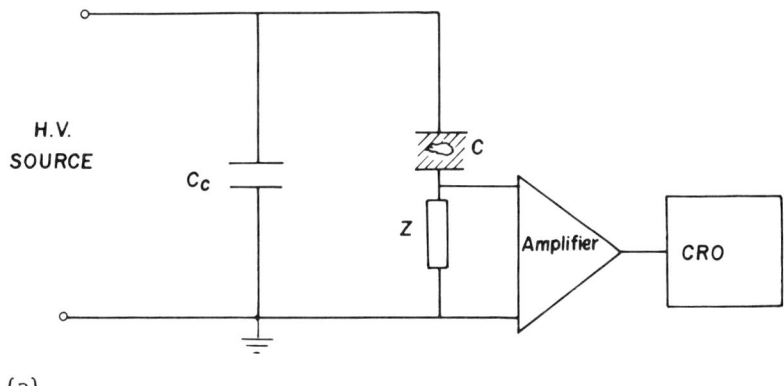

(a)

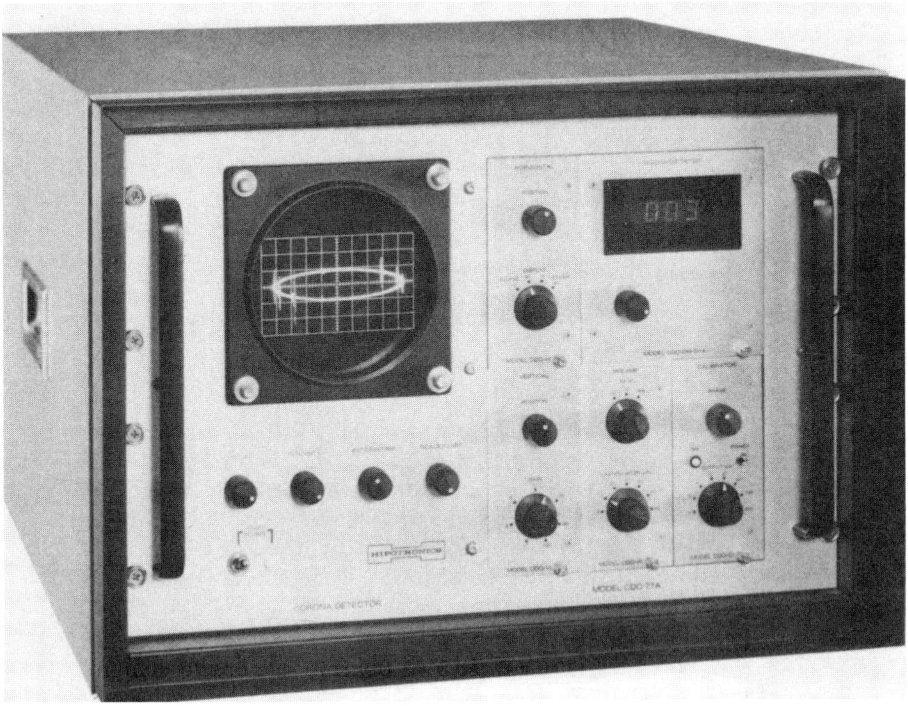

(b)

Fig. 18.4 Partial discharge detection circuit (a) and apparatus (b). (Courtesy of Hypotronics.)

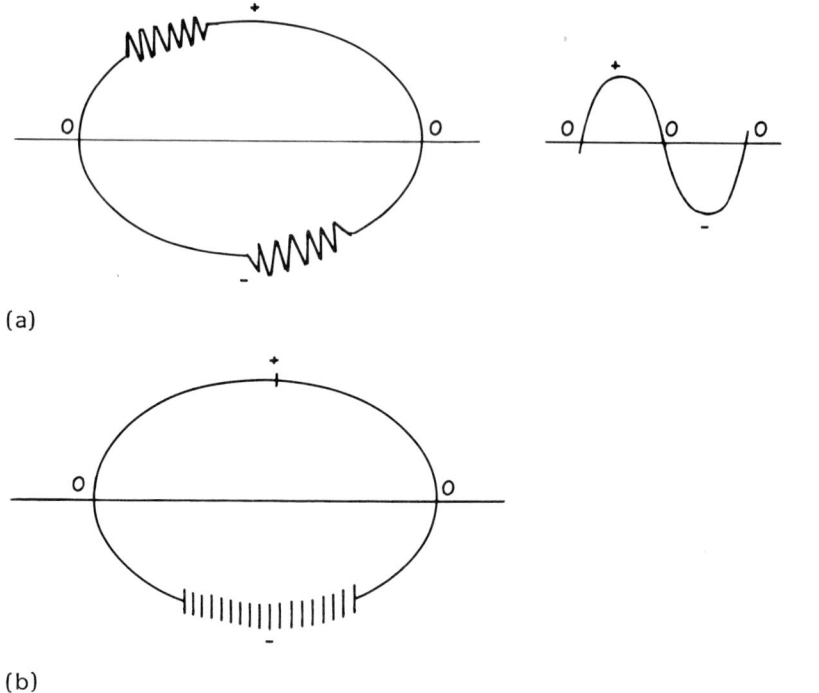

(a)

(b)

Fig. 18.5 Partial discharge display: (a) void, (b) point to plane.

Partial discharge current measurement is difficult and precautions taken to improve the fidelity of displaying the current waveshape has been discussed in detail (Nattrass, 1988). Partial discharge measuring instruments available measure the apparent charge (i.e., the charge transfer that takes place at the terminals of the specimen). The amplified discharge pulses are displayed oscillographically superimposed on a power-frequency elliptic time base, as shown in Figure 18.5. The discharge shown in Figure 18.5a is for a gas void and is characterized by its symmetry around both voltage peaks, while that for a point-to-plane gap is characterized by its regular pattern around the negative voltage peak (Fig. 18.5b).

The dependence of the apparent charge Q_a on the actual charge Q_v in the test specimen can be obtained by considering the discharge model shown in Figure 18.2. When breakdown occurs in the void, charge transfer amounts to

$$Q_v = \left(C_v + \frac{C_a C_b}{C_a + C_b}\right) \delta V_v \tag{18.5}$$

where δV_v is the voltage drop in the void at breakdown (Fig. 18.3).

The capacitance C_a is usually much greater than C_b and Q_v can be approximately expressed as

$$Q_v = (C_v + C_b) \delta V_v \tag{18.6}$$

After the void breakdown the system restores the voltage across the capacitor C_b to its original value, and this requires a charge Q_a to be supplied to C_b. This charge is the apparent charge and is given by

$$Q_a = C_b \delta V_v \tag{18.7}$$

From equations (18.6) and (18.7) the apparent and actual charges are related by the approximate expression

$$Q_a = \frac{C_b}{C_v + C_b} Q_v \tag{18.8}$$

Commercial partial discharge measuring instruments are usually calibrated in terms of apparent charge. Figure 18.6 shows a simplified circuit for PD detection employing a passive RC differentiator to sense the high-frequency signals generated by PD pulses. The RC network acts as a single-pole high-pass filter whose function is analogous to that of the power-frequency separation filter in conventional PD detection equipment. With low-noise amplifiers, the apparatus sensitivity could reach 0.01 pC (Bilodeau et al., 1987; Kreuger, 1989).

Balanced Detection Methods

Balanced detection methods are much more sensitive than straight detection methods. In the straight PD detection methods discharges in any part of the test circuit and not within the test sample may be detected and displayed along with the discharge impulses in the sample. This implies the use of discharge-free high-voltage sources or the provision of filters. Discharges on the high-voltage leads or loose earth connections, although they can be recognized, should be eliminated. In addition, noise may be picked

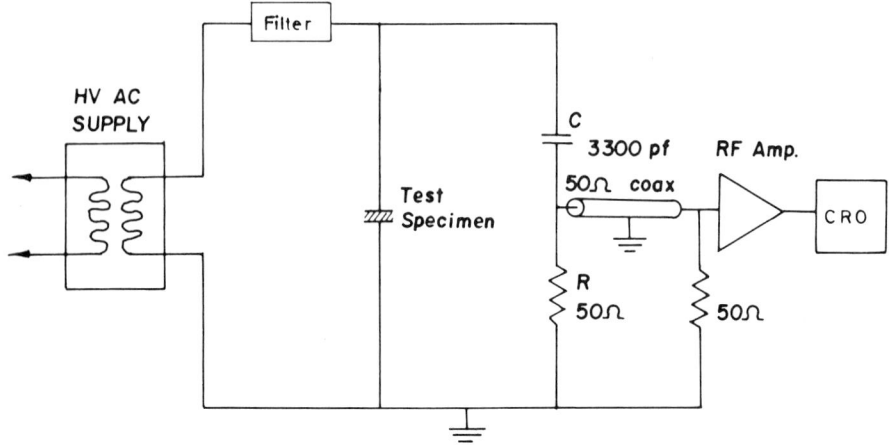

Fig. 18.6 A simplified circuit of a PD measuring system that utilizes a passive RC differentiator as the sensing element.

up from a variety of possible sources having nothing to do with the test setup (e.g., nearby thyristor-controlled machines, ultrasonic generators, and arcing contacts).

The detection of PD with bridges similar to that of Schering has been adopted for more than 50 years. The high-voltage arms of the bridge contain the test sample and a separation capacitor. The low-voltage arms contain balancing resistors and capacitors. The output of the bridge is supplied to and displayed on an oscilloscope through a filter and an amplifier. External disturbances are thus considerably reduced. If the separation capacitor and test sample are of equal capacitance, complete rejection of external interference would be possible. Otherwise, the bridge would be balanced only at one frequency, resulting in less effective rejection of external interference. Screening of the low-voltage arms and good earthing will no doubt substantially decrease external interference.

18.6.2 Radio-Interference Tests

Corona discharges are known to produce radio noise over a considerable portion of the radio-frequency spectrum (Chapter 5). Measurement of radio interference from HV equipment can be carried out by a circuit similar to that shown in Figure 18.6 except that a radio-noise meter is used instead of the PD amplifier and oscilloscope.

18.6.3 Dielectric Loss Measurements

Losses always occur in dielectrics due to conduction, polarization, and ionization (Chapters 7 and 8). Dielectric losses cause certain electrical effects, which can be utilized for nondestructive high-voltage testing (Kind, 1978).

Schering Bridge

The Schering bridge, devised by Schering in 1919, has since been widely used to measure the capacitance and loss angle of high-voltage insulators, capacitors, and cable samples. The bridge comprises two high-voltage (HV) arms and two low-voltage (LV) arms. The HV arms are the test piece and a standard capacitor. This capacitor should have no significant losses over the full working range. A suitable design employs smooth electrodes with corona shields, insulated with compressed gas. The LV arms are adjustable precision capacitor and resistors. Both the LV arms and the null detector are shielded from the high-voltage circuit to eliminate any errors in the measurements caused by the effect of stray capacitances. Proper grounding of the circuit is essential for safety and accuracy. Each of the LV arms is shunted by an overvoltage protective device, which operates at a few tens of volts in case either of the high-voltage arms fails. The capacitance C_2 and loss angle δ of the test piece (Fig. 18.7) could easily be evaluated:

$$C_2 = C_1 \frac{R_4}{R_3} \tag{18.9}$$

$$R_2 = \frac{C_4 R_3}{C_1} \tag{18.10}$$

$$\tan \delta = \omega C_4 R_4 \tag{18.11}$$

Bridge Incorporating Wagner Earth

The Schering bridge is earthed at the low-voltage end of the high-voltage source. The capacitance of the detector screened leads as well as the stray capacitances of branches AB and AD affect the balance conditions. To overcome this drawback, auxiliary arms are used for maintaining points B and D at earth potential under balance. In Figure 18.7 an additional arm N is connected between the low-voltage terminal and earth (Kuffel and Zaengl, 1984). The stray capacitance of the high-voltage terminal to earth is represented

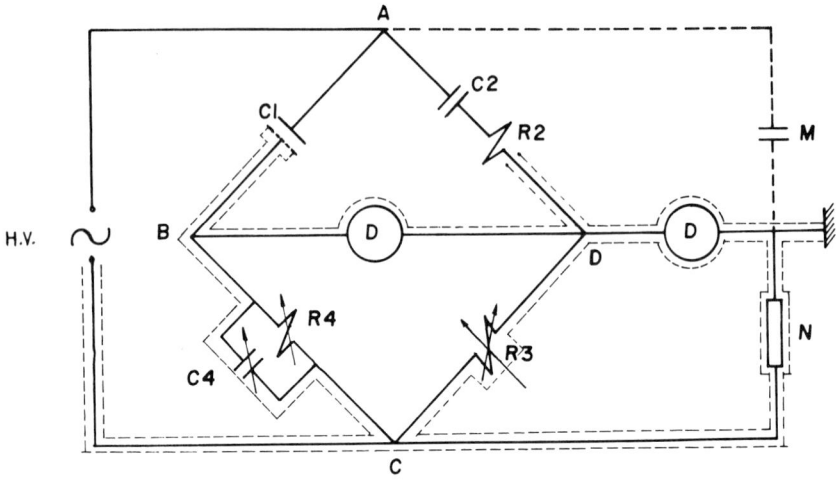

Fig. 18.7 Schering bridge incorporating Wagner earth.

by a capacitance M. The arrangement becomes equivalent to a six-arm bridge. The ratio M/N must be balanced for phase as well as magnitude with the main bridge arms. At balance the terminals of the detector are at earth potential. The capacitances between the leads and screens are in parallel with the impedance N and thus do not affect the balance conditions. An electronic device for automatic balancing of Wagner earth could be developed.

18.6.4 Insulation Resistance Measurement

The magnitude of the resistance of a solid or liquid insulation varies with its temperature, its moisture content, the applied voltage, and whether it is ac or dc (Chapters 7 and 8). It is evident that if the resistance or the loss factor of the insulation in a certain machine, transformer, or cable is measured periodically, under almost identical service conditions, the results would indicate its degree of aging. When the measured insulation resistance starts to decrease considerably below the reading of the previous test, the machine, transformer, or cable should be taken out of service. Its insulation should be more carefully tested and overhauled to avoid sudden failure during operation. The insulation resistance test could be performed using an adjustable high-voltage dc source (IEEE Committee, 1981). The dc voltage applied during the test should be raised very slowly up to no more than 80% of the acceptance test

Testing Techniques 505

values. The initial current component corresponds to charging the circuit capacitance. After several seconds, or minutes, the circuit current decreases to its steady component, which corresponds to conduction through the insulation. This last component is the quantity to be measured, recorded, and compared with the previous readings during similar tests.

18.7 CIRCUIT BREAKER TESTING

Testing of circuit breakers requires highly equipped laboratories and sophisticated testing procedures. Standard specifications give and describe in detail the tests to be carried out on circuit breakers and how they are performed (IEC, 1972).

18.7.1 Mechanical Test

Mechanical failures represent more than 80% of the total failures in circuit breakers. The circuit breaker is subjected to 1000 operating cycles, with no voltage or current in its main circuit. During the test, lubrication of the mechanical parts is allowed according to the manufacturer's instructions. However, mechanical adjustment or replacement of any part is not permissible. After the test a complete check on the circuit breaker is performed to make sure that all mechanical linkages and contacts are in good condition.

18.7.2 Temperature Rise Test

During operation the temperature of any part of the circuit breaker should not exceed the specified limits of temperature rise. These limits depend on each individual part and circuit breaker type. The temperature rise test is carried out with normal rated current flowing in the main circuit and the circuit breaker mounted as under normal service conditions. The maximum observed temperature rise is then compared with the stipulated limits.

18.7.3 Insulation Tests

Dielectric tests are carried out to make sure that the circuit breaker withstands the overvoltage expected within the power system with a reasonable safety factor. The main dielectric tests are impulse and power-frequency voltage tests. The impulse voltage wave shape usually used is the standard wave 1.2/50 µs, with voltage magnitudes as given in the relevant standard specifications (Tables 18.1 and 18.2). Power-frequency voltage tests are applied for 1 minute to indoor and outdoor circuit breakers when dry or wet (i.e., under simulated rain), respectively. The breaker under test should withstand the specified voltage tests, without flashover or puncture.

18.7.4 Short-Circuit Tests

Circuit breakers should be capable of making and breaking the circuit under short-circuit conditions without appreciable deterioration of its components or change in its performance. Short-circuit tests are usually carried out according to test duties that specify the test current, percentage of dc component, and transient and power-frequency recovery voltages. These tests are numerous and it is difficult to discuss them all in detail within the space of this section. However, they are classified as follows:

1. Breaking current tests
2. Making current tests
3. Short-time current tests
4. Operating sequence tests
5. Single-phase short-circuit tests
6. Short-line fault tests
7. Out-of-phase switching tests
8. Capacitor charging-current breaking tests
9. Small inductive current breaking tests

The short-circuit tests need high-power testing plants to meet the ever-increasing circuit breaker ratings, with currents reaching the order of 10 kA or more at voltages ranging up to 750 kV. Such plants are extremely expensive. The synthetic method of cir-

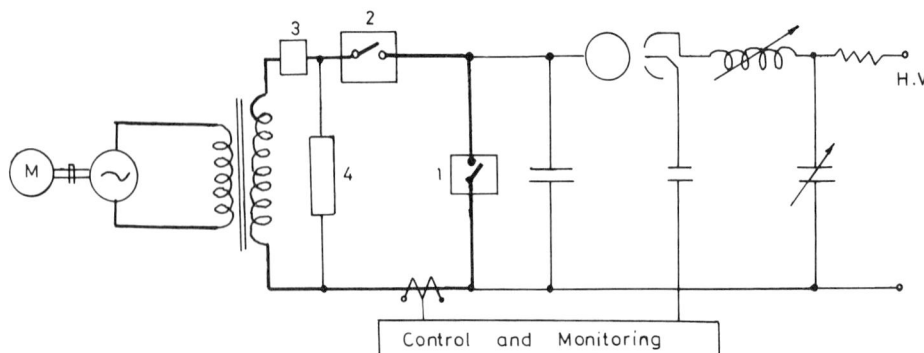

Fig. 18.8 Synthetic testing circuit: (1) breaker under test, (2) master breaker, (3) synchronizing switch, (4) overvoltage protector.

cuit breaker testing provides an alternative solution. In the basic form it consists of two independent voltage and current sources (Fig. 18.8). The current source injects the current into the circuit breaker under test at a relatively reduced voltage, while the voltage source injects a high-voltage transient across the circuit breaker contacts at the moment it interrupts the current. The high-voltage source usually contains a large capacitor bank. Actual synthetic testing circuits are much more elaborate than the simplified one shown in Figure 18.8. They would include elaborate control and instrumentation schemes, in addition to a backup circuit breaker for protecting the sources in case of failure of the breaker under test. The voltage source could be connected either across or in series with the current circuit (Lythall, 1986).

18.8 CABLE TESTING

Cables are subjected to electrical and thermal stresses while in service. They also undergo mechanical stresses during their installation and repair. These mechanical stresses may cause insulation cracking, and hence the leakage current increases, leading to thermal breakdown of the cable insulation.

18.8.1 Conductor Resistance

The conductor resistance of a complete length or a specimen of cable is measured and the result is corrected to a temperature of 20° C and 1 km length in accordance with the formula

$$R_{20} = \frac{R_m}{\ell_c [1 + \alpha_{20} (\theta - 20)]} \quad \Omega/km \qquad (18.12)$$

where the temperature coefficient α_{20} is 3.93×10^{-3} for copper and 4.03×10^{-3} for aluminum, and R_m is the measured resistance in ohms, ℓ_c the sample length in kilometers, and θ the ambient temperature in °C.

18.8.2 High-Voltage Test

The high-voltage test is carried out under power-frequency alternating voltage or direct voltage. Because of the much lower power needed under dc, it is used for commissioning and maintenance tests of complete cable networks. According to the IEC 502 (IEC, 1978), the power-frequency test voltage is $(2.5U_0 + 2)$ kV for cables of rated voltages (U_0) up to and including 6 kV, and $2.5 U_0$ for cables of higher-rated voltages.

18.8.3 Partial Discharge and Loss-Angle Tests

Partial discharge tests are carried out to ensure that the level of the discharge does not exceed a specified limit. Single-core and multicore cables are tested by applying a voltage of 1.25 times the rated voltage between the conductor and the metallic screen. The magnitude of the PD impulses should not exceed 20 pC for cables insulated with butyl rubber, EPR, and XLPE; and 40 pC for PVC cables (IEC, 1987b). Also, the loss factor tan δ is measured at different voltages up to 1.25 times the rated voltage using a Schering bridge.

18.9 TESTING OF POWER TRANSFORMERS

The following tests are required on most power transformers, as recommended by the IEC 76-1,,2 (IEC, 1976c,d), and 76-3-1 (IEC, 1987a).

1. Routine tests
    a. Measurement of winding resistance
    b. Measurement of voltage ratio and check of voltage vector relationship
    c. Measurement of impedance voltage, short-circuit impedance, and load loss
    d. Measurement of no-load loss and current
    e. Tests on tap changers
2. Type tests
    a. Dielectric tests
    b. Temperature rise tests

Some special tests may be required, such as a short-circuit test, measurement of zero-sequence impedance on three-phase transformers, and measurement of acoustic sound level. In this section, only type tests are discussed; other tests are given in the standard specifications.

18.9.1 Insulation Levels and Dielectric Tests

It is essential to specify higher test voltages for the internal insulation of the transformer than for the external insulation of the system. A failure in the non-self-restoring internal insulation would be catastrophic and normally leads to the transformer being withdrawn from service for a long period.

The recommended clearances among the various terminals and windings are referred to the rated withstand voltage of the internal

Testing Techniques

insulation of the transformer (IEC, 1976a, 1987a). For the equipment with rated voltages less than 300 kV the reference voltage for external clearance is the lightning impulse voltage, whereas for equipment of higher ratings, the switching impulse voltage is the one to be employed.

18.9.2 Temperature Rise Test

The temperature rise values for the windings, cores, and oil of transformers designed for operation at normal altitudes are listed in tables according to the cooling medium (IEC, 1976d). For oil-immersed transformers, the temperature rise tests include the determination of top oil temperature rise and of winding temperature rise.

18.10 TESTING OF SURGE ARRESTERS

Surge arresters are identified by their rated voltage, rated frequency, nominal discharge current, and class of long-duration discharge (IEC, 1970). Some of the type tests made on nonlinear resistor-type arresters for ac systems are summarized below.

18.10.1 Power-Frequency Sparkover Tests

Dry and wet tests are made on complete arresters. The voltage initially applied to the arrester should be of a low value, to avoid sparkover of its series gaps. The permissible time in which the applied voltage may exceed the rated voltage of the arrester is in the range 2 to 5 seconds. After sparkover the test voltage is switched off by automatic tripping within 0.5 s. The power-frequency sparkover voltage is the average of five test results.

18.10.2 Impulse Sparkover Tests

These tests are made on the same samples as those used for power-frequency sparkover tests. Five positive and five negative impulses of the standard shape are applied to the test sample, and the series gaps of the arrester must spark over on every impulse. If in either series of five impulses the gaps fail to spark over once only, an additional 10 impulses of that polarity are applied and the gaps must spark over on all these impulses. Arresters with rated voltages exceeding 100 kV are additionally tested under switching impulses.

18.10.3 Impulse Current Withstand Tests

The rated voltage of these tests must be in the range 3 to 6 kV. Each sample is subjected to two current impulses of the standard shape with peak values depending on the arrester class. Before each test, the samples must be at approximately the ambient temperature. Subsequently, the power-frequency sparkover voltage is determined. It should not change by more than 10% (IEC, 1970). Also, no evidence of puncture or flashover of the nonlinear resistors or significant damage to the series gaps or grading circuit should occur.

18.11 TESTING OF INSULATORS

Insulators used on overhead transmission lines and in substations are subjected to type tests, sample tests, and routine tests.

18.11.1 Type Tests

The main type tests include the following:
 1. *Dry lightning impulse withstand test.* The dry insulator is tested under positive and negative $1.2/50$-µs impulse voltage waves. Two test procedures are in common use for the lightning impulse withstand test: the withstand procedure with 15 impulses and the 50% flashover voltage procedure (IEC, 1979).
 2. *Dry switching impulse withstand test.* The insulator is tested under positive and negative $250/2500$-µs impulse voltage waves. The 50% flashover procedure is normally used.
 3. *Dry power-frequency withstand test.* This test is applicable only to insulators for indoor use. The applied test voltage is the specified dry power-frequency withstand voltage adjusted for atmospheric conditions.
 4. *Wet power-frequency withstand test.* This test is applicable only to insulators for outdoor use. The applied test voltage is the specified wet power-frequency withstand voltage adjusted for atmospheric conditions. The characteristics of artificial rain are given in IEC 168 (IEC, 1979). The wet power-frequency withstand voltage is determined by the same procedure as that described in determining the dry power-frequency withstand voltage.
 5. *Artificial pollution tests.* These tests are intended to provide information on the behavior of outdoor insulators under conditions representative of pollution in service. The pollution tests may be made either to determine the maximum degree of pollution withstood by the insulator under a given voltage or to determine its

withstand voltage for a specified degree of pollution. The pollution tests involve application of the pollution and simultaneous or subsequent application of voltage. The pollution tests fall into two categories: the saline fog method and the pre-deposited pollution method (IEC, 1975).

18.11.2 Sample Tests

The sample tests on insulators include the following two principal tests:

1. *Power-frequency puncture test.* The insulator, after having been cleaned and dried, is completely immersed in a tank containing a suitable insulating medium (usually oil) to prevent surface discharges. The test voltage is applied on the insulator and then raised rapidly to the minimum puncture voltage specified. No puncture may occur below the specified minimum puncture voltage (IEC, 1979).

2. *Temperature cycle test.* This test is performed on individual insulator units made of ceramic or toughened glass. The unit is cyclically immersed in hot- and cold-water baths maintained with a temperature difference of 50 K. The insulator is left in each bath for (15 + 0.7m) minutes, with a maximum of 30 minutes, m being the mass of the insulator in kilograms. This heating-cooling cycle is repeated three times in succession. The insulator is then examined It should suffer no cracks or damage and should withstand a subsequent dry flashover test. For insulators made of annealed glass, the test procedure is slightly different (IEC, 1979).

NOTE

From the discussion above it is evident that each item or type of equipment has its specific tests. Tests on pieces of equipment other than those mentioned above are described in detail in their relevant standard specifications. The ideas behind these tests, and the general procedures, are essentially similar to those described above.

REFERENCES

ANSI (1968). *Standard Techniques for Dielectric Tests*, Publication 68.1, American National Standards Institute, New York.

Bilodeau, T. M., Shea, J. J., Fitzpatrick, G. J., and Sarjeant, W. J. (1987). *IEEE Electrical Insulation Magazine*, 3:4.

IEC (1970). *Lightning Arresters, Part 1, Non-linear Resistor Type Arresters for A.C. Systems*, Publication 99-1, International Electrotechnical Commission, Geneva.

IEC (1972). *High Voltage Alternating Current Circuit Breakers*, Publication 56-4, International Electrotechnical Commission, Geneva.

IEC (1973). *High Voltage Test Techniques*, Publication 60-1,2, International Electrotechnical Commission, Geneva.

IEC (1975). *Artificial Pollution Tests for HV Insulators for AC Systems*, Publication 507, International Electrotechnical Commission, Geneva.

IEC (1976a). *Insulation Co-ordination, Part 1, Terms, Definitions, Principles and Rules*, Publication 71-1, International Electrotechnical Commission, Geneva.

IEC (1976b). *Insulation Co-ordination, Part 2, Application Guide*, Publication 71-2, International Electrotechnical Commission, Geneva.

IEC (1976c). *Power Transformers*, Publication 76-1, International Electrotechnical Commission, Geneva.

IEC (1976d). *Power Transformers*, Publication 76-2, International Electrotechnical Commission, Geneva.

IEC (1978). *Extruded Solid Dielectric Insulated Power Cables for Rated Voltages from 1 kV up to 30 kV*, Publication 502, International Electrotechnical Commission, Geneva.

IEC (1979). *Tests on Indoor and Outdoor Post Insulators of Ceramic Material or Glass for Systems with Nominal Voltages Greater Than 1000 V*, Publication 168, International Electrotechnical Commission, Geneva.

IEC (1987a). *Power Transformers*, Publication 76-3-1, International Electrotechnical Commission, Geneva.

IEC (1987b). *Partial Discharge Tests for Cables*, Publication 885-I,II, International Electrotechnical Commission, Geneva.

IEEE Committee (1981). *IEEE Trans.*, PAS-100:3292−3300.

Kind, D. (1978). *Introduction to High Voltage Experimental Technique*, Friedr Vieweg & Sohn Verlagsgesellschaft mbH, Braunschweig, West Germany, p. 52

Kreuger, F. (1989). *Discharge Detection in High Voltage Equipment*, Butterworth & Company (Publishers) Ltd., London.

Kuffel, E. and Zaengl, W. S. (1984). *High-Voltage Engineering Fundamentals*, Pergamon Press Ltd., Oxford.

Lythall, R. T. (1986). *The J & P Switchgear Book*, Butterworth & Company (publishers) Ltd., London.

Nattrass, D. (1988). *IEEE Electrical Insulation Magazine*, 4(3):10−23.

Index

A
Absorption
  coefficient, 79, 88
  range, 81
Acceptance tests (see under Tests)
AC peak voltmeters (see under Voltmeters)
Active dielectrics, 214
Aging of insulation (see under Insulating liquids)
Air-blast circuit breakers (see under Circuit breakers)
Air clearances (see Bus bar clearances)
Air relative density (see Relative air density)
Alpha particles, 80, 81
Apparent charge, 500, 501
Arc, 82, 93, 145–161
  burning ambient, 147, 276, 294, 298
  as a circuit element, 156–158
  erosion, 161, 162
  industrial applications, 93, 162, 163
  interruption, 85, 87, 275
    ac, 159
    dc, 158, 159
  magnetic phenomena, 154
  plasma, 148–151, 154, 159, 162, 163

[Arc]
  regions, 147–151
  reignition, 160, 161
  V-I characteristics, 152–156
    dynamic, 154, 156
    static, 152, 153
  voltage, 152, 153, 157, 159
Arresters (see Surge arresters)
Askarels, 165, 167, 168, 170, 171, 173, 183
Asymmetrical short circuits, 276, 277, 279
Attachment coefficient (see Electron attachment)
Atmospheric conditions, effects on discharges, 87–90 (see also Gas parameters)
Avalanche (see Electron avalanche)

B
Basic impulse insulation levels (BIL, BSL), 396–398
Beta particles, 80, 81
Breakdown of electronegative gases, 106–109
Breakdown field strength, 211
Breakdown in gases
  under ac, 112
  under dc, 97, 109, 111

513

[Breakdown in gases]
  under impulse, 112, 113, 115–117
  in long gaps, 112–117
  at minimum voltage, 101, 102, 115
  in nonuniform fields, 103, 109, 115
  in uniform fields, 97, 99–102, 107
Breakdown in liquids (*see under* Insulating liquids)
Breakdown probability, 113, 114, 385–387
Breakdown in solids (*see under* Solid insulating materials)
Breakers (*see* Circuit breakers)
Bridges (*see* Schering bridge)
Bubble theory (*see under* Insulating liquids, theories of breakdown)
Bundle conductors, 18–20, 134, 137
Burst pulse corona, 124
Bus bars
  arrangements, 220–224
  clearances, 226–231
  current ratings, 231–236
  electrodynamic forces, 238
  insulators, 225, 239
  mechanical stresses, 236–238
Bushings, 38, 47, 253, 255

C
Cables
  constants of, 319, 320
  construction of, 306–310
  current-carrying capacity, 321–323

[Cables]
  electric fields in, 22, 320, 321
  insulation, 310–315
  jointing, 323
  locating faults in, 324–326
  losses, 315–319
  sheath phenomena, 315–319
  superconducting, 326–329
  testing of, 329, 507, 508
  types
    compressed-gas, 256–258, 314
    gas-filled, 256, 314
    oil-filled, 311–313
    paper-insulated, 310
    synthetic insulation, 311
Calculation of electric fields (*see* Field analysis and computation)
Capacitive dividers (*see* Potential dividers)
Cascaded rectifier circuits (*see under* High-voltage dc generators)
Cascaded transformers (*see under* Testing transformers *and* Voltage transformers)
Cathode region of arc (*see* Arc regions)
Cavities in insulation, 190, 487 (*see also* Voids in insulation)
Chemical reactions in oils (*see under* Insulating liquids)
Circuit breakers
  rated quantities, 276–279
  testing, 505, 506
  types
    air, 288, 289
    air-blast, 291, 293, 304
    bulk-oil, 288
    direct current, 302–304

Index

[Circuit breakers]
  [types]
    minimum-oil, 289, 291
    SF_6, 294, 295, 304
    solid-state, 201, 300
    vacuum, 298
Circuit breaking
  asymmetrical short circuits, 281
  capacitive current, 285, 506
  resistance switching, 287, 375
  RRRV (rate of rise of restriking voltage), 287
  short-line faults, 285, 286, 506
  single-phase short circuit, 506
  small inductive currents, 506
  3-phase short circuits, 279, 280
Clearances (see Bus bar clearances)
Coefficient
  of attachment (see Electron attachment)
  of diffusion (see Electron diffusion)
  of ionization (see Ionization coefficient)
Collision (see Particle collision)
Compressed-gas cables (see under Cables)
Compton effect, 78
Conduction through dielectrics, 172, 173, 199
Conductor bundles (see Bundle conductors)
Conductor coating, 265–268
Contamination of insulators (see Insulator pollution)
Corona discharge, 108–110, 112, 123–142
  ac, 51, 128, 130

[Corona discharge]
  applications in industry, 140–142
  impulse, 131, 132
  negative dc, 124, 128
  positive dc, 124, 127
Corona onset, 105, 110, 112, 133–135
  computation, 135, 136
Corona power loss
  computation, 136–138
  effect of conductor bundling, 137, 138
  effect of weather, 137, 138
  empirical formulas, 136
Corona pulse
  negative Trichel, 124, 128, 139
  positive, 124, 127, 139
Corona radio noise, 139, 140
Corona in SF_6, 136
Cosmic rays, 65, 76, 112, 206
Current-carrying capacity of cables (see under Cables)
Current chopping, 160, 283
Current growth in discharges, 98, 112
Current transformers, 253

D
DC circuit breakers (see under Circuit breakers)
DC corona (see under Corona)
Deionization of gases
  by diffusion, 65, 87
  by electron attachment, 71, 85, 86
  by recombination, 52, 65, 83–85
Deltatron circuit (see under High-voltage dc generators)

Dielectric
  breakdown, 109, 174–179, 205–211
  constant (see Permittivity)
  dispersion, 199
  hysteresis, 173, 189
  loss, 173, 199, 200, 207, 210, 314
  loss factor, 173, 183, 201, 202
  recovery (see Arc interruption)
  relaxation time, 197
  solids (see Solid insulating materials)
  susceptibility, 192
Dielectrics
  active, 214
Diffusion coefficient, 87, 89
Dipole method (see Field computation)
Dipole sensor, 484
Discharges in voids, 202, 211, 311, 312 (see also Partial discharges)
Disconnectors (see GIS components)
Disruptive discharge, 182, 487, 489, 493, 494
Dissipation factor (see Dielectric loss factor)
Distortion of waves (see under Traveling waves)
Dividers (see Potential dividers)
Double bus (see Bus bar arrangements)
Driven rods (see Grounding)

E
Earth factor, 365
Earth fault (see Ground fault)
Earthing (see Grounding)
Earthing switch, 248, 249, 252

Earth resistivity, 334, 352–354
  measurement of, 352, 353
Electric fields
  in cable insulation, 320, 321
  nonuniform, 11–59, 109
Electric shock, 331, 349, 350
Electric stress control, 59–61
Electrode effects (see Insulating liquids breakdown)
Electrode optimization, 60, 61
Electrodynamic forces on bus bars (see under Bus bars)
Electrolytic tank, 22, 339
Electromagnetic compatibility, 270
Electromagnetic interference, 485, 502
Electron
  attachment, 71, 85, 86, 88
  avalanche, 93–97, 103, 104, 112, 114, 115, 118
  detachment, 65, 82, 112
  diffusion, 87, 89, 90
  drift velocity, 89, 94, 119
  emission, 78, 127
  energy, 71, 74
  mobility, 71, 89, 119
Electronegative gases, 71, 82, 85, 106, 108, 109, 136 (see also under SF_6)
Electro-optical high-voltage measurement, 485
Electrostatic generators (see under High-voltage dc generators)
Electrostatic precipitators, 140, 141, 409
Electrostatic voltmeters (see under Voltmeters)
Emission (see Electron emission)
Erosion (see Arc erosion)
Esters, 165, 166, 168–171, 181
Excitation energy, 71

Index

External insulation, 239 (see also Insulators)
Extrahigh voltages (EHV), 139, 166, 184, 210, 219, 225, 231, 291, 297, 260, 403

F
Faraday cage, 495
Fibers in insulating liquids (see Insulating liquid impurities)
Field, electric
  analysis, 11-51
    conductor-to-plane, 12-19
    in multidielectric media, 20, 21
  computation, 25-59
    by charge simulation, 30-36, 55-59
    by dipole method, 29, 30
    by finite differences, 36-39
    by finite elements, 40-46, 52-55
    by integral equations, 46-49
    by Monte Carlo method, 49, 50
    in multidielectric media, 20, 21, 39, 40, 46, 49
    by successive imaging, 25-29
Field distribution, 18
Field emission of electrons, 109
Field inside a cavity, 191, 192
Field mapping
  by electrolytic tank, 22
  by resistive mesh analog, 24, 25
  by semiconducting paper analog, 23, 24
Field measurement (see Measurement of electric fields)

Formative time lag (see Time lag to breakdown)
Free path, 71, 73-76, 78, 101
Full-wave rectifiers (see under High-voltage dc generators)

G
Gap factors, 116, 384
Gas bubbles (see Voids in insulation)
Gas density monitor, 256
Gas parameters
  effects on breakdown, 109 (see also under Atmospheric conditions)
Generation of high current impulses, 444-446
  double exponential, 444, 445
  rectangular, 444, 446
Generation of high voltages (see High-voltage generators)
GIS (gas-insulated switchgear)
  components, 251-256
  conductor systems, 261-265
  dimensions, 258-260
  effect of moisture, 262
  enclosure configurations, 245-248
  insulation coordination, 403-408
  insulation strength, 261-267
  layout, 245-248
  maintenance, 272, 273
  materials, 248, 251
  particle contamination, 263-270
  pressures, 250
  spacers, 251, 258, 262
  testing, 270, 272
Ground
  impulse impedance, 341, 342

[Ground]
  resistance, 332–340
  safety (see Safety ground)
Ground faults, 279, 324, 365
Grounding
  of GIS, 347, 348
  grids, 338, 348
  in high-voltage laboratories, 485, 486
  of power system neutral (see Power system neutral grounding)
  for safety (see Safety ground)
  of transmission line towers, 348
  system design, 343–346
Ground modes of traveling waves, 374

H
Hall-effect current transducers, 480
High-current impulse generators (see Generation of high-current impulses)
High-current impulse measurement (see Transient current measurement)
High-frequency breakdown (see under Breakdown)
High-voltage ac generators (see Testing transformers)
High-voltage dc generators, 416–426
  cascaded multiplier circuits, 418–425
  deltatron circuit, 425
  electrostatic, 140, 141, 425, 427
  regulation and stabilization 428

[High-voltage dc generators]
  variable-capacitance generators, 427
High-voltage dividers (see Potential dividers)
High-voltage impulse generators, 429-443
  circuits, 429-437
  control of, 439, 440
  for switching surges, 440–443
  multistage, 433, 438–440
High-voltage laboratory
  grounding, 485, 495
  safety precautions in, 495
High-voltage measurements
  by electro-optical methods, 485
  by electrostatic voltmeter, 459–461
  by peak voltmeter, 458, 459
  by sphere gap, 449–458
  sources of error, 456, 485
  by voltage transformers, 461, 462
Humidity, effect on breakdown, 88, 89, 109, 449

I
Impulse current generators (see Generation of high-current impulses)
Impulse generators (see High-voltage impulse generators)
Impulse ratio, 180
Impulse sparkover, 509
Impulse voltage dividers (see Potential dividers)
Impulse withstand voltages, 490–492

Index

Impurities, effect on copper
    conductivity, 306
    (see also under Insulating
    liquids)
Insulating liquids, 165–184
    additives in, 182, 183
    aging of, 182
    chemical properties, 169–171
    dielectric constant, 174, 177
    dielectric strength, 179–181
    electric conductivity, 172, 173
    electrical properties, 171–177
    impurities, 180, 181
    oxidation, 170, 183
    physical properties, 168, 169
    tests, 183, 184
    theories of breakdown, 174–179
    types
        mineral oil, 166
        organic, 165
        synthetic, 166, 167
Insulating materials, solid (see
    Solid insulating materials
    and Dielectrics)
Insulation, 6
    non-self-restoring, 382, 491
    self-restoring, 93, 382, 491
Insulation coordination, 113,
    381, 392–408
    conventional, 396–398
    statistical, 392–396
    with arresters, 405–407
Insulation level (see Basic insulation levels)
Insulation losses (see Dielectric loss)
Insulator pollution, 239, 240,
    383, 487
Insulators, 239, 510, 511
Insulator surface and volume
    resisitivities, 199, 200
Internal discharges (see Discharges in voids)

Internal overvoltages, 357
Intrinsic breakdown (see under
    Solid insulating materials)
Ionization
    by electron impact, 65, 71–76,
        93
    by metastables, 65, 80
    by nuclear particles, 80, 81
    by photons (see Photoionization)
    thermal, 65, 81, 82
Ionization
    coefficient, 74, 82, 87–89
    cross-section, 72, 73
    potential, 74–76, 78
Ion mobility, 51, 117, 152, 173
Irradiation, effect on gas
    ionization, 76, 80

K
Kinetic theory of gases, 65–70
    interpretation of gas pressure, 65–67
    interpretation of gas temperature, 67

L
Laplace's equation, 11, 12, 22,
    25, 36, 37, 40, 45, 48,
    55
Lattice diagram, 372, 373
Layout of GIS (see under GIS)
Leader channel, 106
Leader streamers, 106, 115,
    116
Leakage current, 119, 200, 212
Leakage resistance, 473
Lightning
    current, 329
    discharge, 358
    impulse, 112, 116
    impulse voltage, 228, 359

[Lightning]
  overvoltage, 357
  stroke, 106, 358, 359, 404, 494
Liquid dielectrics (see Insulating liquids)
Loss angle, 200, 202
Losses in dielectrics (see Dielectric loss)

M
Main stroke (see Lightning stroke)
Marx generator (see High-voltage impulse generators)
Maxwell-Boltzmann distribution, 69
Mean free path, 73–76, 78
Measurement
  of electric fields, 21–25, 480–485
    by electro-optical methods, 485
    by mills, 481, 482
    by probes, 480, 481, 484
  of high voltages (see High-voltage measurements)
Measuring cables, 474–477
Mechanical stresses (see under Bus bars)
Metal oxide (see under Surge arresters)
Metastables, 69, 71, 72, 75, 80, 96
Minimum breakdown voltage (see under Breakdown)
Mobility of electrons (see Electron mobility)
Mobility of ions (see Ion mobility)
Moisture
  in oil, 169, 177, 181, 188

[Moisture]
  in soil, 332, 333
  in solid insulating materials, 204, 211
Molecular speeds, 67–69
Multiplier circuits (see High-voltage dc generators)

N
Negative corona (see under Corona)
Neutral grounding (see Power system neutral grounding)
Nonuniform fields (see under Fields)
Nuclear radiation, 65, 76, 80

O
Oil-filled cables (see under Cables)
Oil, insulating (see Insulating liquids)
Onset voltages, 57, 127, 133–135
Organic oils (see under Insulating liquids)
Oscillatory switching impulses (see under High-voltage impulse generators)
Overvoltages, 113, 115, 357–359, 374–376
  control of, 374–376
  harmonics, 365
  lightning, 357–359
  protective devices, 398–402
  statistical characteristics, 377–379
  switching, 360–364
  temporary, 364, 365
Outdoor insulation (see Insulators)

Index 521

Oxidation of insulating liquids (see under Insulating liquids)

P
Partial discharges, 489, 497–502
  detection methods, 497–502
Particle collision
  elastic, 72
  inelastic, 72
Particles in liquids (see under Insulating liquids)
Paschen's law, 100–102, 109
Peak voltage measurement (see High-voltage measurements)
Peek's formula (see Corona loss)
Permittivity (see under Insulating liquids and Solid insulating materials)
Peterson's formula (see Corona loss)
Photocopying machines, 140
Photoelectric emission (see Electron emission)
Photoexcitation, 76
Photo-ionization, 65, 72, 76-80, 88, 103, 109
Photon absorption coefficient, 79, 88
Poisson's equation, 11, 25, 51, 54, 55, 58
Polarization (see under Solid insulating materials)
Pollution (see Insulator pollution)
Polymers, 201, 211, 214
Potential coefficient, 17, 28, 30, 31, 57
Potential dividers
  for ac and dc, 463–469
  for impulse 469–474

[Potential dividers]
  capacitive, 467, 473
  resistive, 470, 483
  screened, 465, 467, 468
  unscreened, 465–467
Potential transformers (see Voltage transformers)
Power networks, 3, 4, 346
Power system neutral grounding, 346, 347
Pulse generator (see High-voltage impulse generators)

Q
Quality factor, 201

R
Radio interference, 502 (see also under Corona)
Rate of rise of restriking voltage (see under Circuit breaking)
Recombination of ions, 52, 65, 83–85, 112
Refraction of electric field lines, 21
Relative air density (RAD), 76, 109, 133, 451, 453
Relaxation time (see under Dielectrics)
Resins, 204, 214
Resistance switching (see under Circuit breaking)
Resonance radiation, 78
Resonant transformers (see under Testing transformers)
Response time of measuring system, 471–474
Return stroke (see Lightning stroke)

Ring bus (*see under* Bus bar arrangements)
Ripple
  factor, 416, 418, 420, 423, 489
  voltage, 416, 422, 425
Rod gaps, 5, 31, 117, 228, 229, 231
Rogowski's coil, 479

S
Safety ground, 349
Schering bridge, 503, 504
Secondary avalanches, 103, 105, 124
Secondary electrons, 96, 97
Semiconducting paper (*see under* Field mapping)
Self-restoring insulation (*see under* Insulation)
Silicon oils (*see* Insulating liquids)
Single bus (*see under* Bus bar arrangements)
Single-stage generator (*see* High-voltage impulse generators)
Soil, 332, 333 (*see also* Earth *and* Ground)
Solid insulating materials, 213, 214
  breakdown, 205–211
    electrochemical, 210
    intrinsic, 205, 206
    thermal, 207–210
  dielectric constant (*see* Relative permittivity)
  dispersion, 199
  electric conduction, 199
  polarization, 193–197
  relative permittivity, 190–192, 194, 195, 198, 199, 319, 321
  tests, 205
  thermal classes, 202–204
  treeing in, 211–213
Spacers (*see under* GIS)
Spark discharge, 93, 99, 103
Sphere gaps, 449–457, 489
Statistical overvoltage (SOV), 393
Statistical safety factor (SSF), 393, 395
Statistical time lag (*see* Time lag to breakdown TBD)
Statistical withstand voltage (SWV), 393
Step response (*see* Response time)
Step voltage, 345, 350 (*see also* Unit step voltage)
Streamer mechanism of breakdown, 103–106
Sulfur hexafluoride (SF_6), 71, 75–77, 82, 86–89, 93, 96, 106, 108, 243–245 (*see also* Electronegative gases)
  breakdown characteristics, 107, 108, 136
  in circuit breakers, 294, 295, 304
  in GIS, 245–250, 256–269, 273
  properties, 243, 244, 258, 294
Superconducting cables (*see under* Cables)
Surge arresters, 253–255, 299–403, 405–407
  magnetically blown, 399, 400
  metal oxide, 401, 403
  tests, 509, 510
Surge impedance, 359, 372, 375, 404

Index 523

Surge voltage (*see* Impulse voltages)
Switching impulses, 112, 115–117, 375, 377, 384, 398, 399, 409, 488, 489, 492, 494, 509
  generation of, 440–443
Switching overvoltages, 357, 360–364
Symmetrical short circuits, 277, 279
Synthetic insulating materials, 204, 211, 213
Synthetic oils, 165–168
Synthetic testing of circuit breakers, 506

T

Temporary overvoltages (*see under* Overvoltages)
Tesla coil, 443
Test voltages, 490–492
Testing
  nondestructive, 497–505
  under impulse voltages, 491–494
    multiple-level method, 493
    up-and-down method, 493, 494
Testing transformers
  cascaded, 410–412
  resonant, 412–416
  single-unit, 410
Tests
  acceptance, 488
  commissioning, 507
  maintenance, 488, 504, 505, 507
  routine, 508, 510
  sample, 509-511
  type, 508, 510
Thermal breakdown (*see under* Solid insulating materials)
Thermal classes of insulating materials (*see under* Solid insulating materials)
Thermal instability, 206
Thermal ionization (*see under* Ionization)
Thermal rating
  of bus bars, 231, 234–236
  of cables, 321–323
Time lag to breakdown (TBD), 113, 115, 386, 388, 390, 391
  formative, 113, 115
  statistical, 113, 114
Touch voltage, 345
  of GIS, 348
Transfer bus (*see* Bus bar arrangements)
Transformers, testing of, 508, 509
Transient current measurements, 477–480
  by Hall-effect transducers, 480
  by magnetic coupling, 479
  by resistive shunts, 478
Transmission lines, 1–4
  tower grounding, 348
  voltages, 3, 219
Traveling waves, 131, 366-374
  distortion by corona, 131, 374
  reflection, 370–373
  refraction, 372
  velocity, 369, 374, 474, 475
Treeing in plastics (*see under* Solid insulating materials)
Trichel pulses (*see under* Corona)
Trigatron gap, 439, 440

U

Ultrahigh voltages, 2, 219
Uniform-field breakdown (see under Breakdown)
Uniform-field gaps, 94, 97 103, 112, 119
Unit step voltage, 469–471
Up-and-down method (see under Testing)

V

Vacuum circuit breakers (see under Circuit breakers)
Van de Graaff generator (see under High-voltage dc generators)
Variable capacitance generator (see under High-voltage dc generators)
Velocity of traveling waves (see under Traveling waves)
Voids in solid insulation, 174– 176, 192, 211, 213, 497, 498, 500
Voltage dividers (see Potential dividers)
Voltage gradient, 105, 106, 115–117, 133, 135, 139, 150, 469

Voltage multiplier circuits (see under High-voltage dc generators)
Voltage ripple (see Ripple voltage)
Voltage transformers, 252, 461, 462
  cascaded, 462
Voltmeters
  electrostatic, 459–461
  high-resistance, 463
  peak, 458, 459, 491

W

Wagner earth, 503, 504
Water in insulating liquids, 169, 170, 181, 193
Water treeing, 211, 212
Wave shape of impulse, 135, 431, 433, 436, 444, 489
Wet withstand test, 510

X

X-rays, 65, 76–79, 409